Water Treatment: Techniques and Technologies

Water Treatment: Techniques and Technologies

Edited by Aria Sawyer

SYRAWOOD
PUBLISHING HOUSE

New York

Published by Syrawood Publishing House,
750 Third Avenue, 9th Floor,
New York, NY 10017, USA
www.syrawoodpublishinghouse.com

Water Treatment: Techniques and Technologies
Edited by Aria Sawyer

© 2019 Syrawood Publishing House

International Standard Book Number: 978-1-68286-835-5 (Hardback)

Cataloging-in-Publication Data

Water treatment : techniques and technologies / edited by Aria Sawyer.
    p. cm.
Includes bibliographical references and index.
ISBN 978-1-68286-835-5
1. Water--Purification. 2. Water--Purification--Technological innovations. 3. Sewage--Purification. I. Sawyer, Aria.
TD430 .W38 2019
628.162--dc23

TABLE OF CONTENTS

PREFACE

Any process that improves the quality of water and makes it acceptable for specific uses, such as drinking, irrigation, industrial water supply, etc. is termed as water treatment. It involves the removal of contaminants or the reduction in their concentration. The techniques and technologies used for water treatment vary according to use. Some of the processes used in the treatment of municipal drinking water include pre-chlorination, aeration, sedimentation, filtration, disinfection, etc. The principal methods of industrial wastewater treatment are cooling water treatment and boiler water treatment. Water supplied to domestic properties is treated via water softening or ion exchange. This book provides significant information of water treatment techniques and technologies to help develop a good understanding of the management of contaminated water. It is a collective contribution of a renowned group of international experts. It aims to serve as a resource guide for students and experts alike.

Significant researches are present in this book. Intensive efforts have been employed by authors to make this book an outstanding discourse. This book contains the enlightening chapters which have been written on the basis of significant researches done by the experts.

Finally, I would also like to thank all the members involved in this book for being a team and meeting all the deadlines for the submission of their respective works. I would also like to thank my friends and family for being supportive in my efforts.

Editor

Optimization and kinetic evaluation of acid blue 193 degradation by UV/peroxydisulfate oxidation using response surface methodology

Mojtaba Ahmadi[*], Mohmmad Hamed Ardakani, ALi Akbar Zinatizadeh

Chemical Engineering Department, Faculty of Engineering, Razi University, Kermanshah, Iran

Keywords:
Potassium peroxydisulfate
Blue dye
Response surface methodology
Color removal
Optimization

ABSTRACT

The optimization of process conditions for the degradation of Acid Blue 193 by UV/peroxydisulfate was investigated using response surface methodology (RSM). The effects of four parameters namely initial $K_2S_2O_8$ concentration, UV irradiation, temperature, and initial dye concentration on two process responses, color removal and the rate constants of the first-order kinetic equations, were investigated using a second-order polynomial multiple regression model. The analysis of variance (ANOVA) explained a high determination coefficient (R^2) value of 0.927-0.967, which ensures a good fit of the first-order regression model with the experimental data. The central composite design (CCD) was used to optimize the process conditions, which showed that an initial $K_2S_2O_8$ concentration of 5 mM, UV irradiation of 250 W, temperature of 50 °C, and the initial dye concentration of 40 mg/L were the best conditions. Under optimum conditions, the maximum color removal from the wastewater and the rate constants of the first-order kinetic equation were 100% and 0.086 min^{-1}, respectively.

1. Introduction

About 700,000 tons of dyes per year are produced worldwide. Nearly 10-15 percent of these dyes are discharged in effluent from dyeing operations [1]. Therefore, the decolorization of the colored effluent from dye industries is of vital importance before their release into the environment. Many conventional wastewater treatment methods such as chemical or electrochemical precipitation [2], biological oxidation [3, 4], biosorption [5], activated carbon adsorption [6-9], ozonation [10], and coagulation–flocculation [11] have been widely used to remove refractory pollutants from textile wastewaters [12]. Some of these processes are not very efficient because the dye compounds are hardly removed due to their resistance to biodegradation. However, these methods are not destructive and only transform the impurities in the solid state, and therefore the waste must be treated.

Biological oxidation is the most cost-effective process compared to the other treatments [13]. However, wastewater from the textile industry are known to contain harmful substances and/or non-biodegradable pollutants; therefore, conventional biochemical processes to treat solutions containing dyes and soluble biorefractory are inadequate [14].

Among the various treatment methods, advanced oxidation processes (AOPs) are considered as one of the most effective methods to degrade toxic pollutants and non-biodegradable organic pollutants [8, 14]. Among the advanced oxidation processes, the homogeneous AOPs employing peroxydisulfate [15-17] and UV/ peroxydisulfate [18-20] have been found to be very effective in the degradation of dye and pollutants.

Acid Blue 193 dye is widely used for silk and wool as well as for leather dyeing. It was chosen as a model contaminant for our degradation studies because it contains aromatic rings that make it difficult to treat with

*Corresponding author
E-mail address: m_ahmadi@razi.ac.ir

traditional processes [21]. To date, various techniques, including adsorption on bentonite [22] , natural sepiolite [23], red pine sawdust [24], declorization by *Cladosporium cladosporioides* [25], and biosorption using ash prepared from cow dung, mango stone, parthenium leaves, and activated carbon [26] have been reported for sepration and/or degradtion of Acid Blue 193. The research of Vijay kumar et al. was conducted using fungus *Cladosporium cladosporioides* as a decolorizing microorganism for the mineralization of Acid Blue 193 in an aqueous solution. Under the best conditions, the fungus completely decolorized a waste concentration of 100 mg l^{-1} of Acid Blue 193 in 8 days. Rasoulifard et al. studied the decolorization of Acid Blue 25 by $UV/S_2O_8^{2-}$ [27]. Under optimum conditions, the maximum color removal efficiency was 95%. The decolorization and mineralization of synthesized Acid Blue 113 by the UV/persulfate advanced oxidation process was examined by Shu et al. [28]. Also, Shu et al. studied the decolorization of azo dye Acid blue 113 by the UV/Oxone process. Various operating parameters for the removal of dye have been studied [29]. The research of Akay et al. was conducted using the heterogeneous fenton process for mineralization of burazol blue ED [30]. Under optimized reaction conditions, 96% color and 75% COD removals were obtained.

To the best of our knowledge, no previous studies have been conducted on the degradation of acid blue 193 by UV/peroxydisulfate oxidation. This study was conducted to optimize the process parameters (initial dye concentration, $K_2S_2O_8$ concentration, UV irradiation, and temperature) of Acid Blue 193 degradation and its kinetic study by UV/peroxydisulfate oxidation with response surface methodology (RSM). In doing so, the degradation process of the contaminent was enhanced.

2. Materials and Methods

2.1. Chemical

Synthetic colored wastewater containing an organic model pollutant (Acid Blue 193), commonly used as a textile dye, was obtained from the Boyakhsaz Company, Iran. The other chemicals, Potassium peroxydisulfate ($K_2S_2O_8$), H_2SO_4, and NaOH, were of laboratory reagent grade (Merck Co., Germany) and used without further purification.

2.2. Experimental methods

Acid Blue 193 degradation experiments were conducted in a photoreactor (Fig1.). For the UV/peroxydisulfate process, irradiation was carried out with a 125 and 250 W (UV-C) mercury lamp (Philips, the Netherland), which was put above a batch photoreactor. The distance between the solution and UV source was adjusted according to the experimental conditions. The reactor was equipped with a water-flow jacket for regulating the temperature by means of an external circulating flow of a thermostat. Since the photocatalysis was sustained by a ready supply of

dissolved oxygen, air was supplied to the reaction system at a constant flow-rate using a micro-air compressor.

Dilute solutions of sodium hydroxide and sulfuric acid were used for pH adjustment and the initial pH values were measured by a Metrohm 827 pH/LF portable pH/conductivity-meter, Schott Instruments GmbH, Mainz, Germany. The prepared solution was transferred to the reactor and after adjusting the temperature, the UV lamp was switched on to initiate the process. During the experiments, a mild aeration was kept for mixing the content and saturation with O_2. Samples (6 mL) were taken at regular time intervals. A maximum total sampling volume of 24 mL was withdrawn during each experimental run, which is not significant when compared with the solution volume.

2.3. Analysis

The concentration of Acid Blue 193 in solution at different times was obtained by measuring absorbance at a maximum absorption wavelength (574 nm); the calculation of the decolorization decay was obtained from the calibration curve prepared earlier. An UNICO (USA) UV spectrophotometer (model- 2001SUV, USA) was employed for absorbance measurements by using a quartz cell of path length 1 cm. The color removal efficiency (η) was achieved by using the following formula:

$$\eta\ (\%) = \frac{C_o - C_t}{C_t} \times 100 \tag{1}$$

where C_o and C_t are the initial concentration and the concentration of Acid Blue 193 at reaction time t, respectively.

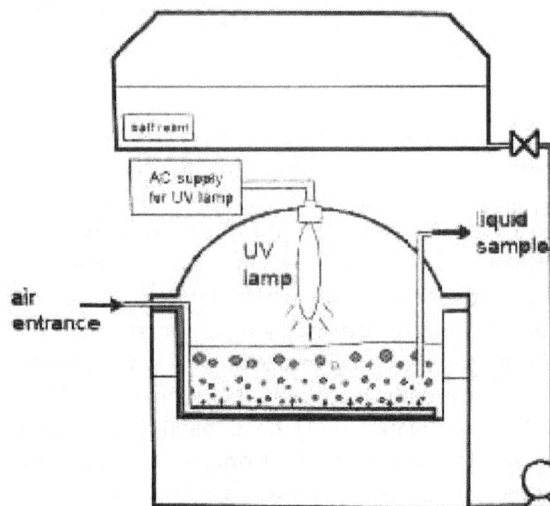

Fig. 1. The schematic view of the photo reactor set-up.

2.4. Kinetic modeling of the process

As a source of sulfate radicals, peroxydisulfate ($S_2O_8^{2-}$) has the advantages of high aqueous solubility and high stability at room temperature [15]. The reactions of peroxydisulfate are generally slow at normal temperatures. The thermal or photochemical activated decomposition of $S_2O_8{}^{2-}$ ion to $SO_4{}^{\bullet-}$ radical has been proposed as a method of accelerating the process [31] and is summarized in the following reactions:

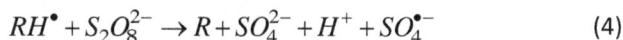

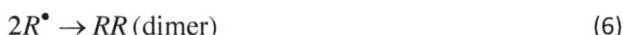

$$S_2O_8^{2-} + h\upsilon \rightarrow 2SO_4^{\bullet-} \tag{2}$$

$$SO_4^{\bullet-} + RH_2 \rightarrow SO_4^{2-} + H^+ + RH^{\bullet} \tag{3}$$

$$RH^{\bullet} + S_2O_8^{2-} \rightarrow R + SO_4^{2-} + H^+ + SO_4^{\bullet} \tag{4}$$

$$SO_4^{\bullet-} + HR \rightarrow R^o + SO_4^{2-} + H^+ \tag{5}$$

$$2R^{\bullet} \rightarrow RR \, (\text{dimer}) \tag{6}$$

(R is an organic matter) In this oxidation process, sulfate ions will be generated as the end-product [32], which leads to a decrease in pH and an increase in salt content in the effluent. The $SO_4{}^{2-}$ is practically inert and is not considered to be a pollutant. The USEPA has listed it under the secondary drinking water standards with a maximum concentration of 250 mg/L (1.43 mM), based on aesthetic reasons such as taste and odor [33].

In the present study, first and second-order reaction kinetics was used to study the decolorization kinetics of Acid Blue 193 by the UV/peroxydisulfate process. The individual expression was presented as below:

First-order reaction kinetics:

$$\frac{dC}{dt} = -k_1 C \tag{7}$$

By integrating the equation (7), the following equations could be obtained:

$$C_t = C_o \cdot \exp(-k_1 t) \tag{8}$$

where,

C_t: The concentration of RY84 at reaction time t (mM)

C_o: The initial concentration of the RY84 (mM)

k_1: The rate constant of the first-order kinetic equation (min^{-1})

2.5. Experimental design and data analysis

A central composite design (CCD) was employed in order to optimize the dye removal. Four factors were considered: initial $K_2S_2O_8$ concentration, UV irradiation, temperature, and initial dye concentration. Table 1 summarizes the levels for each factor involved in the design strategy. Table 2 shows the standard array for three factors and 26 experiments. It also shows the run order and the observed responses. The obtained model was evaluated for each response function and the experimental data (the color removal and the rate constants of the first-order kinetic equations) were analyzed statistically applying an analysis of variance (ANOVA).

The design consists of three series of experiments: (i) a two-level full factorial design 2^4 (all possible combinations of codified values +1 and −1); (ii) two central, replicates of

the central point (0); and (iii) eight axial located at the center and both extreme levels of the experimental models [34].

Table 1. Independent variables and their levels in the central composite design used in the present study.

Parameter name	Code	Low (−1)	High (+1)
Initial $K_2S_2O_8$ concentration, mM	X_1	1	5
UV, W	X_2	0	250
Temperature, °C	X_3	10	50
Initial dye concentration, mg/L	X_4	40	200

$$y = \beta_0 + \sum_{j=1}^{k} \beta_j X_j + \sum_{j=1}^{k} \beta_{jj} X_j^2 + \sum_{j<j}^{k} \sum_{j=2}^{k} \beta_{ij} X_i X_j \tag{9}$$

where β_0, β_i, β_{ii}, and β_{ij} are the regression coefficients for intercept, linear, quadratic, and interaction terms, respectively; and X_i and X_j are the independent variables. The "Design expert" (version 7) was used for regression and graphical analyses of the obtained data.

3. Results and Discussion

3.1. Development of mathematical models

Design-Expert software was used for the analysis of the measured responses and determining the mathematical models with the best fits. The adequacy of the model was tested using the sequential f-test, lack-of-fit test, and the analysis-of-variance (ANOVA) technique using the same software to obtain the best-fit model

3.2. Analysis of color removal

The fit summary for color removal suggests a quadratic relationship where the additional terms are significant and the model is not aliased. The ANOVA table of the quadratic model with other adequacy measures R^2, adjusted R^2, and predicted R^2 and are given in Table 3. The associated p-value of less than 0.05 for the model (i.e., $\alpha=0.05$, or 95% confidence level) indicates that the model terms are statistically significant. The lack-of-fit value of the model indicates non-significance, as this is desirable. The ANOVA result shows that the initial $K_2S_2O_8$ concentration, UV irradiation, temperature, initial dye concentration, and quadratic effect of the UV irradiation along with the interaction effect of the initial $K_2S_2O_8$ concentration and initial dye concentration and UV irradiation and temperature were the significant model terms associated with color removal. The other model terms were not significant, and thus eliminated by a backward elimination process to improve model adequacy.

Table 2. Factorial experimental design for treatment of dye by AOP.

Experiments no.	Initial $K_2S_2O_8$ concentration (X_1)	UV, W (X_2)	Temperature, $^{\circ}$C (X_3)	Initial dye concentration (X_4)	Color removal	K_1 min^{-1}
1	-1	-1	-1	-1	11.76	0.001
2	1	-1	-1	-1	29.44	0.001
3	-1	1	-1	-1	96.12	0.025
4	1	1	-1	-1	99.96	0.076
5	-1	-1	1	-1	38.46	0.001
6	1	-1	1	-1	65.00	0.005
7	-1	1	1	-1	99.57	0.065
8	1	1	1	-1	99.60	0.092
9	-1	-1	-1	1	2.19	0.0001
10	1	-1	-1	1	19.15	0.001
11	-1	1	-1	1	46.47	0.005
12	1	1	-1	1	91.83	0.023
13	-1	-1	1	1	17.14	0.002
14	1	-1	1	1	52.47	0.004
15	-1	1	1	1	78.14	0.011
16	1	1	1	1	95.61	0.034
17	-1	0	0	0	89.10	0.021
18	1	0	0	0	97.20	0.03
19	0	-1	0	0	33.61	0.001
20	0	1	0	0	92.82	0.022
21	0	0	-1	0	91.64	0.02
22	0	0	1	0	97.27	0.025
23	0	0	0	-1	99.63	0.03
24	0	0	0	1	86.60	0.016
25	0	0	0	0	95.13	0.027
26	0	0	0	0	92.10	0.026

The ANOVA table for the reduced quadratic model is shown in Table 3. The reduced model results indicate that the model is significant (p-value less than 0.05). The other adequacy measures R^2, adjusted R^2, and predicted R^2 are in reasonable agreement and are close to 1, which indicate the adequacy of the model. The adequate precision compares the signal-to-noise ratio and a ratio greater than 4 is desirable. The value of the adequate precision ratio of 28.23 indicates an adequate model discrimination. The lack-of-fit f-value of 11.71 implies that the lack-of-fit is not significant relative to the pure errors [35].

The final mathematical models for color removal, which can be used for prediction within the same design space, are given as follows:

In terms of coded factors:

Color removal=93. 58+9.52X_1+29.49X_2+8.59X_3-8.33X_4+4.19 X_1X_4-4.50 X_2X_3-34.18 X_2^2

In terms of actual factors:

Color removal=17.92+1.62X_1+0.84X_2+0.65X_3-0.18X_4+0.026X_1X_4-1.8×10^{-3}X_2X_3-2.2×10^{-3}X_2^2

3.3. Analysis of the rate constants of the first-order kinetic equations

For the rate constants of the first-order kinetic equations, the fit summary recommends the quadratic model where the additional terms are significant and the model is not aliased. Table 4 presents the ANOVA table of the quadratic model with other adequacy measures R^2, adjusted R^2, and predicted R^2. The associated p-value of less than 0.05 (i.e. α=0. 05 or 95% confidence level) indicates that the model terms can be considered as statistically significant. The lack-of-fit value of the model indicates non-significance as desired. The analysis-of-variance result shows that the main effect of the initial $K_2S_2O_8$ concentration, UV irradiation, temperature, and initial dye concentration along with the interaction effect of the initial $K_2S_2O_8$ concentration and UV irradiation, UV irradiation and temperature, UV irradiation, and initial dye concentration are the significant model terms associated with the rate constants of the first-order kinetic equations. The other model terms are not significant and are thus eliminated by a backward elimination process to improve model adequacy.

Table 3. ANOVA analysis of the color removal (after elimination).

Source	Sum of square	df	Mean square	F-value	P-value
Model	26940.47	7	3848.64	75.41	< 0.0001
Residual	918.68	18	51.04		
Lack of Fit	914.08	17	53.77	11.71	0.2263
Pure Error	4.59	1	4.59		
Total	27859.15	25			

R^2 = 96.70%; R^2 (adjust) = 95.42%; Adequate precision = 28.230. Coefficient of variation (CV) = 10.22%

Table 4. ANOVA analysis of the rate constants of the first-order kinetic equation.

Source	Sum of square	df	Mean square	F-value	P-value
Model	0.013117	7	0.001874	32.857	< 0.0001
Residual	0.001027	18	5.7E-05		
Lack of Fit	0.001026	17	6.04E-05	120.709	0.0715
Pure Error	5E-07	1	5E-07		
Total	0.014143	25			

R^2 = 92.74%; R^2 (adjust) = 89.92%; R^2 (predicted) = 83.3%; Adequate precision = 21.94. Coefficient of variation (CV) = 34.81%

The ANOVA table for the reduced quadratic model is presented in Table 4. The reduced model results indicate that the model is significant (p-value less than 0.05). The other adequacy measures R^2, adjusted R^2, and predicted R^2 are in reasonable agreement and are close to 1, which indicate an adequate model. The value of the adequate precision ratio 21.94 indicates adequate model discrimination. The lack-of-fit f-value of 120.71 implies that the lack-of-fit is not significant relative to the pure error [35].

The final mathematical models for the rate constants of the first-order kinetic equations, as determined by the Design Expert software, are given as follows:

In terms of coded factors:

$$K_1 = 0.022 + 7.5 \times 10^{-3} X_1 + 0.019 X_2 + 4.8 \times 10^{-3} X_3 - 0.011 X_4 + 7 \times 10^{-3} X_1 X_2 + 4 \times 10^{-3} X_2 X_3 - 0.012 X_2 X_4$$

In terms of actual factors:

$$K_1 = 4.13 \times 10^{-4} + 2.44 \times 10^{-4} X_1 + 1.55 \times 10^{-4} X_2 + 4.11 \times 10^{-5} X_3 + 5 \times 10^{-6} X_4 + 2.8 \times 10^{-5} X_1 X_2 + 1.6 \times 10^{-6} X_2 X_3 - 1.2 \times 10^{-6} X_2 X_4$$

3.4. Effects of process parameters on the responses
3.4.1. Color removal

Figure 2 is a perturbation plot which illustrates the effect of all the factors at the center point in the design space. It is apparent from this figure that the initial dye concentration has a negative effect on the color removal. This is because the rise in dye concentration induces an inner filter effect, and hence the solution becomes more and more impermeable to UV radiation [36]. It can also be observed from this plot that the initial $K_2S_2O_8$ concentration and temperature both have a minute positive effect on the color removal efficiency. These effects could be attributed to the following reason. An increase in the initial $K_2S_2O_8$ concentration increases the generation of sulfate radicals and hydroxyl radicals simultaneously and improves the decolorization of the dye [18, 19]. The results indicate that the color removal efficiency increases with the increase of UV radiation until it reaches its optimum value; the color removal then starts to decrease slightly as the UV radiation increases near the high limit.

The interaction effects of initial dye concentration and initial $K_2S_2O_8$ concentration on color removal are shown in Fig. 3 (a) and (b). These figures demonstrate that the color removal increases with decreasing initial dye concentration and increasing initial $K_2S_2O_8$ concentration. The experiments for the UV/$S_2O_8^{2-}$ process showed that a 100% reduction of color occurred for the optimum conditions. Chang et al. showed that an overall degradation of 90% was achieved in 30 min for the photooxidation of iopromide by combined UV irradiation and peroxydisulfate [37]. This is due to the fact that an increase in the initial $K_2S_2O_8$ concentration results in an increase of the sulfate and hydroxyl radicals, and hence improves the photooxidation of the dye. The decolorization of the dye in the photooxidation process decreased with the increasing of the initial dye concentration. This was due to the fact that the higher dye concentration causes more absorption of UV light. In fact, an increase in the concentration of dye causes a rise of the internal optical density. Therefore, the solution is more impassable to UV light.

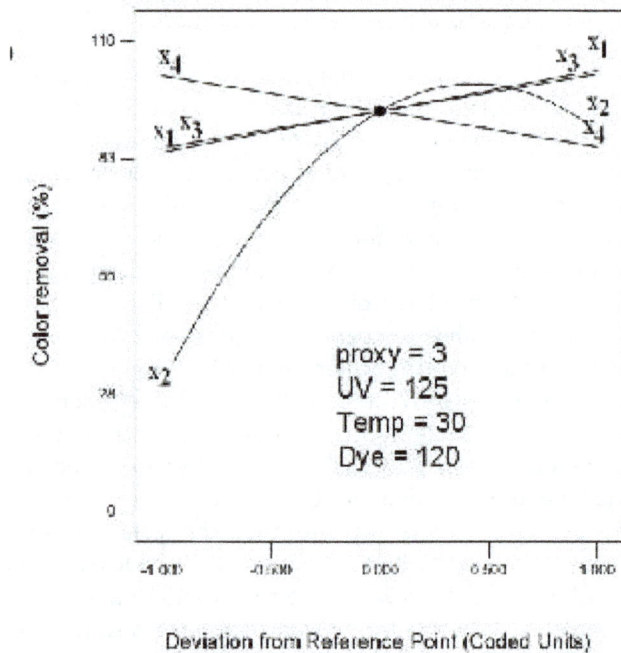

Fig. 2. Perturbation plot showing the effect of all factors on color removal.

Figure 4 (a) and (b) represent the interaction of UV and temperature on color removal. The UV radiation had a positive effect on the color decomposition. In the models for the color removal, x_2 was the major regressor variable affecting the responses (greatest coefficients, β_2=29. 49). In relation to the interaction of these two process parameters, the results indicated that increasing the UV up to the optimum value increased the color removal. Further, an increase of UV decreased the color removal. In this figure, it is evident that as the temperature increased, the color removal was increased.

3.4.2. The rate constants of the first-order kinetic equations

Figure 5 shows a perturbation plot to compare the effect of all the process parameters at the center point in the design space. From this figure, it can be observed that the rate constants of the first-order kinetic equations increases with the initial $K_2S_2O_8$ concentration. This was due to the fact that with an increase in initial $K_2S_2O_8$ concentration, the k_1 is increased. The result showed that the initial dye concentration has a negative effect on k_1, i.e., the decrease in the rate constants with increasing initial dye concentration has been observed. The presumed reason is from the effect of the higher dye concentration which causes more absorption of the UV light. Indeed, an increase in the dye concentration induces a rise in the medium optical density, and the solution therefore becomes more impermeable to the UV radiation. It can be observed from this plot that the rate constant varies positively with the UV radiation. Increasing the UV light

increases the enhanced production of sulfate and hydroxyl radicals. At a low UV radiation, the rate of photolysis of $S_2O_8^{2-}$ is limited. However, at a high UV radiation, the formation of sulfate and hydroxyl radicals increases upon the photodissociation of $S_2O_8^{2-}$, and hence the constant rate of decolorization increases [19, 36]. The temperature contributes a slightly positive effect on the rate constant.

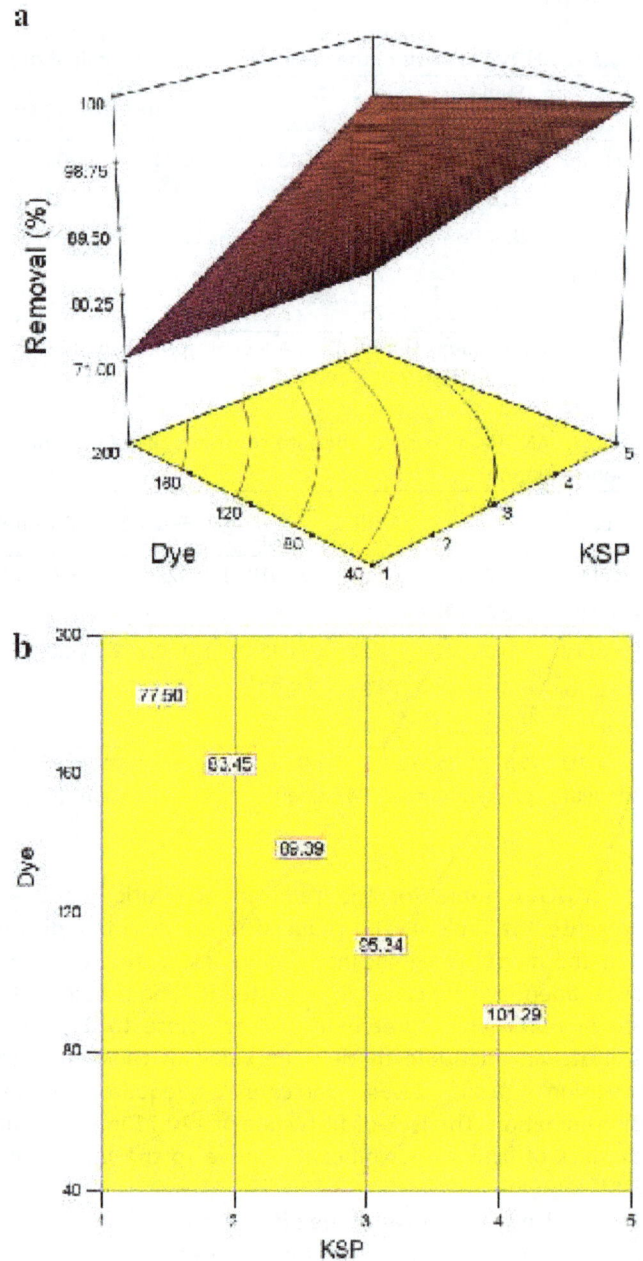

Fig. 3. (a) Response surface plot and (b) contours plot showing the effect of initial dye concentration (mg/l) and initial $K_2S_2O_8$ concentration (mM) on the color removal at UV=125 W and Temperature=30°C.

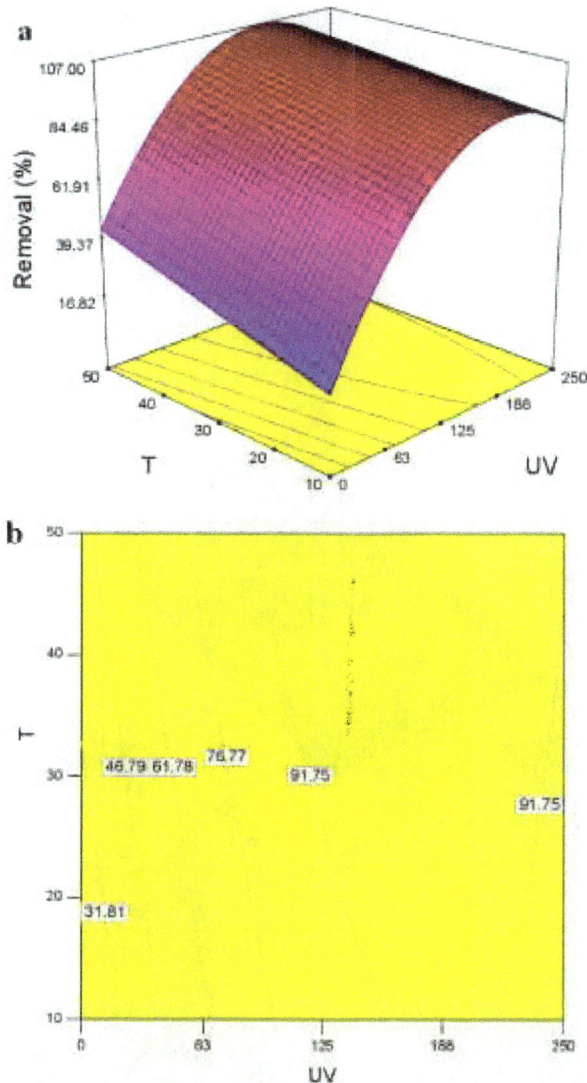

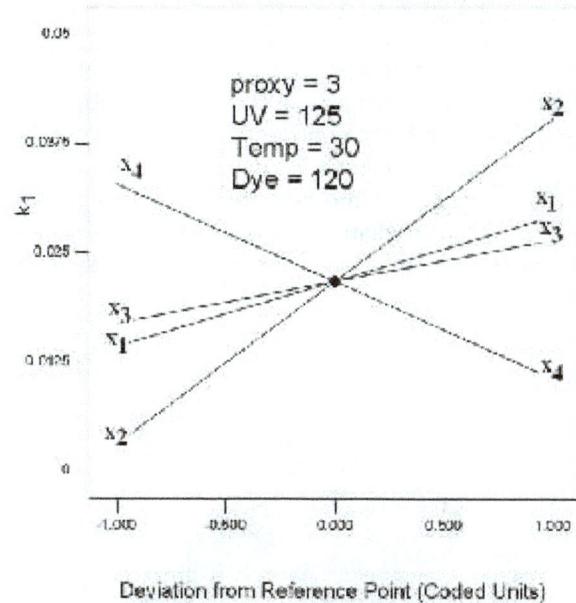

Fig. 5. Perturbation plot showing the effect of all factors on the K_1.

Fig. 4. (a) Response surface plot and (b) contours plot showing the effect UV (W) and temperature (30°C) on the color removal at an initial $K_2S_2O_8$ concentration 3mM and initial dye concentration 120mg/l.

Figure 6 (a) and (b) show the interaction effect of UV and initial $K_2S_2O_8$ concentration on the decolorization rate constant. From this figure, it is evident that as the UV and initial $K_2S_2O_8$ concentration increase, the decolorization rate constant also increases. This observation agrees with the results of other studies. It is reasonable to expect that by increasing the concentration of peroxydisulfate ion, more sulfate and hydroxyl radicals are available to attack the aromatic rings and the decolorization rate constant is increased [38]. Similar results were also obtained by former studies [39]. Khataee et al. investigated the decolorization of Basic Blue by UV/peroxydisulfate treatment. They described that an increase in the decolorization rate constant resulted in an increase in $S_2O_8^{2-}$ [18].

In the decolorization rate constant models, the UV radiation was the major regressor variable affecting the responses (greatest coefficients, $\beta_2=0.019$). Figure 7 depicts the response surface plot showing the effect of UV and temperature on the decolorization rate constant at a fixed initial $K_2S_2O_8$ concentration (3 mM) and an initial dye concentration (120 mg/l). The results indicate that even at high UV levels, an increase in temperature at a fixed UV led to a marked increase in the decolorization rate constant. The present study shows that the decolorization rate constant could be increased with increasing the UV at all the levels of temperature. These results are significant in view of the earlier literature reports [40, 41]. At a higher temperature, the accelerated peroxydisulfate decolorization has been more effective which improved the generation rate of ·OH, and therefore enhanced the decolorization of the dye [42].

The interaction effects of initial dye concentration and UV on the decolorization rate constant is shown in Fig. 8 (a) and (b). These figures demonstrate that the decolorization rate constant increases with decreasing initial dye concentration and increasing UV. The reason is the same as that stated earlier. A similar result was reported by Khatee et al. that found a decrease in photooxidative decolorization rate with increasing initial dye concentration [18].

4. Conclusions

The following conclusions can be drawn from this study based on the range of values of parameters considered.
1. Increasing initial $K_2S_2O_8$ concentration increases the color removal efficiency and the rate constants of the first-

order kinetic equations; whereas, increasing initial dye concentration decreases both the responses.

2. In the case of UV irradiation, the color removal increases with the UV irradiation until it reaches its closest to optimum value, the color removal then starts to drop as the UV irradiation is increased.

3. The rate constants of the first-order kinetic equations increases as the UV irradiation increases.

4. The temperature has a slight positive effect on the color removal efficiency.

5. The temperature contributes positively with a statistically significant effect on the rate constants of the first-order kinetic equations.

6. The optimum conditions for this treatment was achieved by setting the experiment with an initial $K_2S_2O_8$ concentration at 5 mM, the UV irradiation at 250W, and the temperature at 50 $^{\circ}$C while the initial dye concentration level was at 40 mg/L.

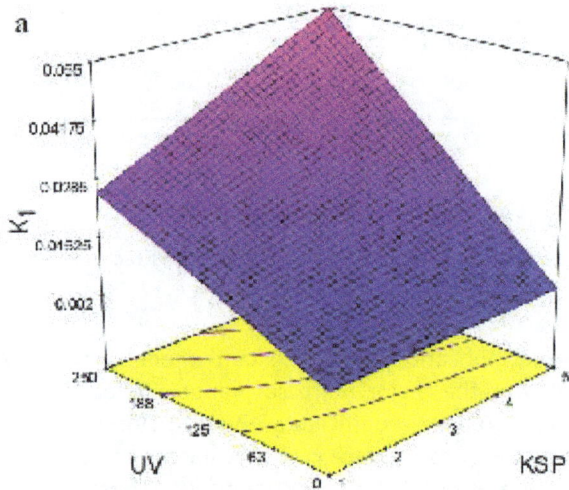

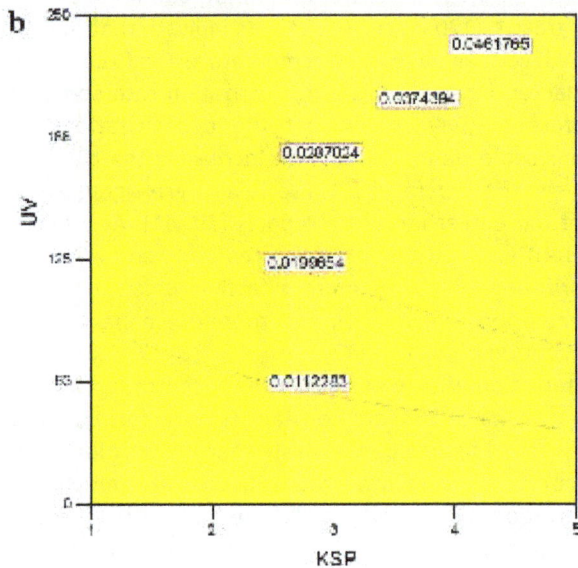

Fig. 6. (a) Response surface plot and (b) contours plot showing the effect of UV (W) and initial $K_2S_2O_8$ concentration (mM) on decolorization rate constant at initial dye concentration=120 mg/l and Temperature=30°C.

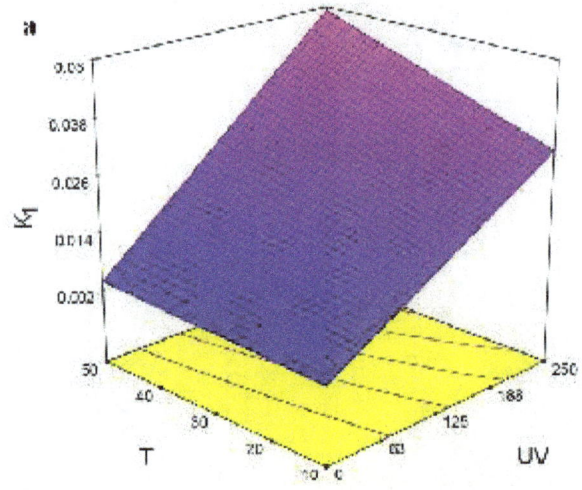

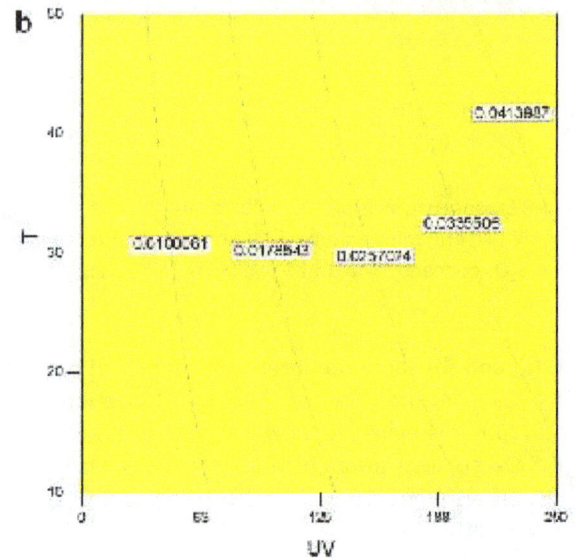

Fig. 7. (a) Response surface plot and (b) contours plot showing the effect of UV (W) and temperature (30°C) on decolorization rate constant at initial $K_2S_2O_8$ concentration=3 mM and initial dye concentration=120 mg/l.

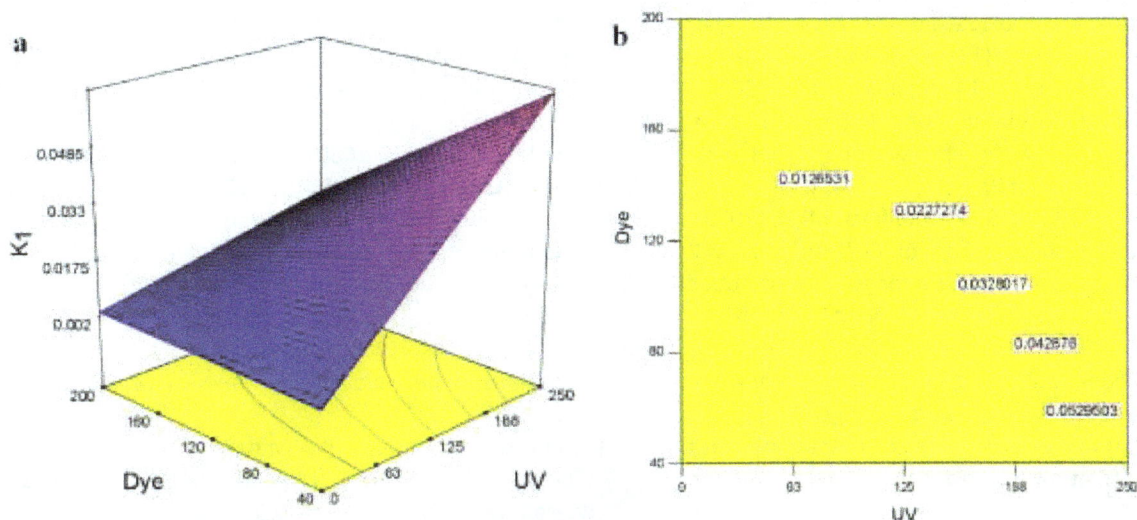

Fig. 8. (a) Response surface plot and (b) contours plot showing the effect of UV (W) and initial dye concentration (mg/L) on decolorization rate constant at initial $K_2S_2O_8$ concentration =3mM and Temperature=300°C.

References

[1] Zollinger, H., (1991). Colour chemistry: Synthesis properties and application of organic dyes and pigments. New York VCH publishers.

[2] Fongsatitkul, P., Elefsiniotis, P., Boonyanitchakul, B. (2006). Treatment of a textile dye wastewater by an electrochemical process. *Journal of environmental science and health part A, 41*(7), 1183-1195.

[3] Işık, M., Sponza, D. T. (2008). Anaerobic/aerobic treatment of a simulated textile wastewater. *Separation and purification technology, 60*(1), 64-72.

[4] Pandey, B. V., & Upadhyay, R. S. (2006). Spectroscopic characterization and identification of pseudomonas fluorescens mediated metabolic products of Acid Yellow-9. *Microbiological research, 161*(4), 311-315.

[5] Ranjusha, V. P., Pundir, R., Kumar, K., Dastidar, M. G., Sreekrishnan, T. R. (2010). Biosorption of remazol Black B dye (Azo dye) by the growing Aspergillus flavus. *Journal of environmental science and health Part A, 45*(10), 1256-1263.

[6] Özacar, M., Şengil, I. A. (2003). Adsorption of reactive dyes on calcined alunite from aqueous solutions. *Journal of hazardous materials, 98*(1), 211-224.

[7] Akkaya, G., Uzun, İ., Güzel, F. (2007). Kinetics of the adsorption of reactive dyes by chitin. *Dyes and pigments, 73*(2), 168-177.

[8] Benkli, Y. E., Can, M. F., Turan, M., Celik, M. S. (2005). Modification of organo-zeolite surface for the removal of reactive azo dyes in fixed-bed reactors. *Water research, 39*(2), 487-493.

[9] Geethakarthi, A., & Phanikumar, B. R. (2011). Adsorption of reactive dyes from aqueous solutions by tannery sludge developed activated carbon: Kinetic and equilibrium studies. *International journal of environmental science & technology, 8*(3), 561-570.

[10] Lin, S. H., Liu, W. Y. (1994). Treatment of textile wastewater by ozonation in a packed-bed reactor. *Environmental technology, 15*(4), 299-311.

[11] Kashefialasl, M., Khosravi, M., Marandi, R., Seyyedi, K. (2006). Treatment of dye solution containing colored index acid yellow 36 by electrocoagulation using iron electrodes. *International journal of environmental science and technology, 2*(4), 365-371.

[12] Daneshvar, N., Sorkhabi, H. A., Kasiri, M. B. (2004). Decolorization of dye solution containing Acid Red 14 by electrocoagulation with a comparative investigation of different electrode connections. *Journal of hazardous materials, 112*(1), 55-62.

[13] Metcalf & Eddy (Empresa comercial). (1991). *Wastewater Engineering: Treatment disposal and reuse.* Irwin Mcgraw Hill.

[14] Ternes, T. A., Hirsch, R. (2000). Occurrence and behavior of X-ray contrast media in sewage facilities and the aquatic environment. *Environmental science & technology, 34*(13), 2741-2748.

[15] Xu, X. R., Li, X. Z. (2010). Degradation of azo dye Orange G in aqueous solutions by persulfate with ferrous ion. *Separation and purification technology, 72*(1), 105-111.

[16] Rastogi, A., Al-Abed, S. R., Dionysiou, D. D. (2009). Sulfate radical-based ferrous–peroxymonosulfate oxidative system for PCBs degradation in aqueous and sediment systems. *Applied catalysis B: environmental, 85*(3), 171-179.

[17] Wang, P., Yang, S., Shan, L., Niu, R., Shao, X. (2011). Involvements of chloride ion in decolorization of Acid Orange 7 by activated peroxydisulfate or

peroxymonosulfate oxidation. *Journal of environmental sciences, 23*(11), 1799-1807.

[18] Khataee, A. R., Mirzajani, O. (2010). UV/peroxydisulfate oxidation of Cl Basic Blue 3: Modeling of key factors by artificial neural network. *Desalination, 251*(1), 64-69.

[19] Salari, D., Niaei, A., Aber, S., Rasoulifard, M. H. (2009). The photooxidative destruction of Cl Basic Yellow 2 using $UV/S_2O_8^{2-}$ process in a rectangular continuous photoreactor. *Journal of hazardous materials, 166*(1), 61-66.

[20] Zhang, H., Zhang, J., Zhang, C., Liu, F., Zhang, D. (2009). Degradation of Cl acid Orange 7 by the advanced Fenton process in combination with ultrasonic irradiation. *Ultrasonics sonochemistry, 16*(3), 325-330.

[21] Panizza, M., Cerisola, G. (2009). Electro-Fenton degradation of synthetic dyes. *Water research, 43*(2), 339-344.

[22] Özcan, A. S., Erdem, B., Özcan, A. (2004). Adsorption of Acid Blue 193 from aqueous solutions onto Na–bentonite and DTMA–bentonite. *Journal of colloid and Interface science, 280*(1), 44-54.

[23] Özcan, A., Öncü, E. M., Özcan, A. S. (2006). Kinetics, isotherm and thermodynamic studies of adsorption of Acid blue 193 from aqueous solutions onto natural sepiolite. *Colloids and surfaces A: Physicochemical and engineering aspects, 277*(1), 90-97.

[24] Can, M. (2015). Investigation of the factors affecting acid blue 256 adsorption from aqueous solutions onto red pine sawdust: equilibrium, kinetics, process design, and spectroscopic analysis. *Desalination and water treatment*, (DOI: 10.1080/19443994.2014.1003974).

[25] Vijaykumar, M. H., Veeranagouda, Y., Neelakanteshwar, K., Karegoudar, T. B. (2006). Decolorization of 1: 2 metal complex dye Acid blue 193 by a newly isolated fungus, Cladosporium cladosporioides. *World journal of microbiology and biotechnology, 22*(2), 157-162.

[26] Purai, A., & Rattan, V. K. (2012). Biosorption of leather Dye (Acid Blue 193) from aqueous solution using ash prepared from cow dung, mango stone, parthenium leaves and activated carbon. *Indian chemical engineer, 54*(3), 190-209.

[27] Rasoulifard, M. H., Fazli, M., Inanlou, M., Ahmadi, R. (2015). Evaluation of the effectiveness of process in removal trace anthraquinone Cl acid blue 25 from wastewater. *Chemical engineering communications, 202*(4), 467-474.

[28] Shu, H. Y., Chang, M. C., Huang, S. W. (2015). UV irradiation catalyzed persulfate advanced oxidation process for decolorization of Acid Blue 113 wastewater. *Desalination and water treatment, 54*(4-5), 1013-1021.

[29] Shu, H. Y., Chang, M. C., Huang, S. W. (2015). Decolorization and mineralization of azo dye Acid Blue 113 by the UV/Oxone process and optimization of operating parameters. *Desalination and Water Treatment*, (DOI: 10.1080/ 19443994.2015.1031188).

[30] Akay, U., Demirtas, E. A. (2015). Degradation of burazol blue ED by heterogeneous fenton process: simultaneous optimization by central composite design. *Desalination and water treatment, 56*(12), 3346-3356.

[31] House, D. A. (1962). Kinetics and mechanism of oxidations by peroxydisulfate. *Chemical reviews, 62*(3), 185-203.

[32] Maurino, V., Calza, P., Minero, C., Pelizzetti, E., Vincenti, M. (1997). Light-assisted 1, 4-dioxane degradation. *Chemosphere, 35*(11), 2675-2688.

[33] Weiner, E. R. (2000). A Dictionary of inorganic water quality parameters and pollutants. *Applications of environmental chemistry, A practical guide for environmental professionals*, 27.

[34] Lapin, L. L. (1997). *Modern engineering statistics*. Duxbury.

[35] Vining, G. G., Kowalski, S. (2010). *Statistical methods for engineers*. Cengage Learning.

[36] Modirshahla, N., Behnajady, M. A., Ghanbary, F. (2007). Decolorization and mineralization of Cl Acid Yellow 23 by Fenton and photo-Fenton processes. *Dyes and pigments, 73*(3), 305-310.

[37] Chan, T. W., Graham, N. J., Chu, W. (2010). Degradation of iopromide by combined UV irradiation and peroxydisulfate. *Journal of hazardous materials, 181*(1), 508-513.

[38] Oh, S. Y., Kim, H. W., Park, J. M., Park, H. S., Yoon, C. (2009). Oxidation of polyvinyl alcohol by persulfate activated with heat, Fe 2+, and zero-valent iron. *Journal of hazardous materials, 168*(1), 346-351.

[39] Sun, S. P., Li, C. J., Sun, J. H., Shi, S. H., Fan, M. H., Zhou, Q. (2009). Decolorization of an azo dye Orange G in aqueous solution by Fenton oxidation process: Effect of system parameters and kinetic study. *Journal of hazardous materials, 161*(2), 1052-1057.

[40] Lau, T. K., Chu, W., Graham, N. J. (2007). The aqueous degradation of butylated hydroxyanisole by $UV/S_2O_8^{2-}$: study of reaction mechanisms via dimerization and mineralization. *Environmental science & technology, 41*(2), 613-619.

[41] Kasiri, M. B., Khataee, A. R. (2011). Photooxidative decolorization of two organic dyes with different chemical structures by UV/H_2O_2 process: experimental design. *Desalination, 270*(1), 151-159.

[42] Sun, S. P., Li, C. J., Sun, J. H., Shi, S. H., Fan, M. H., Zhou, Q. (2009). Decolorization of an azo dye Orange G in aqueous solution by Fenton oxidation process: Effect of system parameters and kinetic study. *Journal of hazardous materials, 161*(2), 1052-1057.

2

Modeling of esterification in a batch reactor coupled with pervaporation for production of ethyl acetate catalyzed by ion-exchange resins

Araz Tofigh Kouzekonani, Majid Mahdavian*

Department of Chemical Engineering, Quchan University of Advanced Technology, Quchan, Iran

Keywords:
Membrane reactor
Pervaporation
Esterification
Ethyl acetate
Modeling
Green chemistry

ABSTRACT

In the chemical industry, process intensification is needed to meet important goals such as sustainable and eco-friendly processes. For esterification reaction the "produce more with less pollution" objective can be achieved by coupling reaction and separation in a so called integrated process. In this work a model for describing the esterification reaction of ethyl acetate in pervaporation membrane reactor using Amberlyst 15 as a heterogeneous catalyst and polydimethylsiloxane (PDMS) membrane was developed. The validity of the model was tested by comparing the calculated results with experimental data reported in the literature. It was shown rate of conversion increased by removing ethyl acetate from the reaction mixture. A parametric study was carried out to evaluate the effects of operating conditions on the performance of the pervaporation membrane reactor. Conversion increased by increasing the temperature, molar ratios of reactants and catalyst concentration. According to the calculation the best conditions for the operation of the reactor in the event of temperature ~343 K, catalyst concentration 10 g, excess amount of acetic acid relative to ethanol 50% were shown.

Nomenclature

a_i	activity of species i
C	concentration (mol/m³)
E_P	activation energy for pervaporation (kJ/mol)
J	permeation flux (mol/ (m²·h))
K_P	permeability coefficient of species i (mol/ (m² s))
k_1	rate constant of the forward esterification reaction(m³/
k_{-1}	reaction rate constant of the reverse esterification
M	molar mass (kg/mol)
N	chemical quantity (mol)
N_0	initial quantity (mol)
P_{i0}	pre-exponential factor (m/h)
Q_i	molar flow rate of species i in the permeate side (mol/s)
R	universal gas constant (8.314 J/ (mol.K))
r	reaction rate (mol/(kg. s))
S	membrane area (m²)
T	temperature (K)
t	time (h)
V_0	initial volume of reaction mixture (m³)

V	volume of reaction mixture (m³)
ρ	density (kg/m³)
γ	activity coefficient (dimensionless)

1. Introduction

The importance of membrane separation processes is well known. Rarely can a chemical process be found in which the purification of raw materials or separation of products from reaction side products is not required. The coupling of reaction with pervaporation is simple and straight forward. This "membrane reactor" shows important benefits in the green synthesis of esters and in water detoxification. For estrification processes were water is produced along with ester as a by-product during the reaction, pervaporation can be used for separating the ester and water from each other [1]. This earth-friendly method with low pollution and high yield is an alternative

*Corresponding author
E-mail address: m.mahdavian@qiet.ac.ir

to hazardous methods because using pervaporation membrane reactor reduces both waste and energy consumption. Pervaporation processes are known as one of the newest and most important separation processes in the field of membrane separation processes. A simultaneous separation of the reaction products in the reversible processing in that reaction can be returned with increasing the product concentration (according to Le Chatelier_Braun principle) enabling the limitations of existing thermodynamic to be overcome, thus causing an increase in the percentage of conversion reaction [2,3]. Several separation methods are performed by creating a second phase and mass transfer between the two phases. Methods such as distillation and crystallization of these are considered. Creating a second phase requires considerable energy. But in pervaporation the enthalpies of vaporization are provided by the feed mixture heat. Because the reaction and purification can be done together, a smaller device will be needed [4]. Thus, lower power consumption, lower investment and higher efficiency products make pervaporation membrane reactor an interesting alternative to conventional processes. This process has been compared with distillation process and it has been found that 75% reduction in energy input can be achieved and a 50% reduction in investment and operational costs [5]. A modeling study on a pervaporation membrane reactor for the esterification of n-butanol with acetic acid was carried out by Li et al [6]. However, the effects of temperature and concentration of water were taken into full consideration. Hydrophilic membrane was used and it was assumed only water permeated through the membrane. A kinetic model describing the reactor-pervaporation hybrid process has also been studied by David et al [7], who did not consider the effects of temperature and variations in feed concentration in the reaction mixtures on the esterification of propionic acid with propanol. A parametric study of a pervaporation membrane reactor was done by Feng et al [8] and the effect of membrane area, membrane permeability on performance of the system was analyzed. Hasanoglu et al. [9,10] studied on the esterification of acetic acid and ethanol catalyzed by both sulphuric acid and Amberlyst 15 using polydimethylsiloxane (PDMS) membranes. Experiments and simulations were conducted. They showed the performance of reactor increases with increasing temperature and increasing the initial molar ratio of acetic acid to ethanol and indicated the conversion of system in the presence of sulfuric acid is higher than amberlyst 15. In this study, a model on the synergistic mechanisms was developed to investigate integration of reaction with membrane separation processes for production of ethyl acetate catalyzed by ion- exchange resin with polydimethylsiloxane (PDMS) membrane due to many positive features: lower power consumption, lower

investment, higher efficiency products and also amberlyst 15, as heterogeneous catalysts play an important role in the catalytic elimination of environmental pollutants. Equation for describing the permeation flux of mixture components was calculated with the available experimental data. Also, the non-ideal thermodynamic effects of the component in the mixture were taken into account using the UNIFAC method. Finally, the effects of various operating parameters on pervaporation membrane reactor are investigated to achieve optimal operating condition.

2. Mathematical modeling

A view into the industrial design and the optimal operating condition of the membrane pervaporation reactor can be achieved by mathematical modeling. Mathematical models using kinetic parameters of the reaction over Amberlyst 15 and permeation data for a polydimethylsiloxane (PDMS) membrane are developed and effects of various operating parameters are investigated. The model takes into account the non-ideal effect by expressing the reaction and permeation rates in terms of the activities. In this work an Arrhenius equation for expressing the relationship between the permeability coefficients and operating temperature were used which has been gained by experimental data of Ayca Hasanoglu et al [9, 10]. This schematic representation of pervaporation membrane reactor is shown in Figure 1.

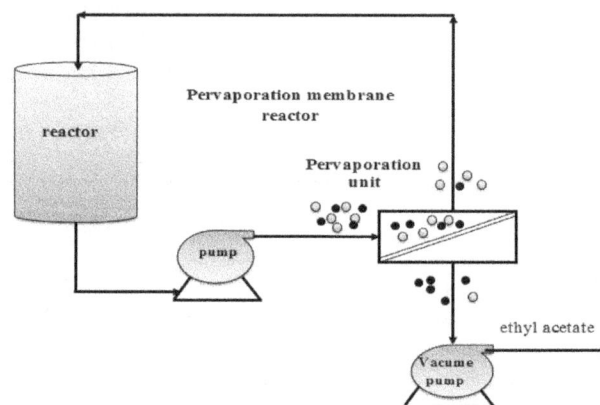

Fig. 1 . schematic diagram of a pervaporation membrane reactor.

2.1. Kinetics of the reaction:

The reaction of the esterification of acetic acid with ethanol can be summarized as follows:

$$CH_3CH_2OH + CH_3COOH \rightarrow CH_3COOCH_2CH_3 + H_2O \qquad (1)$$

Different kinetic models are used to describe esterification reactions catalyzed by ion-exchange resins, in this work to consider non-ideal effect of component and overlook the absorption effect of the components in the reactant medium to simplify kinetic model, the pseudohomogeneous model has been used [11]. The rate model and the kinetic parameters of the reaction over amberlyst15 are expressed as follows:

$$r = \frac{1}{m_{cat}} \frac{1}{v_i} \frac{dn_i}{dt} = -k_1 \left(\alpha_A \alpha_B - \frac{\alpha_C \alpha_D}{K_a} \right) \qquad (2)$$

$$k_e = k^0{}_e * exp \left(\frac{-E_{A,e}}{RT} \right) \qquad (3)$$

$$K_a = \frac{k_1}{k_{-1}} \qquad (4)$$

where m_{cat} is the catalyst mass, v_i the stoichiometric coefficient of the ith component, n_i the number of moles of the ith component, t the time, k_1 the forward reaction rate constant, k_{-1} the backward reaction rate constant, K_a the equilibrium constant a_i the activity of the ith component, T temperature of mixture, $k^0{}_e$ is the pre-exponential factor, $E_{A, e}$ the activation energy and R the gas constant. The kinetic parameters of the reaction over amberlyst-15 are expressed in table 1 The ASOG and UNIFAC procedures are explained in detail in Calvar et al's work [12].

Table 1. Preexponential factors and activation energies of esterification and hydrolysis calculated by ASOG and UNIFAC [12]

	ASOG	UNIFAC
$E_{A,1}$ (kJ mol^{-1})	38.40	28.49
$E_{A,-1}$ (kJ mol^{-1})	35.07	26.70
$k^0{}_{1*} 10^3$ (mol s^{-1}kg^{-1})	161.29	2.811
$k^0{}_{-1*} 10^3$ (mol s^{-1}kg^{-1})	8.34	0.051

2.2. Rates of pervaporation

Hasanoglu et al's experimental data [9, 10] are used to find an accurate equation of component through the Polydimethylsiloxane membrane. Determining diffusion coefficient of species through the membrane is difficult; many parameters can influence it such as temperature, concentration of the component, membrane thickness and so on. By assuming that partial pressures of the component, in permeate side are negligible the following equation is used for describing the component flux through the membrane [13].

$$Q_i = A * kp_i * a_i \qquad (5)$$

Where kp_i is the permeability coefficient, which is related to temperature and a_i is the activity coefficient of each compound. Table 2 summarizes permeability coefficients for the permeation experiments of quaternary mixturesat three temperature levels [9,10].

Table 2. $K_p$experimental permeability coefficients of each component at different temperatures [9, 10]

K_P: Permeability coefficient $g/(m^2 * h)$				
Temperature (K)	Ethyl acetate	Water	ethanol	Acetic acid
323	4963	260	1466	564
333	6495	340	1780	780
343	8157	470	2142	1029

Good relationship between permeability coefficients of each component through the membrane with temperature was obtained with an Arrhenius equation. Ethyl acetate is clearly the dominant component in the total flux because of the high selectivity of the PDMS membrane for ethyl acetate. the expressions are as follows:

$$kp_{ethylnacetate} = 3.23 * 10^7 \, exp(-2840/T) \qquad (6)$$

$$kp_{water} = 8.6 * 10^6 \, exp(-3272/T) \qquad (7)$$

$$kp_{ethanol} = 1.2 * 10^6 \, exp(-2171/T) \qquad (8)$$

$$kp_{acetic \, acid} = 2.95 * 10^7 \, exp(-3428/T) \qquad (9)$$

The variation of permeability coefficient with temperature is shown in Figure 2. Water permeation through the membrane is low, but permeation of ethyl acetate is significantly high. This is due to the tendency of PDMS membrane to ethyl acetate.

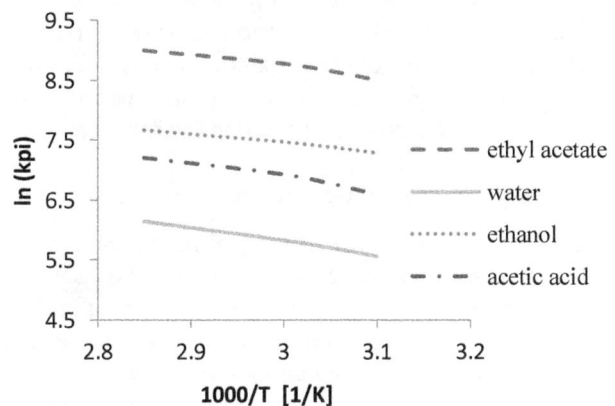

Fig. 2. Arrhenius plot of permeability.

For the non-ideal mixture, concentration term must be corrected to show the departure from the ideal case. Usually a phase model like UNIFAC, UNIQUAC or NRTL models are used to describe this non-ideality factor [14]. In this paper the UNIFAC group contribution method is used to calculate the activity coefficient. The volume, surface, and interaction parameters of all components in the mixture are shown in Tables 3 and 4, data obtained from Dortmund data bank.

Table 3. UNIFAC parameters for the four compounds in the mixture (UNIFAC – VLE Subgroup Parameters)

Main Group	Subgroup	R_k	Q_k
CH_2	CH_3	0.9011	0.8480
CH_2	CH_2	0.6744	0.5400
H_2O	H_2O	0.9200	1.4000
OH	OH	1.0000	1.2000
CCOO	CH_3COO	1.9031	1.7280
COOH	COOH	1.3013	1.2240

Table 4. UNIFAC parameters for the four compounds in the mixture (UNIFAC – VLE Interaction Parameters)

Group	CH_3	CH_2	OH	H_2O	COOC	COOH
CH_3	0.00	0.00	986.5	1318	232.1	507
CH_2	0.00	0.00	986.5	1318	232.1	507
OH	156.4	156.4	0.00	353.5	101.1	267.8
H_2O	300	300	-229.1	0.00	72.87	233.87
COOC	114.8	114.8	245.4	200.8	0.00	-241.8
COOH	329.3	329.3	139.4	124.63	1167	0.00

Activity coefficient changes with time as shown in Figure 3, which shows the non- ideality of the quaternary mixtures. The activity coefficient of ethyl acetate increases as ethyl acetate permeates through the membrane because its content in the mixture decreases during the reaction.

2.3. Modeling of pervaporation membrane reactors

The required equations for modeling of the pervaporation membrane reactor were obtained with mass balance around the reactor by considering production or consumption of reaction components in the reactor and their removal through the pervaporation membrane. The concentration of the reaction components can be written as follows [14]:

$$c_i = c_{io} + reaction - pervaporation$$

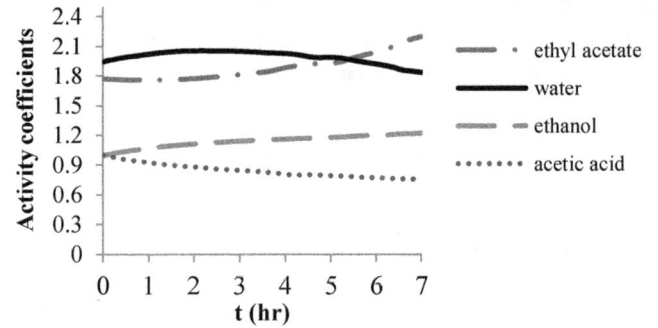

Fig. 3. Activity coefficients of the components in the reaction mixture calculated using the UNIFAC model. Operating conditions: T = 333 K, S/V_0 = 13.25 m⁻¹.

The sets of equations can be summarized as follows:

$$\frac{dN_A}{dt} = c_{cat} * k_1 \left(a_A a_B - \frac{1}{K_a} a_C a_D \right) - S * k p_A \tag{10}$$

$$\frac{dN_B}{dt} = c_{cat} * k_1 \left(a_A a_B - \frac{1}{K_a} a_C a_D \right) - S * k p_B \tag{11}$$

$$\frac{dN_C}{dt} = c_{cat} * k_1 \left(a_A a_B - \frac{1}{K_a} a_C a_D \right) - S * k p_C \tag{12}$$

$$\frac{dN_D}{dt} = c_{cat} * k_1 \left(a_A a_B - \frac{1}{K_a} a_C a_D \right) - S * k p_D \tag{13}$$

$$a_i = x_i * \gamma_i \tag{14}$$

The above equations (Esq. (10) – (14)) are the fundamental equations for describing of modeling of esterification in a batch reactor coupled with pervaporation. The concentration is defined as:

$$C_i = \frac{N_i}{V} \tag{15}$$

$$V == \sum N_i \frac{M_i}{\rho_i} \tag{16}$$

3. Results and discussion

The experimental data and model calculations for the conversion are shown in Fig. 4 (a) and (b) where the dotted line represents the calculated results based on the model of Hasanoglu et al's. [9, 10]. The solid line represents the calculations based on our Work. Both models appear to agree well with the experimental data. However, the reaction conversion is slightly underestimated by the Hasanoglu's model [10] because the non-ideal thermodynamic behavior of the reaction mixtures was not considered in their work.

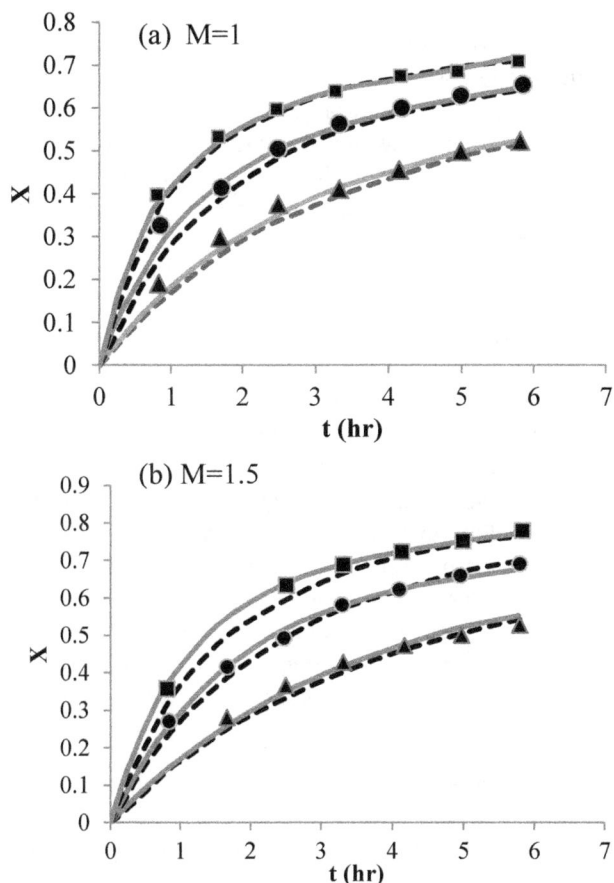

Fig. 4. Conversion versus time: a comparison of model calculations with experimental data with pervaporation (PV) catalyzed by Amberlyst 15 for initial molar ratios, M, of (a) 1 and (b) 1.5 (■:Experimental data 70°C, ●: Experimental data 60°C, ▲: Experimental data 50°C, line 1: 70°C Results of this model, line 2: 70°C Results of Hasanoglu's model [9,10], line 3: 60°C Results of this model, line 4: 60°C Results of Hasanoglu's model [9,10], line 5: 50°C Results of this model, line 6: 50°C Results of Hasanoglu's model [9,10].

Figures 4 (a) and (b) show the effect of initial mole ratio of acetic acid to ethanol, M, on the performance of the pervaporation membrane reactor for esterification of acetic acid with ethanol in the presence of Amberlyst 15 as catalyst under the following conditions: S/V_0 = 13.25 m^{-1}, C_{cat} =2.6g catalyst/100 g acetic acid (g l^{-1}). M was changed from 1 to 1.5 for this reaction. Production was increased by increasing M. An excessive amount of one reactant by progressing the forward reaction leads to an enhancement in conversions. Also a higher yield of ethyl acetate in the reactor was obtained because ethyl acetate formation is increased with the use of an excessive amount of reactant. This caused an increase in permeation flux of ethyl acetate because of the relatively high ethyl acetate concentration at a higher M. However, if this ratio is high the purification of ethyl acetate will be challenged due to the extra volume of residual acetic acid at the end of the reaction. Appropriate value of M is considered 1.5 according to the

experimental work. Thus, in the pervaporation membrane reactor increasing the value of M improves the reactor performance in terms of conversion. Also, it has been shown that a rise in the temperature increases the esterification reaction, because a higher temperature is suitable for removing ethyl acetate through the membrane and increasing the rate of reaction.

Figure 5 shows the effect of increasing temperature on the performance of the pervaporation membrane reactor for esterification of acetic acid with ethanol in the presence of Amberlyst 15 as catalyst under the following conditions: S/V_0 = 13.25 m^{-1}, C_{cat} =2.6g catalyst/100 g acetic acid (g l^{-1}).

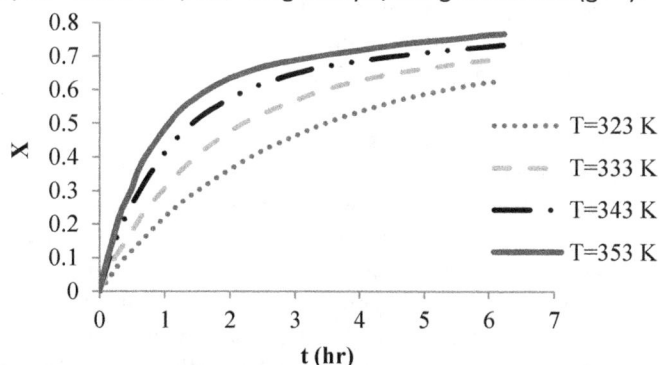

Fig. 5. Effect of temperature on the esterification performance of the pervaporation-coupled reactor. Operating conditions: m = 1.5, S/V_0 = 13.25 m^{-1},C_{cat} = 10 g

The calculation results in Fig. 5 show that at temperatures above 343 K, a further increase in temperature will not enhance the reaction conversion considerably. So, according to the operating costs in connection with heating and cooling of component, a convenient temperature of around 343 K seems suitable for esterification coupled with pervaporation.

Figure 6 shows the effect of catalyst concentration on the esterification performance of the pervaporation membrane reactor. Under the following conditions: S/V_0 = 13.25 m^{-1}, M = 1.5, and T = 343 K.

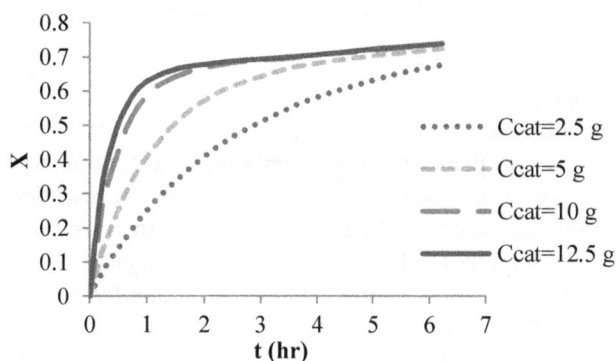

Fig. 6. Effect of catalyst concentration on the esterification performance of the pervaporation -coupled reactor. Operating conditions: S/V_0 = 13.25 m^{-1}, M = 1.5, T = 343 K.

As expected, enhancement of the catalyst concentration promotes the formation of ethyl acetate and also of water in the reactor; an increase in ethyl acetate content increases the rate of ethyl acetate removal. Maximum conversion of acetic acid (X_A) occurs within the first 2 h of the reaction. There is little change in the product yield when the catalyst concentration is above 10 g for a reaction time of 7 h under the conditions studied. This should be considered when selecting a suitable amount of catalyst concentration.

Figure 7 illustrates the effect of S/V_0 on the esterification performance in the pervaporation-coupled reactor under the following conditions: M = 1, C_{cat} = 10 g and T = 323 K. For a given reactor volume, enhancement of the membrane area will increase the rate of ethyl acetate removal and the reaction equilibrium will shift toward the product side, because as soon as ethyl acetate formatted continuously is removed through the membrane, the reactant concentration will increase resulting in an enhancement in the reaction rate. However, selecting a suitable S/V_0 is necessary because an enhancement in surface increases the operating cost. So for selecting the appropriate value of S/V_0, the gain in the enhanced conversion versus the cost associated with the membrane should be taken into account.

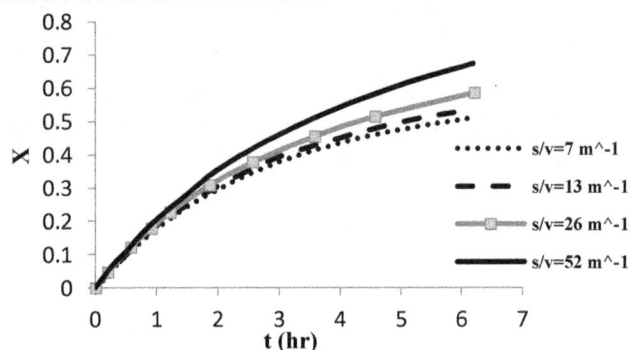

Fig. 7. Effect of (S/V_0) on the esterification performance of the pervaporation - coupled reactor.
Operating conditions: M = 1, C_{cat} = 10 g, T = 323 K.

4. Conclusions

Simulation and selecting the optimal conditions for modeling of pervaporation membrane reactors is very important. The model of estrification of acetic acid with ethanol, catalyzed by Amberlyst 15 has been developed. This model takes into account the non-ideality of the reaction mixtures. It is shown that removal of ethyl acetate through the membrane from the mixture increased the conversion of reaction, because in reversible reactions necessary potential synergies are provided for overcoming the equilibrium limitations. Also Amberlyst 15 as heterogeneous catalyst plays an important role in the catalytic elimination of environmental pollutants and production of clean energy. The development of ion-

exchange resins would definitely contribute to the environmental improvement and efficient utilization of energy resources. The model validation is done with available experimental data. Results showed that, there is good agreement between the model and experimental data. The effect of different parameters such as temperature, catalyst concentration and excess amount of acetic acid relative to ethanol on the reaction conversion rate was investigated and, according to the calculation of the model, the best conditions for the operation of the reactor in the event of temperature ~343 K, catalyst concentration 10 g, excess amount of acetic acid relative to ethanol 50% were shown. Also, with the use of pervaporation membrane reactor, it is possible to achieve 100% conversion. Thus, the prevention of waste can be achieved if the majority of reagents and the solvent are recyclable. For example, ion exchange resin can be regenerated (if needed) and reused in a subsequent run. So, using pervaporation membrane reactor coupled with ion exchange resin can provide the conditions to move towards green chemistry. This method would also help balance environmental concerns with economic development.

Acknowledgements

The authors would like to acknowledge the support and help of 4 th refinery of South Pars Gas Company (SPGC).

References

[1] Lipnizki, F., Field, R. W., Ten, P. K. (1999). Pervaporation-based hybrid process: a review of process design, applications and economics. *Journal of membrane science*, *153*(2), 183-210.

[2] Lipnizki, F., Field, R. W. (2001). Pervaporation-based hybrid processes in treating phenolic wastewater: technical aspects and cost engineering. *Separation science and technology*, *36*(15), 3311-3335.

[3] Datta, R., Tsai, S. P. (1998). *Esterification of fermentation-derived acids via pervaporation* (No. US 5723639). Argonne national laboratory (ANL), Argonne, iL.

[4] Dutta, B. K., Sikdar, S. K. (1991). Separation of azeotropic organic liquid mixtures by pervaporation. *AIChE journal*, *37* (4), 581-588.

[5] Waldburger, R. M., Widmer, F. (1996). Membrane reactors in chemical production processes and the application to the pervaporation-assisted esterification. *Chemical engineering technology*, *19*(2), 117-126.

[6] Li, X., Wang, L. J. (2001). The esterification of acetic acid with n-butanol for the production of n-butyl acetate, Membrane science, 186, 19–24.

[7] David, M. O., Nguyen, T. Q., Neel, J. (1991). Pervaporation-esterification coupling. II: Modelling of the influence of different operating parameters.

Chemical engineering *research & design*, *69*(A4), 341-346.

[8] Feng, X., & Huang, R. Y. (1996). Studies of a membrane reactor: esterification facilitated by pervaporation. *Chemical engineering science*, *51*(20), 4673-4679.

[9] Hasanoğlu, A., Salt, Y., Keleşer, S., Dinçer, S. (2009). The esterification of acetic acid with ethanol in a pervaporation membrane reactor. *Desalination*, *245*(1), 662-669.

[10] Hasanoğlu, A., Dinçer, S. (2011). Modelling of a pervaporation membrane reactor during esterification reaction coupled with separation to produce ethyl acetate. *Desalination and water treatment*, *35*(1-3), 286-294.

[11] Yixin, Q., Shaojun, P. E. N. G., Shui, W., Zhang, Z., & Jidong, W. A. N. G. (2009). Kinetic study of esterification of lactic acid with isobutanol and n-butanol catalyzed by ion-exchange resins. *Chinese journal of chemical engineering*, *17*(5), 773-780.

[12] Calvar, N., Gonzalez, B., Dominguez, A. (2007). Esterification of acetic acid with ethanol: Reaction kinetics and operation in a packed bed reactive distillation column. *Chemical engineering and processing: Process Intensification*, *46*(12), 1317-1323.

[13] Assabumrungrat, S., Phongpatthanapanich, J., Praserthdam, P., Tagawa, T., Goto, S. (2003). Theoretical study on the synthesis of methyl acetate from methanol and acetic acid in pervaporation membrane reactors: effect of continuous-flow modes. *Chemical engineering journal*, *95*(1), 57-65.

[14] Aage, F., Petter, R. (1977). Vapour-liquid equilibria using UNIFAC a Group-contribution method.

[15] Domingues, L., Recasens, F., Larrayoz, M. (1999). Studies of a pervaporation reactor: kinetics and equilibrium shift in benzyl alcohol acetylation. *Chemical engineering science*, *54*(10), 1461-1465.

3

Kinetic study of Pb (II) and Ni (II) adsorption onto MCM-41 amine-functionalized nano particle

Masoumeh Ghorbani*, **Seyyed Mostafa Nowee**
Department of Chemical Engineering, Ferdowsi University of Mashhad, Mashhad, Iran

Keywords:
Pb (II) and Ni (II) adsorption
MCM-41/TMSPDETA
Adsorption kinetic
Pseudo-second order model

ABSTRACT

In the current investigation a novel nano hybrid adsorbent MCM-41/N-(3-trimethoxysilyl)-propyl) diethylenetriamine (MCM-41/TMSPDETA) was prepared and was characterized using DLS (Dynamic Light Scattering), Fourier Transform Infrared (FTIR), X-ray diffraction (XRD), Brunauer–Emmett–Teller (BET) analytical techniques and Transmission electron microscopy (TEM). The synthesized MCM-41/TMSPDETA adsorbent possessed high surface area (867 m^2g^{-1}), narrow pore size distribution (3.6 nm) and pore volume (0.782 cm^3g^{-1}). The nano hybrid adsorbent was applied in batch experiments under different controlling factors by varying pH, contact time and solution temperature of Lead (Pb(II)) and Nickel (Ni(II)) ions. Optimum conditions obtained were 20°C, pH=6 and contact time of 120 min. The maximum capacity of the nano-sorbent was obtained to be 58.823 and 20.921 $mg\,g^{-1}$ for Pb (II) and Ni (II) ions for an initial concentration range 10-70 mgL^{-1}. Pseudo-first order, pseudo-second order and intraparticle diffusion models were used to analyze the kinetic data. Results showed that the pseudo-second order model can well describe the adsorption kinetic data.

1. Introduction

In recent years, Heavy metals pollution has attracted a great deal of attention, especially Lead and Nickel as serious environmental threats. Lead and Nickel pollution results from battery manufacturing, textile dying, petroleum refining, ceramic and glass industries and mining operations [1]. Lead and Nickel may cause retardation, mental disturbance and semi-permanent brain damage [2]. Therefore, the development of clean-up technologies for removing heavy metals from wastewaters is very important.

The removal of toxic metals from wastewaters can be carried out by a number of separation technologies, such as chemical precipitation [3], membrane processes [4], solvent extraction, ion exchange [5], floatation, coagulation and sorption process [6]. Among these methods, sorption is currently considered to be very suitable for wastewater treatment because of its high efficiency, simple operation and cost effectiveness [7]. A number of effective adsorbents have been prepared and reported in recent years [8]. Mesoporous silica materials have received considerable attention due to their unique large surface area, well-defined pore size and pore shape. One of these materials, MCM-41, consists of hexagonal arrays of large and uniform pore size, large surface area, thermal stability and mild acidic property [9]. Mesoporous silica MCM-41 has been functionalized and employed to eliminate traces of toxic heavy metal from wastewater [10]. Amino functional mesoporous silica MCM-41 materials have been prepared to develop efficient adsorbents of heavy metals in wastewater.

The literature studies on the removal efficiency of functionalized mesoporous silica for Pb (II) and Ni (II) are limited. Adsorption of Pb (II) ion with nano MCM-41/TMSPDETA has not been reported.

The structure of the prepared nano adsorbents surface was characterized using DLS (Dynamic Light Scattering), Fourier Transform Infrared (FTIR), X-ray diffraction (XRD), Brunauer–Emmett–Teller (BET) and Transmission electron microscopy (TEM) analytical techniques. The goal of the present paper was the sorption of Pb (II) and Ni (II) on MCM-41/TMSPDETA nano adsorbent. The influence of pH, contact time and solution temperature on the sorption

* Corresponding author
E-mail: Ghorbani.masoumeh@gmail.com

process were investigated. Also, kinetic models (Pseudo-first order, pseudo-second order and intraparticle diffusion) were established.

2. Material and Methods

2.1. Materials

Hexadecyltrimethylammoniumbromide (CTAB); Tetraethoxysilane (TEOS), N-(3-trimethoxysilyl)-propyl) diethylenetriamine (TMSPDETA) were purchased from Aldrich Co. CH_3OH, HCl and $NaHNO_3$ were provided by Merck Co. Sources of metal for sorption experiments were $Pb(NO_3)_2$ and $Ni(NO_3)_2$ of synthesis grade that were supplied from Merck Co. Heavy metal solution (Pb^{2+}) was prepared by dissolving weighed amount of uranyl nitrate (Aldrich) in deionized water. Deionized water was utilized throughout this experiment. Infrared spectra were recorded on a Fourier transform infrared (FT-IR) spectrometer (Shimadzo, FTIR1650 spectrophotometer, Japan) in the range of 400–4000cm^{-1} using spectroscopic quality KBr powder (sample/KBr = 1/100). The residual concentrations of Pb (II) and Ni (II) ions were determined by an inductivity coupled plasma atomic emission spectrophotometer (Shimadzu AA-670).

2.2. Amino-functionalized MCM-41 via direct co-condensation method

The synthesis of functionalized MCM-41 with N silane via the co-condensation method was conducted as follows. CTAB and NaOH were dissolved in deionized water while heating. Afterwards the Si sources (TEOS) were added to the above mixture. Then N-(3-trimethoxysilyl)- propyl) diethylenetriamine (TMSPDETA) was added to the mixture and the solution was stirred 300 rpm for 4 h. The surfactant in the as-synthesized sample was removed by a solvent extraction method; 1.0 g of the as-synthesized sample was stirred in a HCl/MtOH solution (100 ml of methanol containing 5 ml HCl aq. Concentration ca. 35 wt%) at 298 K for 24 h.

2.3. Adsorption experiments

Adsorption experiments were conducted in a batch way. Each experiment was carried out by placing specific amount of nano-adsorbent of MCM-41/ TMSPDETA in aqueous solutions under various operating conditions. The initial Pb (II) and Ni (II) ions concentration of 30 mg/L, contact time of 2h, and adsorbent dose values were varied between 1 and 5 g/l. The effect of contact time was investigated by varying the time from 5 to 120 min, at a temperature of 25 °C. The sorption capacity for heavy metal ions was determined as follows:

$$q_e = \frac{(C_0 - C_e)V}{m} \tag{1}$$

Where C_0 and C_e (mg/L) are the initial and final metal ion concentrations, respectively; V is the volume of solution (L) and m is the weight of adsorbent (g).

3. Results and discussion

3.1. Adsorbent characterization

From DLS (Dynamic Light Scattering), the hydrodynamic particle size of the silica functionalization with amine group, MCM-41/TMSPDETA was estimated to be approximately 627.2nm (Figure 1).

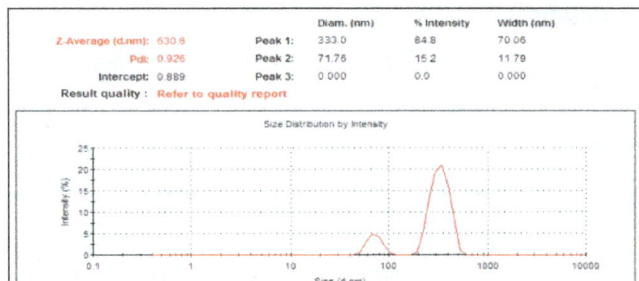

Figure 1. DLS (Dynamic Light Scattering) of the hydrodynamic size.

Infrared technique was used for identification of nano adsorbent. After functionalization with amine group, MCM-41/TMSPDETA showed visible broad absorption bands at 1560cm^{-1} and at 1650cm^{-1} corresponding to the bending vibration of N–H group, while N–H stretching (3200–3500 cm^{-1}) and C–N stretching (1030–1230cm^{-1}) overlap with the broad absorption bands of the silanol group and the Si–O–Si vibrations (Figure 2).

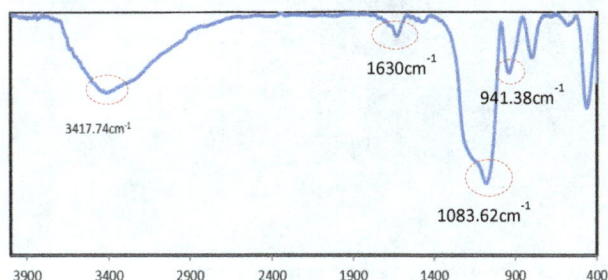

Figure 2. FT-IR spectra of MCM-41/ TMSPDETA.

The XRD pattern of MCM-41/TMSPDETA is shown in Figure 3. As can be seen, the broad peaks indicate the nano-crystalline nature of the particles.

In Table 1. the adsorbed nitrogen volume indicated the mesoporous silica materials functionalized with TMSPDETA have more sorption nitrogen volume, which is attributable to the presence of lower amino groups. The specific surface area of MCM-41 decreased from 1063 to 867 m^2/g when it was functionalized with amine group41/TMSPDETA

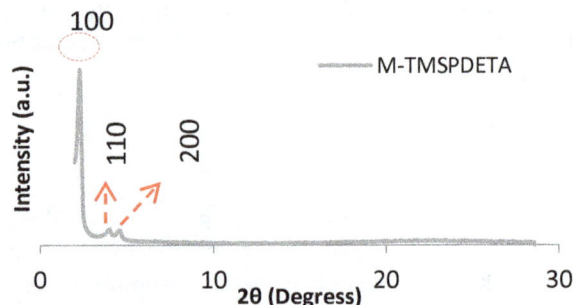

Figure 3. X-ray diffraction patterns of MCM-41 and MCM-41/TMSPDETA.

Table 1. Physicochemical properties of extracted-MCM-41 and MCM-41/TMSPDETA.

Samples	S_{BET} (m²/g)	V_p (cm³/g)	d_{BJH} (nm)	$d_{avg.}$ (nm)
MCM-41	1063	1.082	3.88	4.03
MCM-41/TMSPDETA	867	0.782	3.61	3.81

The determined-field TEM image at high magnification of the MCM-41/TMSPDETA showed a lamellar mesostructure with a well-defined hexagonal arrangement of uniform pores. The pore size was estimated to be ~3.4 nm (Figure 4).

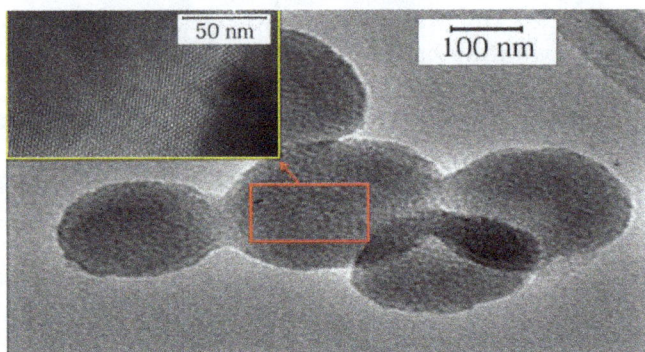

Figure 4. Transmission electron microscopy image of TMSPDETA/MCM-41.

3.2. Effect of contact time

The effect of contact time on the sorption of Pb (II) and Ni (II) ions onto the MCM-41/TMSPDETA nano hybrid adsorbent was indicated in Figure 5. As can be seen, a major portion of the total lead adsorption (i.e. more than 80%) was achieved within 10 min. The intial sorption rate of Ni (II) species has reached the equilibrium about 60 min after the beginning of stirring and remained constant until the end of the experimentation.

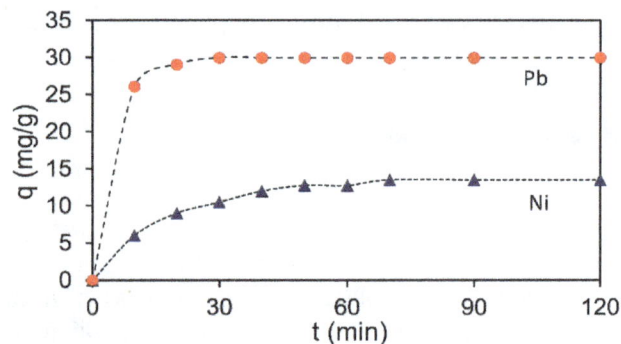

Figure 5. The effect of contact time on the uptake of Pb (II) and Ni (II).

3.3. Effect of temperature

In order to investigate the effect of temperature on the sorption capacity of Pb (II) and Ni (II) ions onto MCM-41/TMSPDETA nano hybrid adsorbent, the experiments were performed at different temperatures (20–50°C) and the temperature impact on Ni (II) sorption was less than Pb (II). The variation in the adsorption might have been the result of an increased tendency of Pb (II) and Ni (II) species to escape from the sorbent surface as the temperature of the solution rises [11]. Notably, for both cases the sorption capacity decreased with temperature and it was concluded that sorption of Pb (II) and Ni (II) species onto the MCM-41/TMSPDETA sorbent was an exothermic process.

3.4. Sorption kinetics

The experimental data were analyzed using three kinetic models including pseudo-first-order, pseudo-second-order and intraparticle diffusion.

Equations studied are defined as [12]:

$$\ln\left(q_e - q_t\right) = \ln q_e - k_1 t \tag{2}$$

$$\frac{t}{q_t} = \frac{1}{k_2 q_e^2} + \frac{t}{q_e} \tag{3}$$

$$q_t = k_{ipd} t^{1/2} + c \tag{4}$$

The parameters including rate constants k_1, k_2, k_{ipd}, c and correlation coefficients were calculated and the results are listed in Table 2.

This indicated that the pseudo-first order kinetic model coincided poorly to the adsorption processes of MCM-41/TMSPDETA for both Pb (II) and Ni (II) ions. However, in case of pseudo-second-order kinetic equation, the correlation coefficients were found to be greater than 0.99, and the corresponding equilibrium adsorption capacities fit well with the experimental data.

Table 1. K_p experimental permeability coefficients of each component at different temperatures [9, 10].

K_P: Permeability coefficient $g/(m^2*h)$				
Temperature (K)	Ethyl acetate	Water	ethanol	Acetic acid
323	4963	260	1466	564
333	6495	340	1780	780
343	8157	470	2142	1029

The R^2 values of intraparticle diffusion model for both heavy metals were estimated much less than pseudo-second-order model which signified that diffusion was not the rate controlling step of the sorption process.

4. Conclusion

A novel MCM-41/TMSPDETA nano adsorbent was used for sorption of Ni (II) and Pb (II) ions from aqueous solutions. The FTIR spectra indicated that MCM-41/TMSPDETA nano adsorbent was functionalized by amine group. The kinetic data was best fitted by pseudo-second-order model kinetic with high regression coefficient ($R^2>0.99$). Contact time of 2h, temperature of 25°C and pH=6, respectively were optimum conditions determined for sorption process in a batch system.

References

[1] Freitas, O. M., Martins, R. J., Delerue-Matos, C. M., Boaventura, R. A. (2008). Removal of Cd (II), Zn (II) and Pb (II) from aqueous solutions by brown marine macro algae: kinetic modelling. *Journal of hazardous materials*, *153*(1), 493-501.

[2] Paulino, A. T., Santos, L. B., Nozaki, J. (2008). Removal of Pb 2+, Cu 2+, and Fe 3+ from battery manufacture wastewater by chitosan produced from silkworm chrysalides as a low-cost adsorbent. *Reactive and functional polymers*, *68*(2), 634-642.

[3] Mellah, A., Chegrouche, S., Barkat, M. (2007). The precipitation of ammonium uranyl carbonate (AUC): thermodynamic and kinetic investigations. *Hydrometallurgy*, *85*(2), 163-171.

[4] Kornilovich, B. Y., Koval'chuk, I. A., Pshinko, G. N., Tsapyuk, E. A., Krivoruchko, A. P. (2000). Water Treatment and Demineralization Technology-Water purification of uranium by the method of ultrafiltration. *Journal of Water chemistry and technology*, *22*(1), 43-47.

[5] Shoushtari, A. M., Zargaran, M., Abdouss, M. (2006). Preparation and characterization of high efficiency ion-exchange crosslinked acrylic fibers. *Journal of applied polymer science*, *101*(4), 2202-2209.

[6] de Pablo, L., Chávez, M. L., Abatal, M. (2011). Adsorption of heavy metals in acid to alkaline environments by montmorillonite and Ca-montmorillonite. *Chemical engineering journal*, *171*(3), 1276-1286.

[7] Babel, S., Kurniawan, T. A. (2003). Low-cost adsorbents for heavy metals uptake from contaminated water: a review. *Journal of hazardous materials*,*97*(1), 219-243.

[8] Hawari, A. H., Mulligan, C. N. (2006). Biosorption of lead (II), cadmium (II), copper (II) and nickel (II) by anaerobic granular biomass. *Bioresource technology*, *97*(4), 692-700.

[9] Zolfaghari, G., Esmaili-Sari, A., Anbia, M., Younesi, H., Amirmahmoodi, S., Ghafari-Nazari, A. (2011). Taguchi optimization approach for Pb (II) and Hg (II) removal from aqueous solutions using modified mesoporous carbon.*Journal of hazardous materials*, *192*(3), 1046-1055.

[10] Thomas, J. M. (1994). The chemistry of crystalline sponges. *Nature*, *368*(6469), 289-290.

[11] Zhou, L., Wang, Y., Liu, Z., Huang, Q. (2009). Characteristics of equilibrium, kinetics studies for adsorption of Hg (II), Cu (II), and Ni (II) ions by thiourea-modified magnetic chitosan microspheres. *Journal of hazardous materials*, *161*(2), 995-1002.

[12] Alkan, M., Demirbaş, Ö., Doğan, M. (2007). Adsorption kinetics and thermodynamics of an anionic dye onto sepiolite. *Microporous and mesoporous materials*, *101*(3), 388-396.

4

Efficient treatment of baker's yeast wastewater using aerobic membrane bioreactor

Mohammad J. Nosratpour[1], Morteza Sadeghi[1,*], Keikhosro Karimi[1,2], Saied Ghesmati[1]

[1]Department of Chemical Engineering, Isfahan University of Technology, Isfahan 84156-83111, Iran
[2]Industrial Biotechnology Group, Institute of Biotechnology and Bioengineering, Isfahan University of Technology, Isfahan 84156-83111, Iran

Keywords:
Baker's yeast wastewater
Hollow fiber membrane
Membrane bioreactor
Membrane fouling
Wastewater treatment

ABSTRACT

A 0.15 µm dead-end immersed hollow fiber membrane and filamentous fungus *Aspergillus oryzae* were used in a membrane bioreactor (MBR) for treatment of baker's yeast wastewater. The fungus was adapted to the wastewater in the bioreactor for two weeks before continuous process. Average organic loading rate of 4.2 kg COD/m^3.d was entered the bioreactor. COD and BOD$_5$ of the wastewater were reduced to 488 and 70 mg/L, respectively, over a period of 45 days, while the turbidity of the wastewater reduced from 134-282 NTU to less than 2.5 NTU in the permeate stream. Critical flux and a suitable operating flux were determined as 6.7 and 5 L/m^2 h, respectively. The system was able to efficiently reduce the turbidity and suspended solid by 99.4% and 98.3%, respectively, resulting in a clear effluent.

1. Introduction

Baker's yeast is nowadays widely produced and used in breading of bread industries worldwide [1]. The wastewater of the baker's yeast factory is one of the high strength wastewaters that contain high concentrations of slowly biodegradable organic materials [2]. High load of organic compounds such as sugars, carbohydrates, and fermented products is one of the main characteristics of the wastewater [3, 4]. Molasses, a by-product of sugar manufacturing factories, is the main raw material for baker's yeast production [5]. It contains 45-50% residual sugars that is easily consumed by yeast, while the rest of molasses is non sugary compounds (15-20% non-sugar organic substances, 10-15% ash that is mineral compounds and around 20% water) [4]. A major part of non-sugary compounds are not converted by yeast and released to the wastewater [3]. Thus, the wastewater contains the non-sugar compounds, yeast metabolites, and chemicals added during the fermentation [3]. Antioxidant properties of melanoidines cause their recalcitrance to biodegradation [5]. Various processes such as carbon adsorption [6], electrocoagulation [7] and biological degradation with ozonation [8] have been investigated on molasses wastewater.

Membrane technology, nowadays, are used in many water and wastewater treatment processes [9, 10]. Membrane bioreactor is one of the efficient technologies for wastewater treatment that received significant attentions in the last two decades [11-15]. The combination of common activated sludge wastewater treatment and physical filtration by membrane called membrane bioreactor. Unlike common biological treatments, membrane bioreactors do not need a downstream stage [16]. Membrane bioreactor has some advantages in comparison with common biological treatment, e.g., less sludge production, less footprint, higher quality of effluent, treatment with different hydraulic retention and sludge retention times, and high performance in disinfection [17, 18].

The purpose of this study is to investigate treatment of baker's yeast wastewater by *Aspergillus orayze* in a membrane bioreactor (MBR). *A. orayze* is a non-pathogenic

*Corresponding author
E-mail address: m-sadeghi@cc.iut.ac.ir

fungus widely used in many food industry wastewater treatment processes because of appropriate cultural status and absence of harmful by-product [19, 20]. During 45 days of experiment changes in mixed liquor suspended solid (MLSS), chemical oxygen demand (COD), biological oxygen demand (BOD) and turbidity was investigated, also membrane operational factors such as, critical flux and membrane resistance distribution was determined.

2. Material and methods

A baker's yeast wastewater was obtained from Shahr-e-kord baker's yeast factory (Shahrekord, Iran). Main characteristics of wastewater are given in Table 1.

Table 1. Wastewater characteristics

Parameter	Unit	Value
COD	mg/L	2550-5070
BOD$_5$	mg/L	460-1460
Turbidity	NTU	134-282
SS	mg/L	500-1200
pH	-	5.4-6.7

A membrane bioreactor (Plexiglas) with 50 cm long, 15 cm wide, 50 cm high, and 30 L working volume was used in all treatments (Pars Polymer, Isfahan, Iran). The wastewater was continuously fed to the reactor using a diaphragm pump (ROUltraTec, USA). Continuous aeration was provided by an air pump (Aco-208, Hailea, china) through the spargers located at the bottom of the reactor. A circular sparger was contrived exactly at the beneath of the membrane module (Fig. 1).

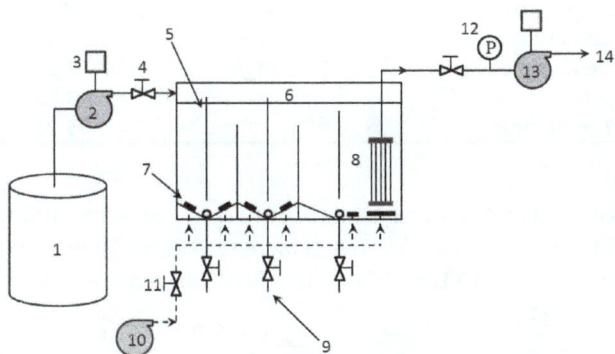

Fig. 1. Experimental membrane reactor setup. The numbers on the figure indicate: (1) feed tank, (2) feed pump, (3) feed pump controller, (4) feed valve, (5) Compartment wall, (6) bioreactor, (7) sparger, (8) membrane module, (9) sludge drain valve, (10) aeration pump, (11) air regulation valve, (12) pressure indicator, (13) effluent pump, (14) permeate stream

Besides providing the oxygen for the system, aeration had a role in agitation and cleaning of the membrane surface [21, 22]. The bottom of reactor was designed gradient to ease the sludge discharge. Hollow fiber, dead-end membrane module was used in the MBR that submerged in the bioreactor. Membrane characteristics are presented in Table 2.

A peristaltic pump was used for MBR effluent (Etaron DS, Italy). After each 5.5 minute of operation, effluent suction pump turned off for 30 seconds. This have done to mitigate cake layer formation and membrane fouling. In case of drastic fouling that occurred time to time, membrane module was removed from the reactor and physically washed by tap water for few minutes. Physical washing effectively removed the cake layer from membrane surface. Occasionally when physical washing was not enough, membrane was soaked in 8% solution of NaOCl for 4 h. After chemical washing, permeability of membrane was mostly recovered. Membrane before and after chemical washing are demonstrated in Fig. 2.

Table 2. Membrane module characteristics

Membrane substance	polypropylene
Fiber length(cm)	12.5
Fiber outer diameter(mm)	0.35
Fiber inner diameter(mm)	0.25
Mean pore size(μm)	0.15
Membrane total area(m^2)	0.3

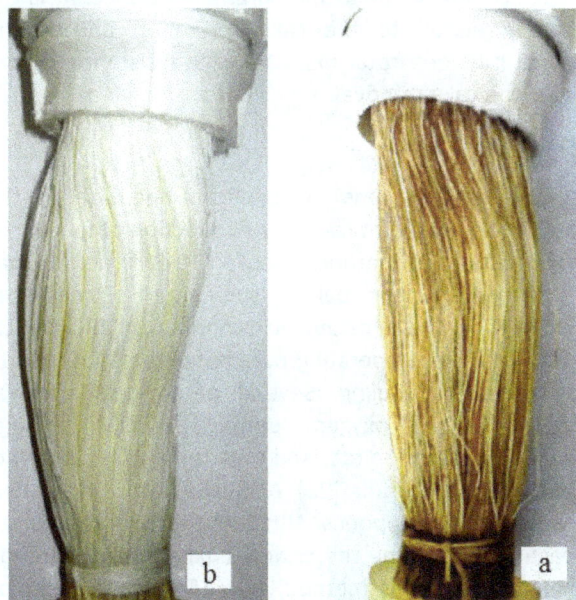

Fig. 2. Membrane before (a) and after (b) chemical washing

2.1. Micro-organism preparation

Filamentous fungus *Aspergillus orayze* PTCC 5163 purchased from Persian Type Culture Collection was used in all treatments. The fungi was cultivated on a solid medium containing 20 g/L malt extract, 20 g/L glucose, 1 g/L peptone and 20 g/L agar. Then, the fungal biomass was prepared in a liquid medium containing 50 g/L glucose for 4 days. The fungus was adapted to the wastewater for two weeks. In the beginning of adaptation, glucose was given to microorganism as a carbon source. During the adaptation, the carbon source of the feed was gradually substituted by the wastewater in a two weeks process.

2.2. Factors affecting the membrane flux

Flux of filtration, an important factor that determines the performance of filtration, depends on different factors such as operational conditions and membrane fouling. Factors affecting the fouling are membrane characteristics such as material, hydrophobicity, and porosity, biomass characteristics such as mixed liquor suspended solid (MLSS), soluble microbial product (SMP), extracellular polymeric substances (EPS), floc size and structure, and operating conditions such as configuration, solid retention time (SRT), hydraulic retention time (HRT), and trans-membrane pressure (TMP) [23-26]. Equation 1 shows the relationship between TMP and flux according to Darcy's law [17]:

$$J = \frac{TMP}{\mu R_t} \tag{1}$$

$$R_t = R_m + R_c + R_f \tag{2}$$

Where R_m is the membrane resistance when the membrane is clean. R_m is a constant depending on the membrane characteristics. R_f is the fouling resistance that is resistance after cake removing minus membrane resistance. R_c is the parameter related to the cake formation and is the difference between total mass transfer resistance before and after the cake removal.

2.3. Critical flux

Membrane fouling leads to significant increase in the hydraulic resistance that causes the reduction of flux in the case of constant TMP or increment in TMP if the flux was kept constant. Principally, submerged membrane bioreactor works in constant flux conditions [27]. Critical flux is an important operating parameter that can help to avoid membrane fouling. Several parameters such as soluble chemical oxygen demand (COD), sludge concentration, aeration rate and even initial flux can affect critical flux of membrane [28]. At fluxes below the critical flux, fouling do not happens. Although there is no accurate approach to determine the critical flux, step flux method can be useful. In this method, two TMP value must be measured that is TMP after sudden increase in flux (TMP$_i$) and TMP at the end of step (TMP$_f$). TMP$_i$ usually measured after 30 seconds from flux change. Initial TMP increment (ΔP_0) represents difference in TMP caused by step increase in flux [29].

$$\Delta P_0 = TMP_i^n - TMP_i^{n-1} \tag{3}$$

Rate of TMP increase and average TMP presented in equations 4 and 5, respectively [29]

$$\frac{dP}{dt} = \frac{TMP_f^n - TMP_i^n}{t_f^n - t_i^n} \tag{4}$$

$$P_{ave} = \frac{TMP_f^n + TMP_i^n}{2} \tag{5}$$

Permeability can be calculated as [29]

$$K = \frac{J}{P_{ave}} \tag{6}$$

2.4. Sampling and analytical methods

Duration the experiments, liquid samples were taken daily from the bioreactor influent, mixed liquor, and membrane effluent at the same time. All samples were taken at least twice and the averages of the results were reported. COD, BOD, turbidity, TSS, and MLSS concentrations were analyzed. COD was determined according to standard methods [30] using a COD reactor (HACH, Germany) and a UV-Vis spectrophotometer (6305, JENWEY, UK). Turbidity of the samples was analyzed using a turbidimeter (AN-2100, HACH, Germany). For DO analysis, a DO meter was used (YSI-55, YSI, USA). Hydraulic retention time (HRT) was calculated theoretically. Dissolved oxygen concentration stabilized in 2 to 4 mg/L. Table 3 represents other operating conditions.

Table 3. MBR operating conditions

Factor	Unit	Value
HRT	h	22
MLSS	mg/L	4100-9100
Organic mean load	kg COD/m³.day	4.2
Permeate flux	l/m².h	5
Effluent rate	l/h	1.5
Dissolved oxygen	mg/L	1.7-3.7
pH	-	7.3-8.4
Temperature	⁰C	25-28

Step flux method was used for critical flux determination. Initial flux was 3.3 L/m².h and increased every 30 min until it reached to 10 L/m².h. At each flux, TMP change was recorded.

3. Results and discussion

3.1. Critical flux

The rate of fouling and permeability are represented in Fig. 3. The permeability decreased by increasing the flux. Furthermore, the higher TMP was required for the higher flux, accompanying with higher fouling. The reason for higher fouling was accumulation of suspended solid on the membrane surface. TMP was increased as result of pore size reduction during the constant flux operation.

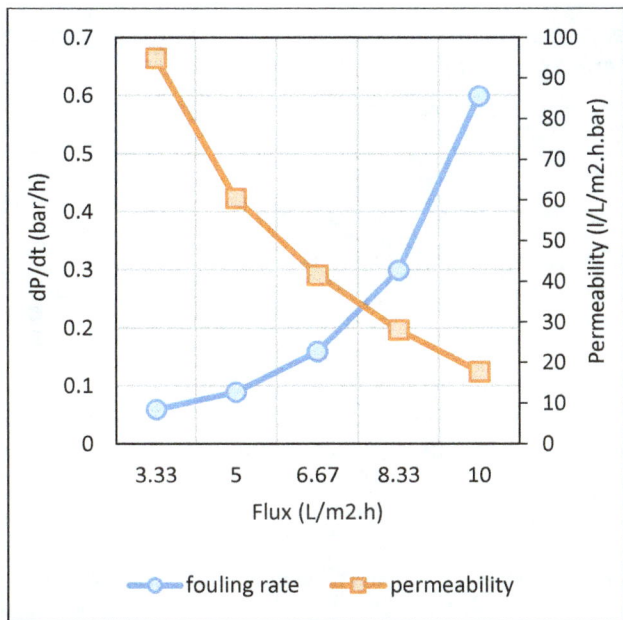

Fig. 3. Fouling rate and permeability of the membrane as a function of the flux rate

According to Fig. 3, critical flux was 6.7 L/m². h. Both TMP and fouling were extremely increased in fluxes higher than 6.7 L/m² h. When the flux was increased from 6.7 to 8.3 L/m² h, the fouling rate was increased from 0.16 to 0.3bar/h, whereas reduction of flux from 6.7 to 5 L/m² h led to reduction of fouling rate from 0.16 to 0.09 bar/h. On the other hand, at fluxes equals to 5, 6.7, and 8.3 L/m² h the average TMP was 82.5, 160, and 295 mbar, respectively, that represented a significant increase in the average TMP by increasing the flux from 6.7 to 8.3 L/m² h. According to the results, the flux of 5 L/m² h had a low fouling rate and at the same time high permeability.

3.2. Membrane resistance distribution

Several resistances contributed in the membrane mass transfer. The cake resistance (R_c) had a major contribution in the overall resistance while the clean membrane resistance (R_m) showed the lowest impact (Table 4). This indicates that cake layer formation is the major mechanism of fouling. Significant effect of physical cleaning on fouling indicates that cake layer was attached to membrane quite loosely.

Table 4. Membrane resistances

Resistance	value$\times 10^{-12}$ (m^{-1})
Overall resistance	8.333
Clean membrane resistance	0.837
Internal fouling resistance	1.674
Cake resistance	5.822

3.3. MBR effluent quality

3.3.1. COD removal

Fig. 4 represents the COD of the bioreactor influent, mixed liquor, and membrane effluent during the process.

Operation was done for 45 days, and the results of COD removal in the bioreactor and overall process are summarized in Fig. 5.

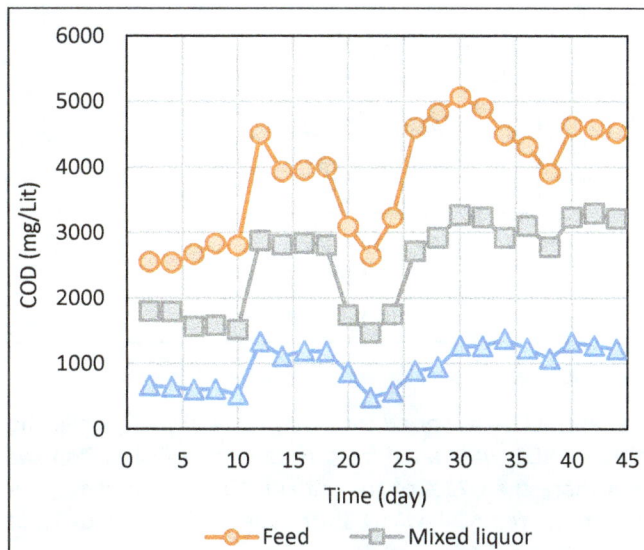

Fig. 4. COD of influent, mixed liquor, and effluent in different days of experiment

As can be seen in Fig. 4, the COD concentration of the influent at different times varied in the range of 2546 to 5074 mg/L, while the effluent had a concentration in the range of 488 to 1380 mg/L. In this situation, the maximum and minimum removal of COD were 82% and 69%, respectively. The concentration of COD in the effluent stream depended on the bioreactor influent concentration and biofilm characteristics formed on the membrane. Biofilm thickness was a factor that could affect the COD removal. The thicker the biofilm layer, the smaller particle were removed.

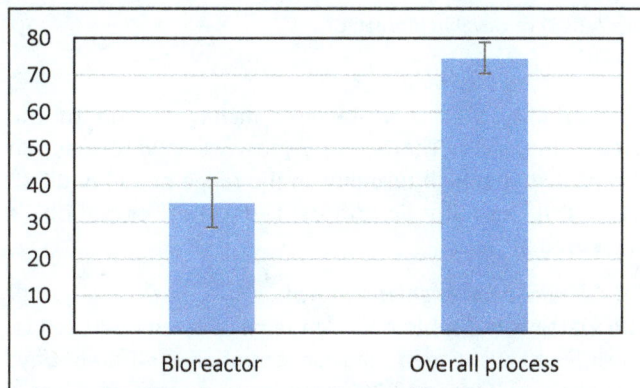

Fig. 5. COD removal in the bioreactor and overall process

3.3.2. BOD₅ removal

One of the important factors of wastewater quality is BOD₅. The BOD of bioreactor influent and membrane effluent was monitored during the experiments and is presented in Fig. 6.

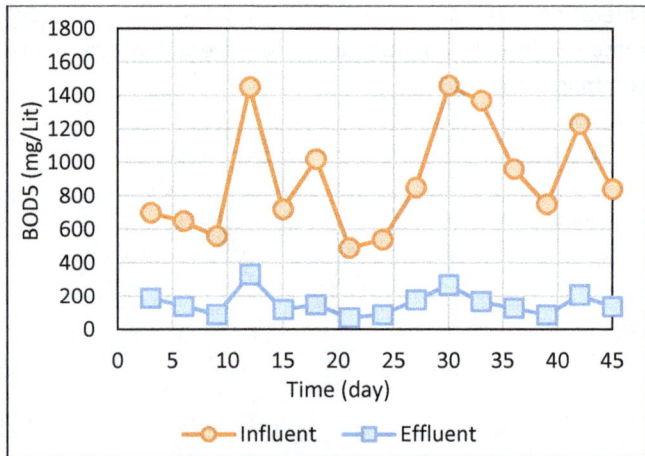

Fig. 6. BOD$_5$ of reactor influent and effluent

Influent BOD was varied from 540 to 1460 mg/L while the effluent BOD was varied from 70 to 330 mg/L. Fig. 7 shows that more than 72% of the initial BOD was removed. The maximum removal was 88% obtained after 39 days of operation.

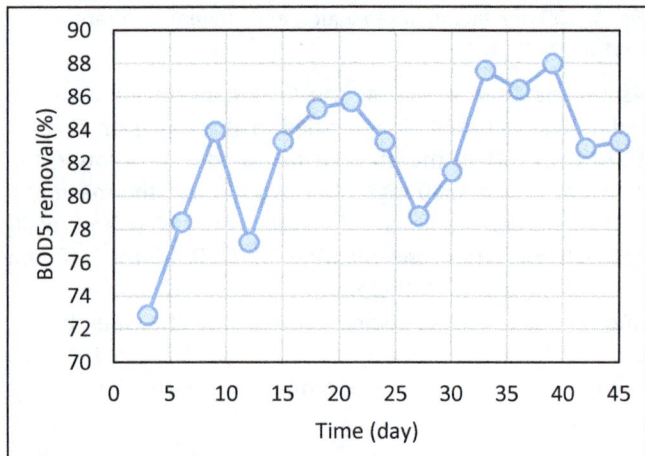

Fig. 7. BOD removal in bioreactor

3.3.3. Turbidity reduction

The turbidity of the wastewater mainly related to the organic materials. The backer's yeast wastewater used in this study had a high turbidity, in the range of 134 and 282 NTU. The turbidity in different days of operation is presented in Fig. 8

3.3.4. Suspended solid removal

The wastewater contained high level of suspended solids. The influent suspended solid concentration in different days was between 500 and 1200 mg/L whereas that of effluent was less than 30 mg/L and the least value was 5 mg/L (Fig. 10). 98.43±0.547 %of the suspended solid was removed during the experiment in the current work.

Effluent turbidity was reduced to less than 2.5 NTU, which was elimination of more than 98% of influent turbidity. Removal of the turbidity depends on the membrane pore

size. Fig. 9 represents the turbidity removal percent at different days of operation.

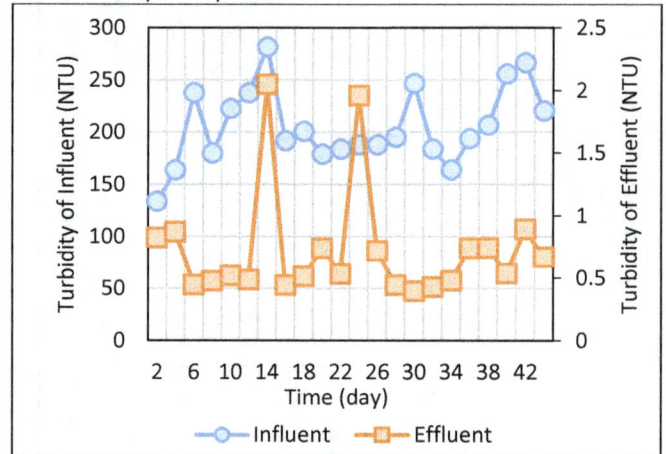

Fig. 8. Turbidity in influent and effluent during the experiment

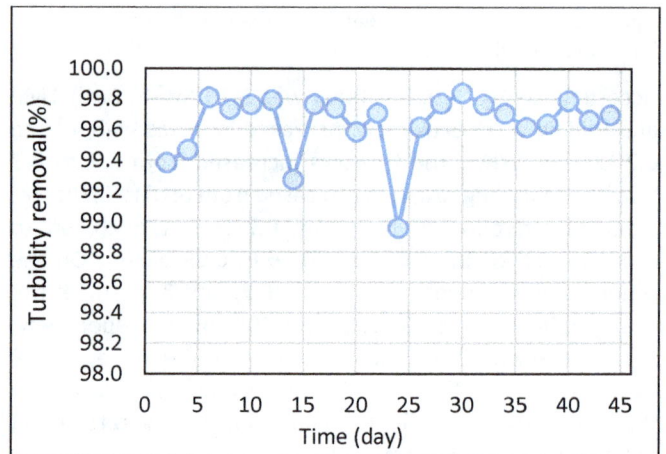

Fig. 9. Turbidity removal in bioreactor

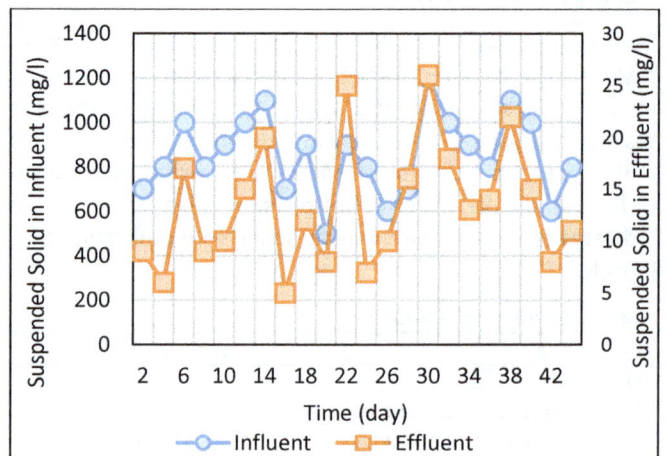

Fig. 10. Suspended solid in the Influent and Effluent

3.3.5. Mixed liquor suspended solid

MLSS concentration in the bioreactor features the active biomass concentration that consumes wastewater organic materials. Greater MLSS concentration accompanied with better treatment of wastewater pollutants. Fig. 11 shows that MLSS in the first days of process was low while it sharply increased in the latter phase of the treatment. At the beginning of experiment, MLSS concentration was 4100 mg/L, while it increased to 7000 mg/L in 8th day of operation. The MLSS concentration increased up to 9100 mg/L. The changes in MLSS concentration at different days of the process was due to the changes in the feed and the fungal biomass concentrations.

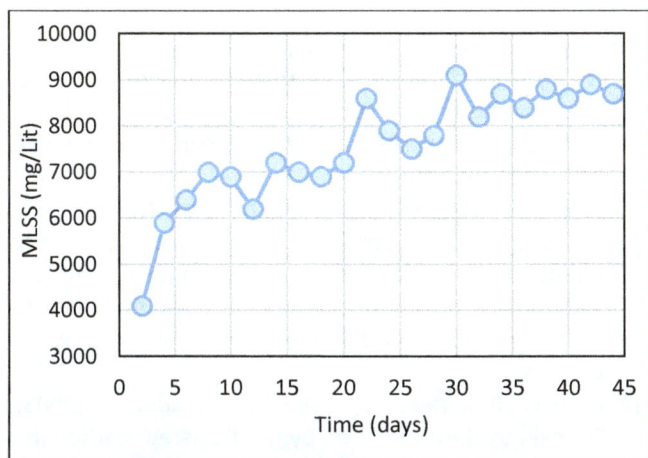

Fig. 11. MLSS concentration in bioreactor

3.3.6. Dissolved oxygen and pH
The bioreactor was continuously sparged with air to provide the aerobic conditions, and the DO concentration was monitored by a DO meter located in the bioreactor (Fig. 12). The maximum and minimum concentration of oxygen during the process were 3.7 and 1.7 mg/L, respectively.

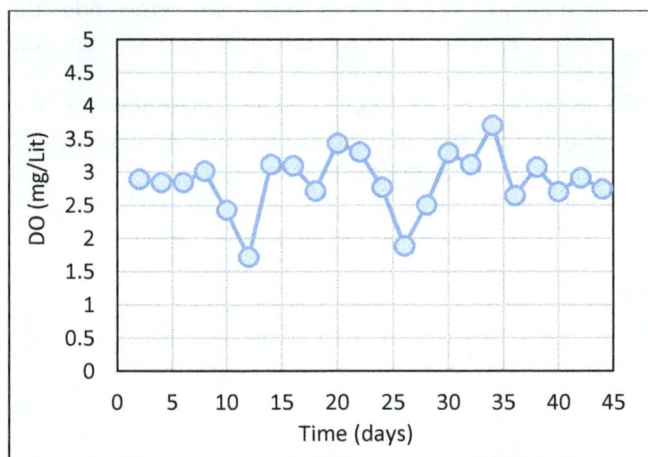

Fig. 12. Dissolved oxygen concentration in bioreactor

4. Conclusions

MBR technology using *A. oryzae* demonstrated a reasonable performance for baker's yeast wastewater treatment. Significant reduction in COD, BOD_5 and turbidity of

bioreactor effluent were observed. Removing 98% of turbidity and reducing effluent to 488 mg/L in the best operation conditions, MBR demonstrates that it is a good solution for treatment of food industries wastewater. The cake resistance had the major part in membrane resistance. Sufficient back-washing protocol and aeration on membrane surface can mitigate cake resistance and fouling of membrane.

References

[1] Takagi, H., Shima, J. (2015). Stress Tolerance of Baker's Yeast During Bread-Making Processes. In *Stress Biology of Yeasts and Fungi* (pp. 23-42). Springer Japan.

[2] Pirsaheb, M., Rostamifar, M., Mansouri, A. M., Zinatizadeh, A. A. L., Sharafi, K. (2015). Performance of an anaerobic baffled reactor (ABR) treating high strength baker's yeast manufacturing wastewater. *Journal of the Taiwan institute of chemical engineers*, 47, 137-148.

[3] Kobya, M., Delipinar, S. (2008). Treatment of the baker's yeast wastewater by electrocoagulation. *Journal of hazardous materials*, 154(1), 1133-1140.

[4] Mischopoulou, M., Naidis, P., Kalamaras, S., Kotsopoulos, T. A., Samaras, P. (2016). Effect of ultrasonic and ozonation pretreatment on methane production potential of raw molasses wastewater. *Renewable energy*, 96, 1078-1085.

[5] Liang, Z., Wang, Y., Zhou, Y., Liu, H., Wu, Z. (2009). Variables affecting melanoidins removal from coagulation/flocculation. *Separation and purification technology*, 68(3), 382-389.

[6] Liakos, T. I., Lazaridis, N. K. (2016). Melanoidin removal from molasses effluents by dsorption. *Journal of water process engineering*, 10, 156-164.

[7 Tsioptsias, C., Petridis, D., Athanasakis, N., Lemonidis, I., Deligiannis, A., Samaras, P. (2015). Post-treatment of molasses wastewater by electrocoagulation and process optimization through response surface analysis. *Journal of environmental management*, 164, 104-113.

[8] Tsioptsias, C., Banti, D. C., Samaras, P. (2015). Experimental study of degradation of molasses wastewater by biological treatment combined with ozonation. *Journal of chemical technology and biotechnology*, 91(4), 857–864.

[9] Maher, A., Sadeghi, M., Moheb, A. (2014). Heavy metal elimination from drinking water using nanofiltration membrane technology and process optimization using response surface methodology. *Desalination*, 352, 166-173.

[10] Sadeghian, M., Sadeghi, M., Hesampour, M., Moheb, A. (2015). Application of response surface methodology (RSM) to optimize operating conditions during

ultrafiltration of oil-in-water emulsion. *Desalination and water treatment, 55*(3), 615-623.

[11] Guglielmi, G., Chiarani, D., Judd, S. J., Andreottola, G. (2007). Flux criticality and sustainability in a hollow fibre submerged membrane bioreactor for municipal wastewater treatment. *journal of membrane science, 289*(1), 241-248.

[12] Hosseinzadeh, M., Bidhendi, G. N., Torabian, A., Mehrdadi, N. (2013). Evaluation of membrane bioreactor for advanced treatment of industrial wastewater and reverse osmosis pretreatment. *Journal of environmental health science and engineering, 11*(1), 1-8.

[13] Kim, I., Choi, D. C., Lee, J., Chae, H. R., Jang, J. H., Lee, C. H., Won, Y. J. (2015). Preparation and application of patterned hollow-fiber membranes to membrane bioreactor for wastewater treatment. *Journal of membrane science, 490*, 190-196.

[14] Basu, S., Kaushik, A., Saranya, P., Batra, V. S., Balakrishnan, M. (2016). High strength distillery wastewater treatment by a PAC-MBR with low PAC dosage. *Water science and technology, 73*(5), 1104-1111.

[15] Deowan, S. A., Galiano, F., Hoinkis, J., Johnson, D., Altinkaya, S. A., Gabriele, B., Figoli, A. (2016). Novel low-fouling membrane bioreactor (MBR) for industrial wastewater treatment. *Journal of membrane science, 510*, 524-532.

[16] Judd, S. (2008). The status of membrane bioreactor technology. *Trends in biotechnology, 26*(2), 109-116.

[17] Judd, S. (2010). *The MBR book: principles and applications of membrane bioreactors for water and wastewater treatment*. Elsevier.

[18] Neoh, C. H., Noor, Z. Z., Mutamim, N. S. A., Lim, C. K. (2016). Green technology in wastewater treatment technologies: Integration of membrane bioreactor with various wastewater treatment systems. *Chemical engineering journal, 283*, 582-594.

[19] Sankaran, S., Khanal, S. K., Jasti, N., Jin, B., Pometto III, A. L., Van Leeuwen, J. H. (2010). Use of filamentous fungi for wastewater treatment and production of high value fungal byproducts: a review. *Critical reviews in environmental science and technology, 40*(5), 400-449.

[20] Machida, M., Yamada, O., Gomi, K. (2008). Genomics of Aspergillus oryzae: learning from the history of Koji mold and exploration of its future. *DNA research, 15*(4), 173-183.

[21] Meng, F., Yang, F., Shi, B., Zhang, H. (2008). A comprehensive study on membrane fouling in submerged membrane bioreactors operated under different aeration intensities. *Separation and purification technology, 59*(1), 91-100.

[22] Amiraftabi, M. S., Mostoufi, N., Hosseinzadeh, M., Mehrnia, M. R. (2014). Reduction of membrane fouling by innovative method (injection of air jet). *Journal of environmental health science and engineering*, 12(1), 1-8.

[23] Huang, Z., Ong, S. L., Ng, H. Y. (2011). Submerged anaerobic membrane bioreactor for low-strength wastewater treatment: effect of HRT and SRT on treatment performance and membrane fouling. *Water research, 45*(2), 705-713.

[24] Liang, S., Liu, C., Song, L. (2007). Soluble microbial products in membrane bioreactor operation: behaviors, characteristics, and fouling potential. *Water research, 41*(1), 95-101.

[25] Meng, F., Zhang, H., Yang, F., Zhang, S., Li, Y., Zhang, X. (2006). Identification of activated sludge properties affecting membrane fouling in submerged membrane bioreactors. *Separation and purification technology, 51*(1), 95-103.

[26] Pourabdollah, M., Torkian, A., Hashemian, S. J., Bakhshi, B. (2014). A triple fouling layers perspective on evaluation of membrane fouling under different scenarios of membrane bioreactor operation. *Journal of environmental health science and engineering, 12*(1), 1-10.

[27] Le Clech, P., Jefferson, B., Chang, I. S., Judd, S. J. (2003). Critical flux determination by the flux-step method in a submerged membrane bioreactor. *Journal of membrane science, 227*(1), 81-93.

[28] Wu, Z., Wang, Z., Huang, S., Mai, S., Yang, C., Wang, X., Zhou, Z. (2008). Effects of various factors on critical flux in submerged membrane bioreactors for municipal wastewater treatment. *Separation and purification technology, 62*(1), 56-63.

[29] Mutamim, N. S. A., Noor, Z. Z., Hassan, M. A. A., Olsson, G. (2012). Application of membrane bioreactor technology in treating high strength industrial wastewater: a performance review. *Desalination, 305*, 1-11.

[30] Wef, A. A. (1998). Standard methods for the examination of water and wastewater. *American public health association, Washington, DC*.

5

Sorption of Cu(II), Zn(II) and Ni(II) from aqueous solution using activated carbon prepared from olive stone waste

Gehan E.Sharaf El-Deen

Radioactive Waste Management Department, Hot Laboratories Center, Atomic Energy Authority, Egypt.

Keywords:
Sorption
Cu(II)
Zn(II)
Ni(II)
Physically activation Olive stone

ABSTRACT

The performance of olive stone activated carbon (OSAC) for the sorption of Cu^{2+}, Zn^{2+} and Ni^{2+} ions was investigated via batch technique. OSAC materials were prepared under different physical activation conditions. Olive stone waste was activated with N_2 and steam at 900°C at a 3.5h hold time (OSAC-3). The characterization for OSAC-3 was performed under BET-surface area, SEM, density and FTIR-spectrum. The optimum adsorption conditions were specified as a function of agitation time, initial metal concentration, pH, and temperature. The kinetic results were found to be fast and described well by the pseudo-second order model. The adsorption capacities were 25.38mg/g (Cu^{2+}), 16.95mg/g (Zn^{2+}) and 14.65mg/g (Ni^{2+}) which followed the sequence Cu^{2+} > Zn^{2+} > Ni^{2+}. The spontaneous adsorption for all the studied cations, endothermic nature for both Zn^{2+} and Ni^{2+} ions and exothermic nature for Cu^{2+} ions were obtained. The results showed that OSAC-3 is an economically feasible material for Cu^{2+}, Zn^{2+} and Ni^{2+} remediation from weak acidic contaminated effluents.

1. Introduction

Environmental pollution by heavy metals like As, Cd, Co, Cr, Cu, Hg, Ni, Pb, Sn, Zn, etc. is a matter of ever-growing concern because of their toxicity. Even relatively low concentrations can have long term health effects on humans [1]. Heavy metals appear in wastewater discharged from different industries, including smelting, metal plating, Cd–Ni batteries, electronic industry, electroplating, metal finishing plants, phytopharmaceutical plants, phosphate fertilizer, mining pigments, stabilizer alloy manufacturing, and many others [2]. Trace amounts of copper, zinc, and nickel are essential for humans, microorganisms, and animals. However, an excessive intake of copper will cause stomach upset, ulcers, mental retardation as well as liver and brain damage in humans [1]. The consumption of nickel that exceeds permissible levels can cause various diseases like lung and nasal cancers, pulmonary fibrosis, renal edema, lung cancer, skin dermatitis, diarrhea, nausea and vomiting [3]. Elevated concentrations of zinc can cause

several health problems, e.g., arteriosclerosis, pancreas damage, vertigo, and disharmony [4]. The US Environmental Protection Agency (EPA) requires the level of copper, zinc, and nickel in drinking water not to exceed 1.3, 5, and 0.04 mg/L, respectively [5]. Also, excessive amounts of these elements are harmful for the environment. Various treatment technologies such as flotation, electrochemical methods, coagulation, filtration, precipitation, adsorption, ion exchanges, reverse osmosis and membrane technologies have been employed to remove hazardous heavy metals from aqueous solutions [6]. However, most of these processes have some drawbacks such as high operating costs, the inability to remove toxic elements from a wide range of wastewaters, and their ineffectiveness in lower concentrations [7]. Adsorption on activated carbons is considered as one of the most practical, easy to operate, environmentally friendly, and economical approach for water treatment. The utilization of low-cost agro-industrial waste as a precursor for the production of activated carbon supports the economic feasibility of this adsorbent. It has

*Corresponding author
E-mail address: gsharaf2000@gmail.com

been shown that specific surface area and porosity are not the only determinant parameters in activated carbon. Surface chemistry is also an effective parameter in the process of metal ions adsorption. Accordingly, activated carbon surfaces can be improved by acidic treatment, oxidation, heating or ammonization to enhance metal species removal [8]. Activated carbons (AC) are used as adsorbent materials because of their large surface areas, high degree of surface activates, microporous structures and high adsorption capacities [9]. As of today, the use of commercially produced activated carbons is still limited due to the high cost of raw materials such as coal and non-renewable materials. Efforts are being made with numerous researchers to produce more effective, cheaper, and environmental friendly activated carbons [10-14]. Therefore, various precursors such as agricultural and agro-industrial by-products materials have been used as precursor materials for activated carbon production by using various physical and chemical activation processes for preparations. One objective of this study is to prepare a more effective, economical, and environmental friendly activated carbon. Olive stone waste as a raw material for the production of activated carbon could be considered as a better choice among agro-industrial wastes because it is abundant and inexpensive, especially in Mediterranean countries. Olive stone waste, a by-product generated from olive oil extraction, are available in large amounts in Mediterranean countries like Egypt, where about 13,500 x130 tones per year are produced. This olive waste which is generated in huge quantities in a short period of time from November to March creates an environmental problem for Egypt [15]. The main use of this biomass is the production of energy. In addition to utilizing olive stone by-product as activated carbon, other uses that are been investigated include furfural production, biosorbent, abrasive, cosmetic, plastic filler, animal feed, or resin formation [16]. Another objective of the present work is to investigate the potential effectiveness of olive stone activated carbon (OSAC) on the removal of Cu^{2+}, Zn^{2+}, and Ni^{2+} ions from contaminated water at different conditions of physical activation in batch technique. At optimum conditions for OSAC-preparation, the effects of agitation time, initial concentration, initial solution pH, and thermodynamic parameters on the sorption capacity of OSAC were studied.

2. Materials and methods

2.1. Chemicals, Materials and Equipments

To perform the sorption experiments, 1000mg/L of copper, zinc and nickel standard solutions were prepared from $CuCl_2.2H_2O$, $ZnSO_4.7H_2O$ and $Ni(NO_3)_2.6H_2O$ (Merck) using bi-distilled water. All chemicals used were of analytical grade purity. The olive stone waste was obtained from the by-product of olive oil producing factories in Wahet Sewa, Egypt. Ultra-pure nitrogen gas was employed for the activated carbon production. A high temperature tube furnace (model CD-1700G, Chida, China) was used for preparing the OSAC-samples, while an Atomic absorption spectrophotometer (Hitachi model Z-8100, Germany), a Fourier Transform Infrared Spectrophotometer (NICOLET 380-FTIR, Thermo-scientific, UK), and a Scanning Electron Microscope were used for measuring. Nitrogen adsorption/desorption isotherms at 77 K on an automatic adsorption instrument (Nova 3200 BET instrument, Quanta chrome, Corporation, USA) were used for measuring surface area, average pore diameter, and total pore volume.

2.2. Preparation of sorbent:

The olive stone waste was supplied by an olive oil factory in the Wahet Sewa region of Egypt. This agro-industrial olive stone waste was dried in the sun, crushed and sieved. The granules of olive stone waste with diameter fractions from 0.7 to 2mm were used as a precursor for the production of activated carbon by a physical activation process at different conditions. Carbonization and activation are the two main steps in the physical activation process. These steps were followed to prepare different samples of OSAC and are summarized in Table (1). The preparation conditions for one of these samples can be explained as follows: carbonized olive stone waste was obtained by heating 100 gram of the clean dry crushed olive stone waste (as the original weight) to a specified temperature of 850°C at a rate of 50°C/10min. for a one hour hold time under flowing N_2 gas at 10psi. in a tube furnace (with inclined position of angle 70°); then, it was cooled to room temperature. In the activation step, the obtained samples from the last step were placed in the reactor, which was situated in the hot zone of the tubular furnace. Next, pure steam was introduced through the sample. The temperature was raised gradually (100°C/10min.) up to 450°C to allow free evolution of volatiles. The heating continued until 900°C for a 3.5h. hold time, then it was cooled and the weight determined. The granular activated carbon final product was crushed until its particle diameter was lower than 0.7mm, this was determined by sieving. After that, the granular activated carbon was kept in a closed bottle for further tests.

3. Characterization of olive stone activated carbon

The physical characterizations (specific surface area, carbon yield percent, weight loss, activation burn off, density), chemical characterizations (surface pH and the percent of oxides in ash-residue obtained by EDX), FTIR, and SEM were analyzed.

3.1. Physical characterizations of OSAC-3 product

Most of the physical properties for all the prepared OSAC were measured and are shown in Table 1. The one which had a higher surface area (OSAC-3) was selected for further investigations, and its physical properties were studied in detail. The calculated specific surface area (S_{BET}=850 m^2/g), total pore volume (V_t=0.47cc/g) and average pore radius (r`=11A°) were determined by the nitrogen adsorption–desorption isotherms.

The activation burn-off, yield percent and the weight loss due to activation were found to be 48.37%, 12.22% and 76.33g, respectively. The OSAC-3 reported good bulk density (B_d=0.67g/ml) and an apparent density (A_d) of 0.42g/ml. An adequate bulk density can help to improve the rate of filtration.

3.2. Chemical characterizations of OSAC-3 product

For pH determination of the OSAC-3 active surface, 1.0g of the dry carbon sample was stirred with 100ml of bi-distilled water for 5h., then soaked for 3 days for equilibrate in a stoppered glass bottle. At end of this period, the pH of the carbon slurry was recorded after 3min. in order for the pH probe to reach equilibrium. It was observed that a neutral OSAC-3 surface with a pH of 7.2 was obtained. Therefore, its pH does not affect the aqueous medium during the sorption process. An EDX-analysis was important for determining the main oxides included in the ash content, shown in Table 2, in which K$^+$ and Ca^{2+} are the major alkaline ingredients of the ash residue of OSAC-3.

Table 1. Preparation conditions for synthesis of different OSAC-materials.

Sample no.	1. Carbonization step 2. Activation step	S_{BET}, m^2/g	V_t, cc/g	r`, A°	B_d, g/cm^3
OSAC-1	(1) 50°C/10min. at 850°C for 2h. (h.t). (2) 100°C/10min. with CO$_2$ gas at 825°C/2h. (h.t).	530	0.283	21.3	0.837
OSAC-2	(1) 50°C/10min. at 850°C for 2h. (h.t). (2) 100°C/10min. with steam gas at 825°C/3h. (h.t).	780	0.533	15.50	0.677
OSAC-3	(1) 50°C/10min. with N$_2$-gas at 850°C for 1h. (h.t). (2) 100°C/10min. with steam gas at 900°C/3.5h. (h.t).	850	0.470	11.00	0.671
OSAC-4	(1) 50°C/10min. with N$_2$-gas at 850°C for 2h. (h.t). (2) 100°C/10min. with steam gas at 950°C/3.5h. (h.t).	300	0.180	11.70	0.731
OSAC-5	(1) 50°C/10min. at 850°C for 1h. (h.t). (2) 100°C/10min. with steam gas at 900°C/3.5h. (h.t).	458	0.235	10.26	0.779

(h.t= hold time)

Table 2. Main Oxides composition of OSAC-3 from EDX analysis

Units	Mg	Al	Si	P	S	Cl	K	Ca	Fe
%	1.9	1.1	3.6	3.4	8.3	7.1	56.1	16.7	1.8

3.3. Scanning electron microscope analysis (SEM)

The surface physical morphology of OSAC-3 activated carbon was examined using a scanning electron microscopy (S-2150, Hitachi High-Technologies) with 1600X magnification; it is shown in Fig. 1. The SEM micrograph clearly revealed that the dark cavities signified pores and the greyish areas were due to the carbon matrix. The reason for the formation of the cavities on the OSAC-3 sample may occur because during pyrolysis (carbonization process) the cellulosic structure loses small molecules as volatile materials such as water and carbon dioxide together with a complexity of aliphatic acids, carbonyls, alcohols, etc. These desired activations don't occur at a single decomposition temperature, but over a range of temperatures. Small molecules are removed from the original macromolecule of the network and the resultant porosity created [16].

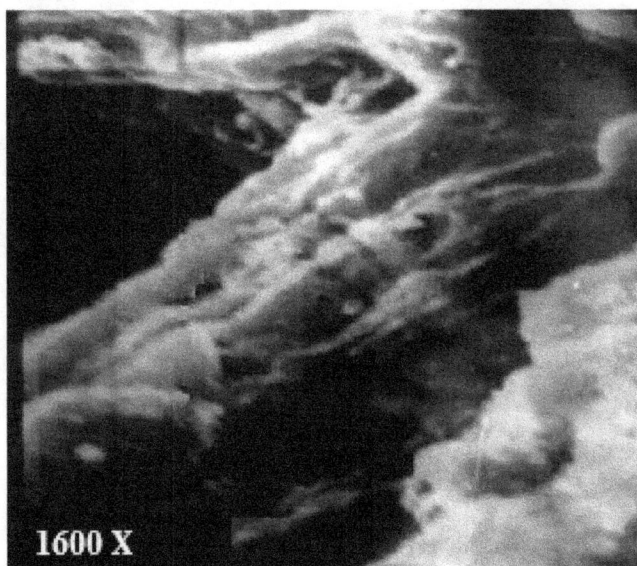

Fig. 1. SEM of OSAC-3 activated carbon

As a result, a new lattice is continuously created with a composition of high C/H and C/O ratios. In regard to the carbonization of cellulosic materials, the obtained char and coal is microporous, but the micro pores may become filled or partially blocked with tars and other decomposed materials. The process used to create high porosity from low-porosity-carbonized material is known as "activation. This is usually completed by reaction with steam or carbon dioxide above 800°C, where the gas molecules penetrate the interior of the char particle to remove carbon atoms and the tar matter [17].

3.4. Fourier-Transform Infrared Spectroscopy (FTIR)

The FTIR spectrum of the olive stone waste before activation is displayed in Fig.2(a). A broad absorption band at a wavenumber ranging from 3500 to 3200cm^{-1} with a maximum absorption at about 3384cm^{-1} means there is O—H stretching vibration owing to intermolecular hydrogen bonding. This band is observed in the spectra of carboxyl, phenols or alcohols groups and adsorbed water. An absorption band at 2925cm^{-1} is due to the aliphatic –CH group. The stretching –C=O group in normal ester was obtained at 1735cm^{-1}, the conjugation interfered with possible resonance with the carbonyl group that led to an increase in the absorption frequency for the C=O band to appear at the 1738cm^{-1} band. A weak band at 1670 to 1640cm^{-1} was assigned to –C=O stretching vibrations of amides (-C=O-N). A strong band at 1035cm^{-1} with a shoulder at 1259cm^{-1} was associated to aliphatic ether (-C-O); an ether group was obtained from alcohol (R–OH) or ester (–C=O) groups. The FTIR spectrum of OSAC-3 after activation, Fig.2 (b), showed that in comparison to the non-activated raw material, more peaks appeared at the OSAC-3 and some disappeared due to the cracking of some bonds by heating; the settling of broad bands at 3713–3683 cm^{-1} could be due to the NH$_2$ stretching vibration of the nitrile functional groups. A broad band at 3448cm^{-1} was assigned to the stretching vibration of hydrogen bonded hydroxyl groups (-OH). A band appeared at 2367- 2369 cm^{-1} and was perhaps due to the -C≡C group. A band at 2337 cm^{-1} may be due to a weak nitrile group attached to the aliphatic chain [18]. Bands around 1705 cm^{-1} may be due to ketone or ketene. A band at 1466cm^{-1} was assigned to stretching vibration of aliphatic –CH$_2$.

4. Results and discussion

4.1. Batch sorption procedure

Sorption experiments were carried out at 22±1°C by shaking a fixed amount (0.03g) of olive stone activated carbon in 50ml of metal solution in a thermostatic water bath mechanical shaker. The water bath was used to maintain a constant temperature. After equilibrium, the solid was filtered and the final heavy metal concentration as well as in the initial solution was measured by flame atomic absorption spectrometer (FAAS) (Hitachi Z-8100). The sorbed cations were determined from the difference between the initial and final concentration in the solution. The capacity of metal ions as sorbates was calculated as:

$$q_e = \frac{V\,(C_o - C_e)}{m} \qquad (1)$$

where m (g) is the weight of olive stone activated carbon, V (L) is the solution volume, and C_o and C_e (mg/L) are the initial and equilibrium bulk ion concentrations, respectively. Initially, a preliminary test was performed to select the best OSAC-adsorbent for removing the investigated cations at a pH of 6. The results presented in Fig.3 indicate that OSAC-3 was the best adsorbent for the removal of the studied cations. Thus, OSAC-3 was selected to complete the remaining tests to study the optimum conditions for removing the investigated cations.

4.2. Effect of pH

The removal of heavy metals as pollutants from wastewaters by adsorption is highly dependent on the pH of the solution, which affects the adsorbent surface charge, the degree of ionization and speciation of the adsorbate. The effect of solution pH on the sorption of heavy metals was investigated by using 0.03 g of OSAC-3 and 20 mg/L of metal ion concentration at initial pH values ranging from 2.5 to 9.2 for a 3 h shaking time at 295 K; the results are shown in Fig. 4. At a pH from 2.5 to 4, the sorption of metals onto the activated carbon was found to be low. This could be due to increasing the competition between the studied cations with H_3O^+ ions on active sites at a lower pH.

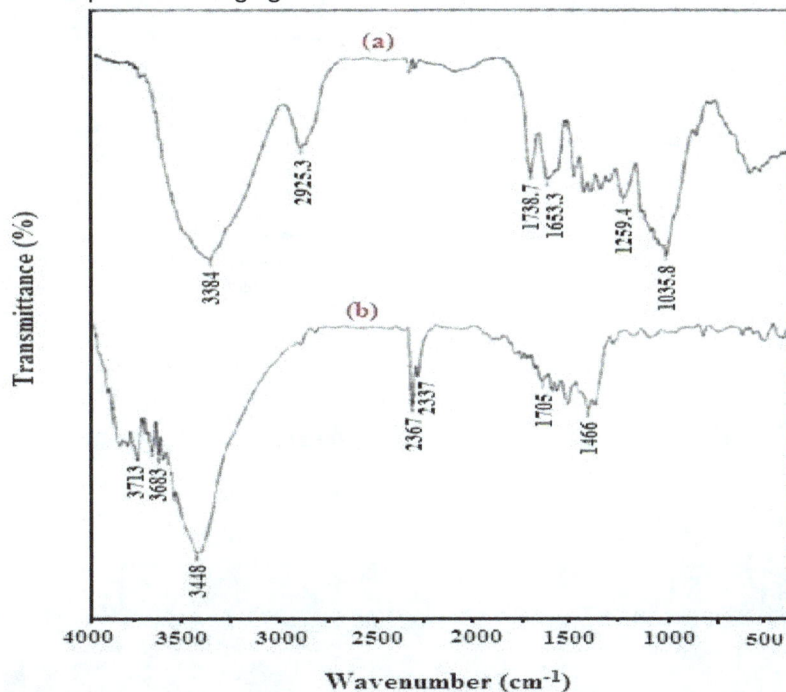

Fig. 2. FTIR spectra for original olive stone waste (a) and for OSAC-sample (b).

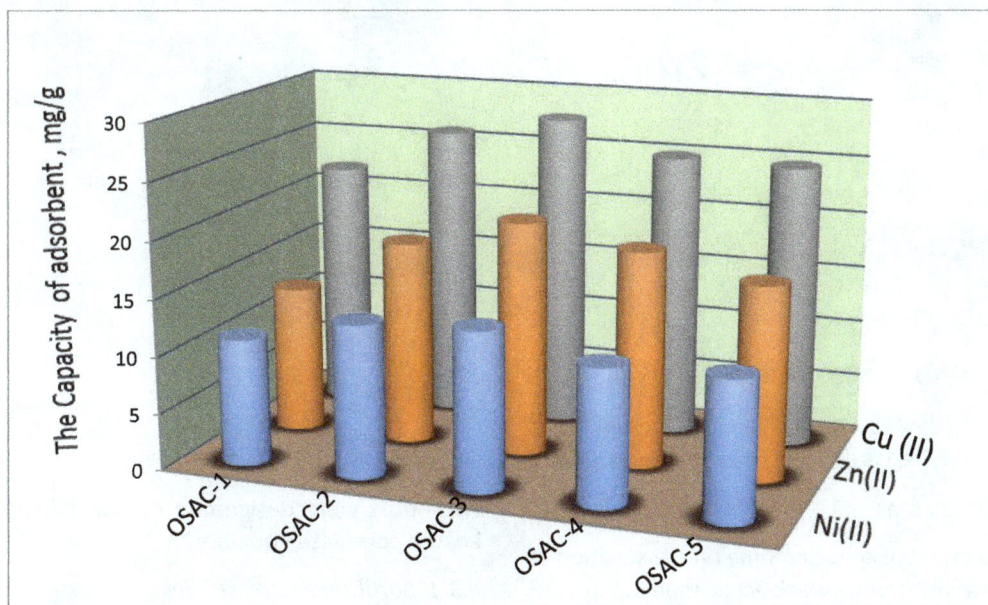

Fig. 3. Preliminary test results to select the best adsorbent to remove Cu(II), Zn(II) and Ni(II)-ions.

Metal uptake increased gradually with increasing pH from 4 to 7.2 for Cu^{2+} ions, and from 4 to 8 for Zn^{2+} and Ni^{2+} ions. This may be attributed to an increase in pH. More active sites with negative charges were expected to be exposed and this would attract the positively charged Cu(II), Zn(II) and Ni(II) ions for binding. The predominant species were Cu^{2+}, $Cu(OH)^+$, Zn^{2+}, and Ni^{2+} at a pH lower than 7.2, as shown in Fig. 6(a,b,c). Then, with increasing of the basicity of the solution, the efficiency of the sorption process reached a steady state for Cu^{2+} ions after a pH of 7.2. But the sorption capacities for Zn^{2+} and Ni^{2+} ions drastically increased with increasing pH values from 7.5 to 9.2. This may be attributed to reducing the solubility and starting precipitation of the metal ions at higher pH-values [19] through the formation of $Cu(OH)_2$, $Zn(OH)_2$ and $Ni(OH)_2$ as precipitate at a pH higher than 7.5 and is shown in Fig.6

(a,b,c). Therefore, metal ions may accumulate inside the activated carbon porous or cracks by a mechanism known as combined sorption-micro precipitation [20]. Hence, the next tests were carried out at an initial pH value of 5.7 to insure that no precipitation occurred for the studied cations.

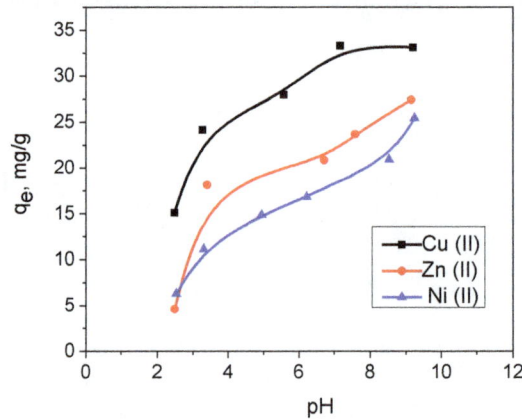

Fig. 4. Effect of pH on sorption of Cu^{2+}, Zn^{2+} and Ni^{2+}-ions on OSAC-3

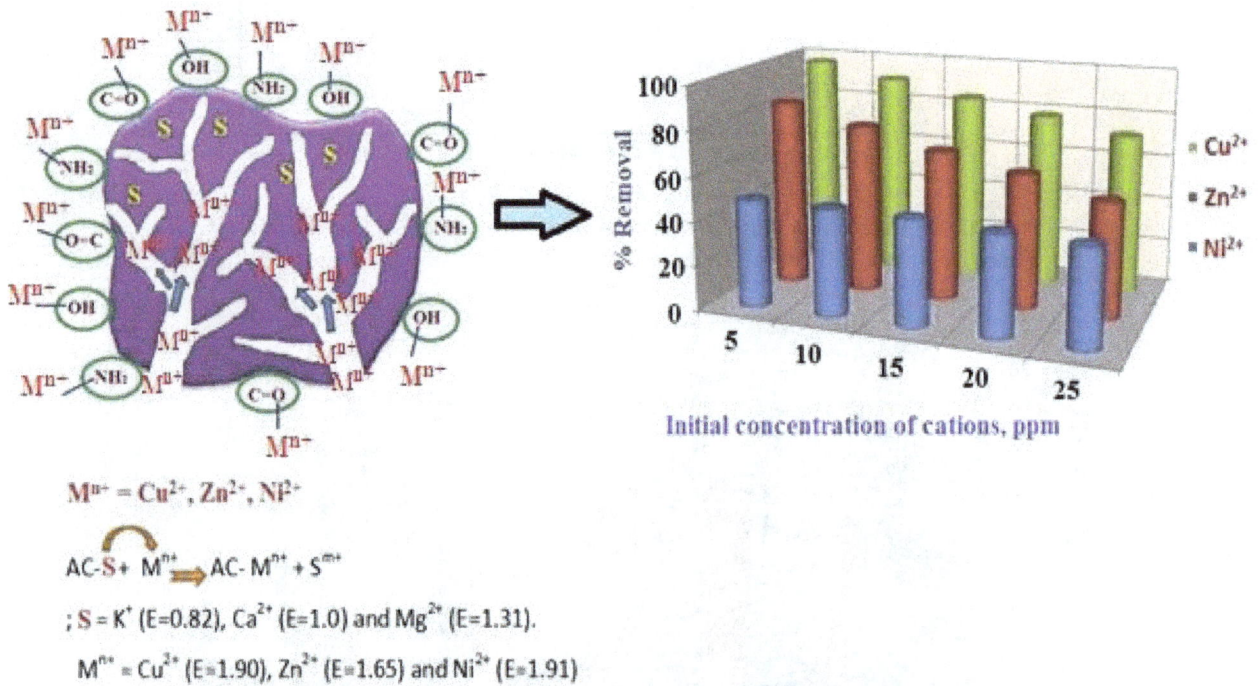

$M^{n+} = Cu^{2+}, Zn^{2+}, Ni^{2+}$

$AC\text{-}S + M^{n+} \rightleftharpoons AC\text{-}M^{n+} + S^{m+}$

$; S = K^+ (E=0.82), Ca^{2+} (E=1.0) \text{ and } Mg^{2+} (E=1.31).$

$M^{n+} = Cu^{2+} (E=1.90), Zn^{2+} (E=1.65) \text{ and } Ni^{2+} (E=1.91)$

Fig. 5. Scheme of sorption mechanism and the % removal of Cu^{2+}, Zn^{2+} and Ni^{2+} ions.

4.3. Effect of agitation time

To investigate the effect of agitation time on the sorption of the investigated cations from a weak acidic medium at a pH of 5.7, the experiments were conducted with a constant concentration of salt solution (20ppm). The plots, Fig. 7, show that the sorption equilibrium of the studied cations were reached after 30min. (for Cu^{2+} ions), less than 60min. (for Zn^{2+}) and at 120min. (for Ni^{2+} ions) of agitation time. All the studied cations were fast removed in the first 5min., this may be attributed to the pores of OSAC-3 that were nearly completely blocked by the ions in the first 5min. Therefore,

two hours were designated for the subsequent studies to ensure complete equilibrium.

4.3.1. Sorption kinetic studies

In order to investigate the behavior of the sorbent and also to determine the rate controlling mechanism of the adsorption process, three sorption kinetic models are used: pseudo-first order, pseudo second order, and the intraparticle diffusion models. The Lagergren pseudo-first order model [21] is described by the following equation:

$$\log (q_e - q_t) = \log q_e - \frac{k_{ad}}{2.303}t \tag{2}$$

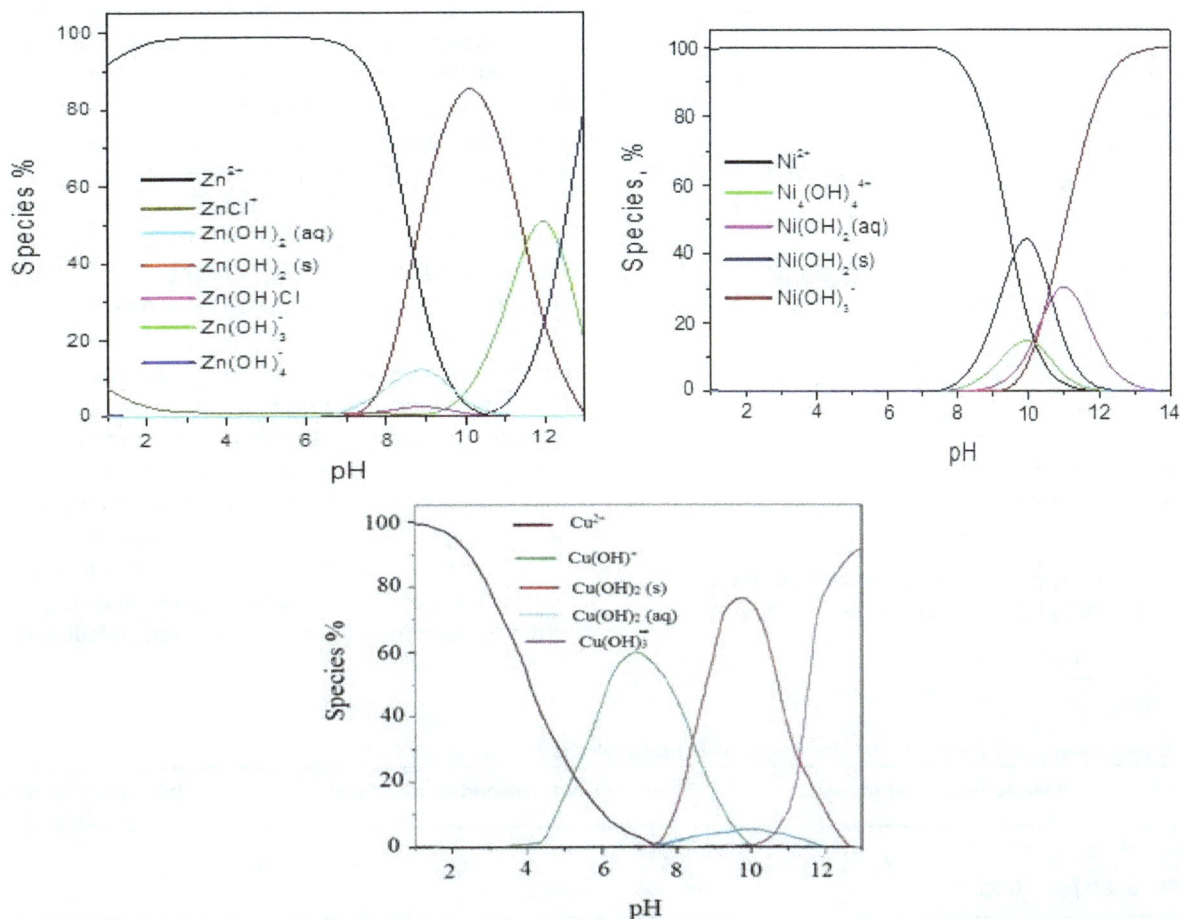

Fig. 6 (a,b,c). Speciation of Cu^{2+}, Zn^{2+} and Ni^{2+} in water obtained using the PHREEQC model according to metal concentration 20mg/L.

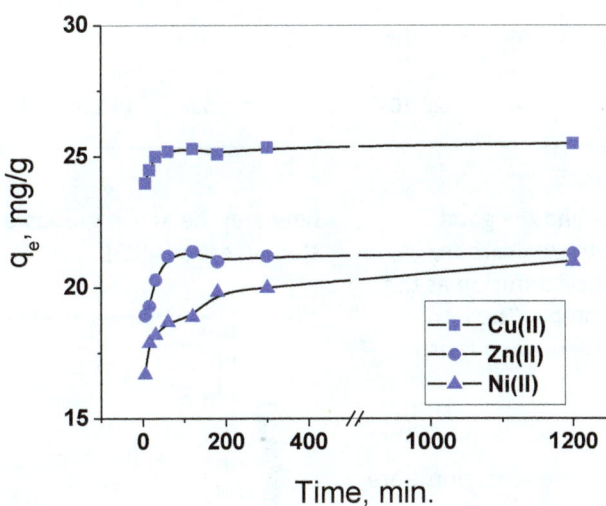

Fig. 7. Effect of agitation time sorption of Cu^{2+}, Zn^{2+} and Ni^{2+} on OSAC-3

where q_e (i.e. $q_{e.cal.}$) (mg/g) and q_t are the amounts of adsorbed cations on the surface of OSAC sorbent at equilibrium and at time t, respectively; k_{ad} is the Lagergren rate constant (min⁻¹). The plot of $\log{(q_e - q_t)}$ versus t (min.) in Figure (8) shows that straight lines are obtained for the studied cations. The first order rate constant, k_{ad}, and q_e are determined and listed in Table 3.In addition, the

experimental uptake (q_e, exp.) and the correlation coefficient (R^2) are also shown. It was noticed that the calculated uptake ($q_{e,calc.}$) values for all the investigated cations are not a match with the experimental data ($q_{e,exp.}$). The wide variation referring to the pseudo-first-order kinetic model did not fit the experimental data for the sorption of the studied cations by OSAC-3.

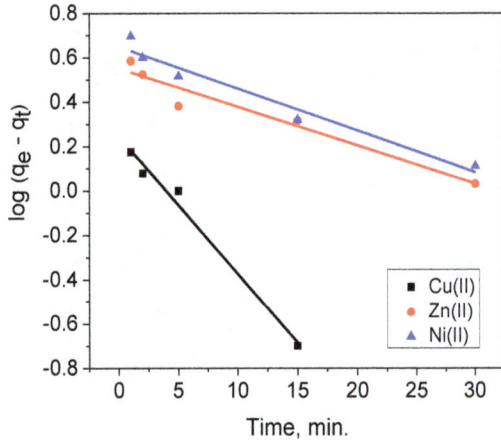

Fig. 8. Pseudo first order kinetic model for sorption of Cu^{2+}, Zn^{2+} and Ni^{2+}-ions on OSAC-3

The chemisorption pseudo-second order model proposed by Hall et al. (1966) [22] is described by equation Eq. (3):

$$\frac{t}{q_t} = \frac{1}{k_2 q_e^2} + \frac{1}{q_e} t = \frac{1}{h} + \frac{1}{q_e} t \qquad (3)$$

Where k_2 (g mg^{-1} min^{-1}) is the rate constant of the second-order model and h $(k_2 q_e^2)$ is the initial adsorption rate constant. When the experimental data were applied into the second-order law, as shown in Figure 9, straight lines were obtained for all the studied cations. Different variables as the calculated and the experimental equilibrium uptake ($q_{e,cal.}$ and $q_{e,exp.}$), the second-order rate constants (k_2), and the correlation coefficient (R^2) were calculated and are shown in Table 3. As can be seen from Table 3, the calculated $q_{e,cal.}$-values are in agreement with the experimental $q_{e,exp.}$-data and they reflect higher correlation coefficients obtained in comparison with that obtained from the pseudo-first-order equation. This denotes that the sorption process follows the pseudo-second-order kinetic model and the chemisorption process may be the rate-limiting step, which involves valence forces through sharing or exchange of electrons between sorbent and sorbates [23]. In addition , Table 3 shows that the sorption rate constant, k_2, of Cu^{2+} ions is the highest value in comparison with the other investigated cations, which follow the order: $Cu^{2+} > Zn^{2+} > Ni^{2+}$.

Table 3. Comparison of the kinetic models for sorption of copper, zinc and nickel ions

Metal	Pseudo-first-order model				Pseudo-second-order model					Intra-particle diffusion		
	$q_{e,exp.}$ (mg/g)	$q_{e,calc.}$ (mg/g)	K_{ad}	R^2	$q_{e,exp.}$ (mg/g)	$q_{e,calc.}$ (mg/g)	k_2	h	R^2	K_i	C_i	R^2
Cu^{2+}	25.00	1.758	0.143	0.98	25.00	24.95	0.164	102.25	0.99	1.04	22.19	0.99
Zn^{2+}	21.38	3.585	0.040	0.93	21.38	21.22	0.078	35.109	0.99	0.74	16.94	0.98
Ni^{2+}	20.10	4.47	0.043	0.94	20.10	20.04	0.038	15.196	0.99	0.99	14.41	0.96

In order to the higher porous structure and the good surface area (850m^2/g) for granular activated carbon, there is a possibility of using intra-particle diffusion as the rate-limiting step for these sorption systems. The intra-particle diffusion model described by Weber and Morris (1963) [24] is expressed as:

$$q_t = K_i t^{0.5} + C_i \qquad (4)$$

Where K_i and C_i are the intra-particle diffusion rate constant (mg/g min$^{0.5}$) and a constant, respectively. In Figure 10, the plot of q_t versus $t^{0.5}$ gives a straight line at q_t before the equilibrium case with slope, K_i, and intercept, C_i, for all the studied cations. The values of K_i and C_i are listed in Table 3, where the intercept, C_i , is the portion of the extent of boundary layer thickness [2]. This indicates that the pore diffusion is not the only rate limiting step for sorption of cations onto OSAC-3. The higher K_i-values for all the cases indicate the enhancement of the sorption rate and best sorption mechanism, which is due to good bonding

between the active groups on the sorbent surface and the studied cations [25].

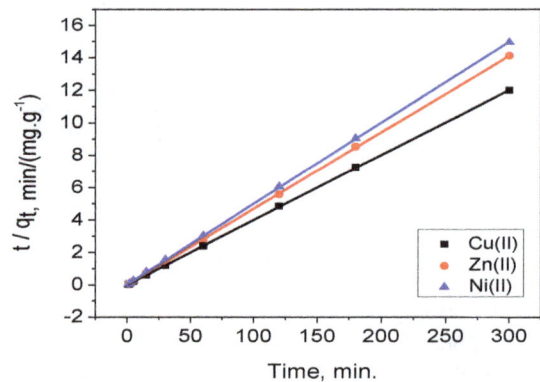

Fig. 9. Pseudo second order kinetic model for sorption of Cu^{2+}, Zn^{2+} and Ni^{2+}-ions on OSAC-3

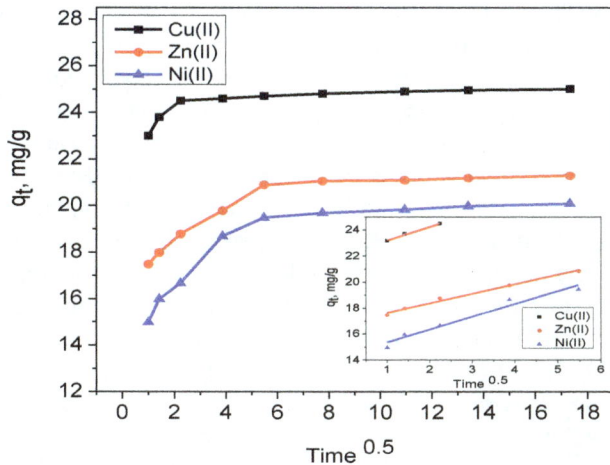

Fig. 10. Intra-particle diffusion plots for sorption of Cu^{2+}, Zn^{2+} and Ni^{2+}-ions on OSAC-3

After comparing the correlation coefficients (R^2) of the three kinetic models, the pseudo second order kinetic model is found to be the best model to fit the experiment data for all the investigated cations.

4.4. Equilibrium concentrations and sorption isotherms

The properties of sorption process depend not only on the properties of the sorbents, but also on the concentration of the metal ion solution. The initial metal ion concentration provides an important driving force to overcome all mass transfer resistances of the cations between aqueous solution and solid phase [26]. The effect of the initial concentration of heavy metal ions was studied at a pH of 5.7 and a shaking time of 2 hours. In Fig.5, it was noticed that the percentage removal efficiency values decreased from 97.7 % to 72.2% for Copper, 84% to 52% for Zinc, and 48.48% to 44% for Nickel, respectively, with an increase in the metal ion concentration from 5 to 25 mg/l by keeping all other parameters constant. Such a high efficiency of OSAC-3 toward the investigated ions make it applicable for the removal of Cu(II), Zn(II) and Ni(II) ions from low concentrated solutions such as those generated from painting/plating processes by Copper, Zinc, and Nickel metals. The effect of initial concentrations of heavy metal ions was studied and the isotherms are illustrated in Fig. 11. An adsorption isotherm is a basic representation showing the relationship between the amount adsorbed by a unit weight of adsorbent and the amount of adsorbate remaining in a test solution at equilibrium, Fig.11. It means it is the distribution of solute between the aqueous and solid phases at various equilibrium concentrations.

The obtained equilibrium sorption experimental data were tested using the commonly used Langmuir and Freundlich isotherm models. The Langmuir isotherm is based on the assumptions that: (i) all sites are equivalent; (ii) adsorption of solutes produces in monolayer coverage; (iii) the adsorbate molecule is adsorbed on a site independent of

the neighboring adsorbed molecules; (iv) coverage is independent of binding energy [27]. It is shown as follows:

$$\frac{1}{q_e} = \frac{1}{Q^o} + \frac{1}{(bQ^o)}\left(\frac{1}{C_e}\right) \qquad (5)$$

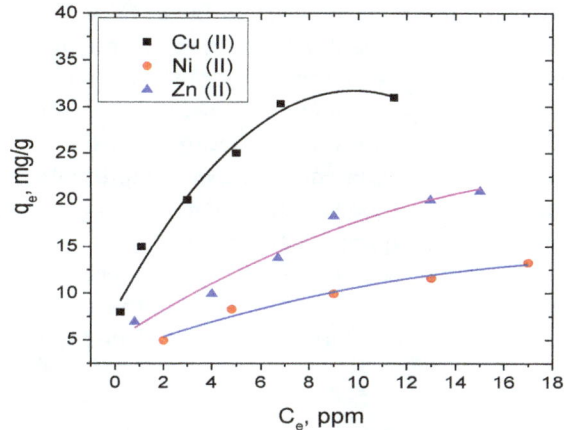

Fig. 11. Effect of initial solute concentration on sorption of Cu^{2+}, Zn^{2+} and Ni^{2+} on OSAC-3.

where q_e is the amount of metal ion sorbed per gram of the adsorbent at equilibrium in mg/g; C_e is the equilibrium concentration of metal ions left in solution at equilibrium in mg/l; Q^o is the maximum adsorption capacity in mg/g to form a complete monolayer coverage on the solid surface; and b is the Langmuir constant associated to the adsorption energy. Q^o and b have been calculated from the intercept, $1/Q^o$, and the slope, $1/(bQ^o)$, of plotting $1/q_e$ versus $1/C_e$, which give straight lines for all the investigated ions. The results are illustrated in Figure 12 (a). The essential characteristics and the feasibility of the Langmuir isotherm can be expressed by Hall et al. (1966) [22] in terms of a dimensionless constant separation factor R_L, which is defined as Eq. (6):

$$R_L = \frac{1}{1 + bC_o} \qquad (6)$$

where C_o is the initial metal ion concentration (mg/l) and b is the Langmuir constant. The value of R_L is announced to the status of the adsorption isotherm to be either unfavorable ($R_L > 1$), linear ($R_L = 1$), favorable ($0 < R_L < 1$), or irreversible ($R_L = 0$). The R_L values are calculated for the adsorption of Cu^{2+}, Zn^{2+} and Ni^{2+} ions, and they are found to be 0.0243, 0.0571 and 0.5880, respectively. This indicates that the adsorption of Cu^{2+}, Zn^{2+} and Ni^{2+} on OSAC-3 are favorable. The Freundlich model is an empirical model which is based on the sorption on a heterogeneous surface with exponential variation active site energies. The linear form of the Freundlich isotherm model is expressed by the following equation [28]:

$$\log q_e = \log k_f + \frac{1}{n}\log C_e \qquad (7)$$

Where k_f and n are empirical constants that indicate respectively adsorption capacity and adsorption intensity. Figure 12 (b) shows the linear plot of $\log q_e$ against $\log C_e$ for the adsorption of Cu^{2+}, Zn^{2+} and Ni^{2+} onto OSAC-3. The intercept and slope were used to calculate k_f and n-values, respectively. The Langmuir and Freundlich constants in addition with R^2-values are given in Table (4). According to the regression coefficient values, R^2, in Table 4, it can be determined that the adsorption isotherms can be well described by the Langmuir model rather than the Freundlich one for Cu^{2+} and Ni^{2+} ions. The opposite was shown for Zn^{2+} ions, where the experimental data were represented well by the Freundlich model rather than the Langmuir one. Furthermore, the experimental data for all the studied ions were represented by the Langmuir and Freundlich models because of the good R^2-values for both of the investigated isotherm models. Therefore, the sorption of the cations was attributed to the mixed mechanisms of ion-exchange as well as to the adsorption process.

According to Figure 13, an aqueous alkaline medium was obtained during the adsorption process, indicating that the ion exchange was the binding mechanism [29]. During the ion-exchange process, metal ions slipped through the pores and channels of the adsorbent material and replaced the elements in the OSAC-3 structure (potassium, calcium and magnesium, as shown in Table 2). In this work, the diffusion of ions was faster through the micro-pores of the OSAC-3 material; the ionic radii of the metal ions Cu^{2+}, Zn^{2+} and Ni^{2+} (0.72A°, 0.74 A° and 0.69 A°, respectively) are smaller than the pores radii of the OSAC material (average OSAC-3 pore radius is 11 A°). According to the uptake amount and the maximum adsorption capacities, Q_o, the selectivity sequence of the studied metal ions by OSAC-3 can be given as $Cu^{2+} > Zn^{2+} > Ni^{2+}$. This is usually attributed to the differences in metal characteristics and the resultant affinity for sorption sites.

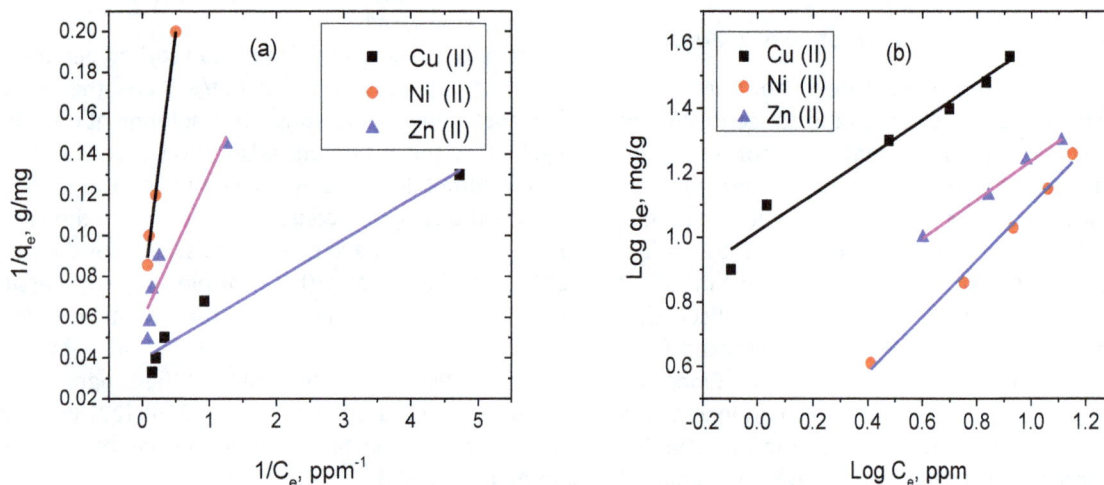

Fig. 12 (a,b). Langmuir and Freundlich plots of sorption of Cu^{2+}, Zn^{2+} and Ni^{2+} on OSAC-3.

Table 4. Langmuir and Freundlich constants with linear fit correlation coefficient (R^2).

Metal ions	Langmuir constants			Freundlich constants		
	Q_o	b	R^2	K_f	n	R^2
Cu^{2+}	25.381	2.010	0.980	12.830	1.740	0.951
Zn^{2+}	18.950	0.825	0.955	4.318	1.658	0.996
Ni^{2+}	14.650	0.259	0.993	3.809	2.252	0.983

Many investigators have tried to explain the action of the physicochemical properties of metal ions, as ionic radius, electron configuration and electronegativity on the mechanism of adsorption [30]. In this study, there was a linear relation between the metal ion uptake and the ionic radius of the ions. The ionic radius (r`) of metal ions considered through this investigation takes the following order: $K^+(r`=1.38A°) > Ca^{2+}(r`=1.02A°) > Mg^{2+} (r`=0.86Ao) > Zn^{2+} (r`=0.74A°) > Cu^{2+} (r`=0.72A°) > Ni^{2+}(r`=0.69A°) > H^+ (r`=0.154A°)$ [31]. Thus, all these heavy metal ions have a smaller ionic radius than K^+, Ca^{2+} and Mg^{2+} ions and the heavy metals can be incorporated in the molecular structure of OSAC-3 replacing K^+, Ca^{2+} and Mg^{2+} ions from there, this is according to the cation exchange theory. A preferential sorption of Cu^{2+} ions against Zn^{2+} ions can be explained by the difference in the electronegativity (E), which was higher for Cu^{2+} ions (E=1.90) than for Zn^{2+} ions (E=1.65); also, the Pauling electronegativity for all the cations Cu^{2+} (E=1.90), Zn^{2+} (E=1.65) and Ni^{2+} (E=1.91) were higher than that for K^+ (E=0.82), Ca^{2+} (E=1.0) and Mg^{2+}

(E=1.31). This indicated that all the investigated cations can replace K^+, Ca^{2+} and Mg^{2+} ions [32], as shown in Fig.5. When taking the electronic configuration of Cu^{2+} ions into consideration, it was noticed that Cu^{2+} ions have one unpaired electron at the 3d-orbital as shown in Table (5); this means it is a paramagnetic metal. Subsequently, Cu^{2+} ions can be attracted by the magnetic field, possibly in the adsorbent in which the ash contained in the OSAC-3 has iron oxide. The main oxides content as tested by EDX is reported in Table 2. OSAC-3 had about 8 % ash content. On the other hand, the electronic configuration of Ni^{2+} and Zn^{2+} ions is stable, with no unpaired electrons at the 3d-orbital as shown in Table (5) and they will be repelled by a OSAC-3 magnetic field. This is perhaps one of the reasons for the lower adsorption capacity of OSAC-3 adsorbent for nickel and zinc ions in comparison to the adsorption of copper ions. The electronegativities (E) and ionic radii of Ni^{2+}, and Zn^{2+} are 1.91, 1.65 and 0.69, 0.74A°, respectively. The average electric dipole polarizabilities (D) of Ni, and Zn atoms are 6.8×10^{-24} cm^3 and 7.1×0^{-24} cm^3, respectively [33-34]. Puls and Bohn (1988) [34] explained the metal sorption

capacity using the concept of the conventional hard–soft acid–base (HSAB) principle.

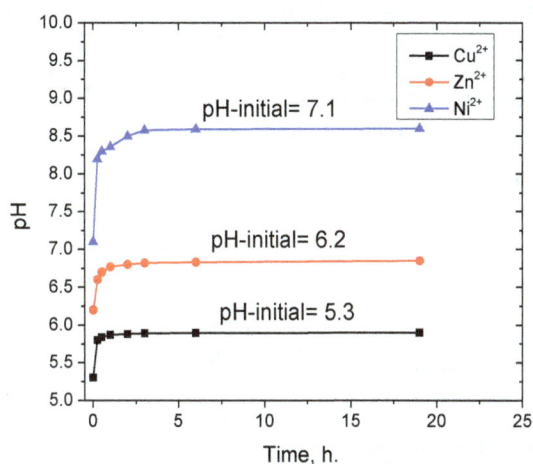

Fig. 13. Effect of time on changing of the initial pH of Cu^{2+}, Zn^{2+} and Ni^{2+} solution adsorbed on OSAC-3.

Table 5. Physicochemical properties of the cations

Character\ Ion	Cu^{2+}	Zn^{2+}	Ni^{2+}	K^+	Ca^{2+}	Mg^{2+}	H^+
Ionic radius, A°	0.720	0.740	0.690	1.380	1.020	0.860	0.154
Electron configuration	$3d^9 4s^2$	$3d^{10}$	$3d^8 4s^2$	$4s^1$	$4s^2$	$3s^2$	$1s^1$
Pauling electronegativity	1.90	1.65	1.91	0.82	1.00	1.31	---

According to the HSAB principle, hard Lewis acids choose to form complexes with hard Lewis bases, and soft acids prefer to make complexes with soft bases. The word "hard" means high electronegativity, low polarizability, and small ionic size, while vise-versa for "soft" ions. The sorption capacity of Ni^{2+} and Zn^{2+} by OSAC-3 follows the order of increasing ionic radii and polarizability as well as decreasing electronegativity and thus decreasing hardness. In this work, the OSAC-3 give a high adsorption of Zn^{2+} as softer ions compared with the less soft ions of Ni^{2+}. The OSAC-3 had a pH of 7.2, which means an alkaline surface of granular activated carbon OSAC-3. Therefore, OSAC-3 was used as a relatively soft Lewis base. The active sites of the adsorbent appeared to form the most stable complexes with the softer cations (as soft acid). Therefore, the adsorbent surface has a high ability for complexation towards the soft acids (as Zn^{2+} ions) [35]. Hence, the sequence of metal sorption capacities in this research was found as $Cu^{2+}>Zn^{2+}>Ni^{2+}$. A comparison of the maximum sorption capacity of Cu(II), Zn(II) and Ni(II) and the surface area of the investigated OSAC-3 with other activated carbons derived from other precursors with different activation methods is shown in Table 6. Generally, OSAC-3 has a good surface area with a higher sorption capacity for Cu(II), Zn(II) and Ni(II) than other precursors; thus, it was recommended to use OSAC-3 for the removal of C(II), Zn(II) and Ni(II)-ions from aqueous solutions.

4.5. Thermodynamics study

Temperature is one of the important parameters affecting the rate of the sorption process. To study the effect of the temperature (285, 298, 308 and 323 K) on the adsorption of Cu(II), Zn(II) and Ni(II) ions, the experiments were carried out at constant concentrations of 20 mg/L for all the investigated cations. The data showed that by increasing the temperature from 285K to 323K, the uptake decreased from 29.43mg/g to 24.55mg/g for Cu(II) ions. And the adsorption of Zn(II) and Ni(II) increased from 18.52mg/g to 20.70mg/g and from 16.05mg/g to 20.18mg/g, respectively. To study the nature of the adsorption process, the thermodynamic parameters of free energy change (ΔG°), enthalpy change (ΔH°), and entropy change (ΔS°) were calculated using the following equations:

$$\Delta G^o = -RT \, lnK_c \quad (8)$$

$$logK_c = \left(\frac{\Delta S^o}{2.303}\right) - \left(\frac{\Delta H^o}{2.303R}\right)\left(\frac{1}{T}\right) \quad (9)$$

$$\Delta G^o = \Delta H^o - T\,\Delta S^o \quad (10)$$

where R is the gas constant (8.3143 JK^{-1} mol^{-1}), T is the absolute temperature in Kelvin and K_c is the equilibrium constant. K_c-values are calculated as the distribution of the cations between solid surface and solution ($K_c = {q_e}/{C_e}$).

The values of ΔG^o were calculated from Eq. (8) at different temperatures and the ΔH^o-values were obtained from the slope of a plot $logK_c$ versus 1/T (Fig. 14) according to Eq. (9) as linear regression analysis. The ΔS^o-values were calculated from Eq. (10). The values of the studied thermodynamic parameters are given in Table 7. The enthalpy of the adsorption, $\Delta H°$, is a measure of the energy barrier that must be overcome by the reacting cations [50].

The positive value of $\Delta H°$ for Zn^{2+} and Ni^{2+} ions suggests that the adsorption reactions onto the adsorbent are endothermic in nature, but for Cu^{2+} ions, the negative value of $\Delta H°$ means an exothermic reaction on the adsorbent surface. Therefore, this means that increasing temperature will favor the adsorption of Zn^{2+} and Ni^{2+} ions and hinder the adsorption for Cu^{2+} ions onto the adsorbent.

Table 6. Surface areas and maximum metal ions (Ni(II), Cu(II) and Cd(II)) adsorption capacity of the activated carbons derived from different carbon precursors by different activation methods.

Carbon precursor	Activating agent	Surface area (m^2/g)	Q_{max} (mg/g)			Reference
			Cu(II)	Zn(II)	Ni(II)	
Phragmites australis	H_3PO_4	894.5	5.421		22.88	[34]
	$(NH_3)_3PO_4$	444.9	6.982		34.04	[34]
	$(NH_3)_2HPO_4$	495.7	6.429		31.81	[34]
	$NH_3H_2PO_4$	408.5	6.213		31.40	[34]
Commercial GAC (F 400)	---	960			3.12	[35]
Commercial GAC (F 200)	---	790	24.10			[36]
Ceiba pentandra	Steam	521	20.8			[37]
Commercial carbon	Tannic acid	325.1	2.23			[38]
Apricot	K_2CO_3	770			32.36	[39]
Briquette	Steam	719			18.79	[40]
Almond husk	H_2SO_4				37.17	[41]
Hazelnut shell	H_2SO_4	441			11.64	[42]
Local wood	--	1400		20.52		[43]
coconut shell (ACABPEX.)	Potassium ethyl xanthate with 4 M nitric acid solution	804.0		17.544		[44]
Residual from solvent extracted olive pulp (ACOP)	steam/nitrogen mixture	364		32.68		[45]
Olive stone (ACO)	steam/N_2 mixture	474		16.08		[45]
Apricot stone (ACA)	steam/N_2 mixture	486		13.175		[45]
Peach stone (ACP)	steam/N_2 mixture	660		6.370		[45]
apple pulp	$ZnCl_2$/ N_2 gas	1067.1		11.72		[46]
olive oil industrial solid waste	N_2/steam gas	850	25.381	18.950	14.650	The present work

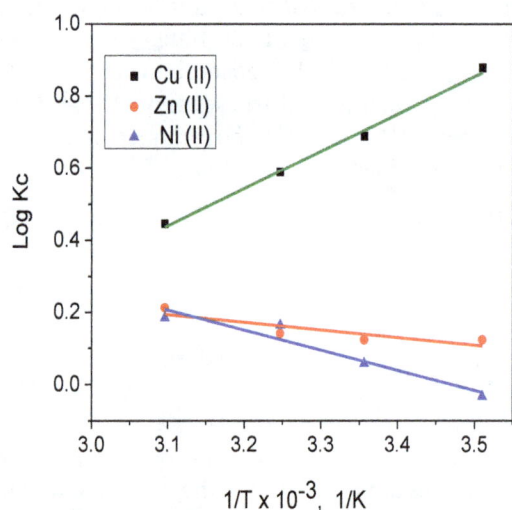

Fig. 14. Effect of temperature on the distribution coefficient

The positive entropy change ($\Delta S°$) values for all the studied cation at all investigated temperatures, except for Ni^{2+} ions at 285K, showed an increase in the randomness at the solid/solution interface during the sorption process. On the other hand, the negative entropy change value for Ni^{2+} ions at 285K corresponded to a decrease in the degree of freedom of the adsorbed nickel species. The degree of freedom decreased with a decrease in temperature for Zn^{2+} and Ni^{2+} ions, the opposite was noticed for the sorption of Cu^{2+}-ions.

Negative $\Delta G°$-values indicated a spontaneous nature of sorption process for all the studied cations at all the investigated temperatures except at a temperature of 285K for Ni^{2+}-ions. The negativity of $\Delta G°$-values decreased with decreasing temperature for Zn^{2+} and Ni^{2+} ions, the opposite was noticed for Cu^{2+} ions. On the other hand, positive or small negative values of Free energy change ($\Delta G°$) could suggest that the reaction required a small amount of energy. It can be seen from Table 7 that the $\Delta G°$ are positive for Ni^{2+} at 285K, while a low negative-value under other temperatures for Ni^{2+} and Zn^{2+} meant that small external energy input was needed for the adsorption process of Ni^{2+} and Zn^2 ions on OSAC-3. This was confirmed by the less

positive values of $\Delta H°$. On the other hand, high negative values of $\Delta G°$ for Cu^{2+} ions suggested a spontaneous adsorption reaction with no need for external energy, and this was supported by the high negative values of $\Delta H°$.

Table 7. Thermodynamic parameters for the cations adsorption on OSAC-3.

T, K.	Cu^{2+}			Zn^{2+}			Ni^{2+}		
	$\Delta G°$	$\Delta H°$	$\Delta S°$	$\Delta G°$	$\Delta H°$	$\Delta S°$	$\Delta G°$	$\Delta H°$	$\Delta S°$
285	-4.78		16.72	-0.53		1.86	0.18		-0.58
298	-3.93		13.12	-0.71		2.39	-0.34		1.19
308	-3.48	-19.77	11.23	-0.84	4.06	2.75	-0.97	10.63	3.19
232	-2.76		8.48	-1.33		4.12	-1.15		3.6

5. Conclusions

In the present study, the removal of Cu^{2+}, Zn^{2+} and Ni^{2+} ions from aqueous solution by adsorption on olive stone activated carbon via batch technique was investigated. Microporous OSAC-3 material with a high surface area of $850 m^2/g$ was obtained from physical activated olive stone waste. The results showed that the removal of the three cations was favorable at a higher pH, but the kinetic, isotherms, and thermodynamic studies were operated at a pH of 5.7 to ensure that no precipitation occurred for the studied cations. It was observed that during adsorption of the three cations, an increase in the solution pH was noticed with agitation time. This indicated that there was an ion-exchange mechanism occurring simultaneous with the adsorption of Cu^{2+}, Zn^{2+} and Ni^{2+} ions on the active sites of OSAC-3 surface. The rate of adsorption for the three elements was very fast in the first 5min., reaching equilibrium after 30min. for Cu^{2+}-ions, after 60min. for Zn^{2+}, and after 120min. for Ni^{2+} ions adsorption. The experimental data were successfully described by the pseudo-second order model. The intra-particle diffusion model showed that the adsorption mechanism of Cu^{2+}, Zn^{2+} and Ni^{2+} ions was a complex mechanism, where the intra-particle diffusion participated in the overall rate of the adsorption process, and it was not the only rate-determining step. The equilibrium adsorption data was described by both Langmuir and Freundlich models and the adsorption capacity of OSAC-3 decreased in the following sequence: $25.381 \ mgCu^{2+}/g > 18.950 \ mgZn^{2+}/g > 14.650 \ mgNi^{2+}/g$. The thermodynamic parameters showed that the negative values of ΔG for the three metals (except for adsorption of Ni^{2+} ions at temperature 285K) indicated spontaneous adsorption of the cations onto OSAC-material. The positive ΔH-values of Zn^{2+} and Ni^{2+} ions confirmed the endothermic adsorption process while the negative value of ΔH referred to the exothermic nature for the sorption process for Cu^{2+} ions. Therefore, the temperature increased

the adsorption of Zn^{2+} and Ni^{2+} ions at equilibrium and hindered the adsorption of Cu^{2+} ions onto OSAC-3. The positive values of ΔS for all the studied cations (except for adsorption of Ni^{2+} ions at 285K) clarified an increase in the randomness at solid/solution interface during the adsorption process. The obtained results indicated that OSAC-3 has the potential to be used as a low-cost material for the sorption of Cu^{2+}, Zn^{2+} and Ni^{2+} ions from aqueous media.

Acknowledgment

The author would like to thank the chairman of the olive oil factory in the Wahet Sewa region in Egypt for supplying the olive oil industrial solid waste.

References

[1] Jayakumar, R., Rajasimman, M., Karthikeyan, C. (2015). Optimization, equilibrium, kinetic, thermodynamic and desorption studies on the sorption of Cu (II) from an aqueous solution using marine green algae: Halimeda gracilis. *Ecotoxicology and environmental safety, 121*, 199-210.

[2] Kula, I., Uğurlu, M., Karaoğlu, H., Celik, A. (2008). Adsorption of Cd (II) ions from aqueous solutions using activated carbon prepared from olive stone by ZnCl 2 activation. *Bioresource technology, 99*(3), 492-501.

[3] Meena, A. K., Mishra, G. K., Rai, P. K., Rajagopal, C., Nagar, P. N. (2005). Removal of heavy metal ions from aqueous solutions using carbon aerogel as an adsorbent. *Journal of hazardous materials, 122*(1), 161-170.

[4] Depci, T., Kul, A. R., Önal, Y. (2012). Competitive adsorption of lead and zinc from aqueous solution on activated carbon prepared from Van apple pulp: study in single-and multi-solute systems. *Chemical engineering journal, 200*, 224-236.

[5] Koplan, J. P. (1999). Toxicological profile for Chlorophenols. Agency for Toxic Substances and

Disease Registry (ATSDR), US Department of Health and Human Services. *Public Health Service*.

[6] Bohli, T., Ouederni, A., Fiol, N., Villaescusa, I. (2015). Evaluation of an activated carbon from olive stones used as an adsorbent for heavy metal removal from aqueous phases. *Comptes rendus chimie, 18*(1), 88-99.

[7] Hafshejani, L. D., Nasab, S. B., Gholami, R. M., Moradzadeh, M., Izadpanah, Z., Hafshejani, S. B., Bhatnagar, A. (2015). Removal of zinc and lead from aqueous solution by nanostructured cedar leaf ash as biosorbent. *Journal of molecular liquids, 211*, 448-456.

[8] Rivera-Utrilla, J., Sánchez-Polo, M., Gómez-Serrano, V., Alvarez, P. M., Alvim-Ferraz, M. C. M., Dias, J. M. (2011). Activated carbon modifications to enhance its water treatment applications. An overview. *Journal of hazardous materials, 187*(1), 1-23.

[9] Alslaibi, T. M., Abustan, I., Ahmad, M. A., Abu Foul, A. (2013). Effect of different olive stone particle size on the yield and surface area of activated carbon production. In *Advanced materials research* (Vol. 626, pp. 126-130). Trans tech publications.

[10] Awwad, N. S., El-Zahhar, A. A., Fouda, A. M., Ibrahium, H. A. (2013). Removal of heavy metal ions from ground and surface water samples using carbons derived from date pits. *Journal of environmental chemical engineering, 1*(3), 416-423.

[11] Alslaibi, T. M., Abustan, I., Ahmad, M. A., Foul, A. A. (2013). Review: Comparison of agricultural by-products activated carbon production methods using surface area response. *CJASR, 2*, 18-27.

[12] Alslaibi, T. M., Abustan, I., Ahmad, M. A., Foul, A. A. (2013). A review: production of activated carbon from agricultural byproducts via conventional and microwave heating. *Journal of chemical technology and biotechnology, 88*(7), 1183-1190.

[13] Alslaibi, T. M., Abustan, I., Ahmad, M. A., Foul, A. A. (2014). Kinetics and equilibrium adsorption of iron (II), lead (II), and copper (II) onto activated carbon prepared from olive stone waste. *Desalination and water treatment, 52*(40-42), 7887-7897.

[14] Lin, L., Zhai, S. R., Xiao, Z. Y., Song, Y., An, Q. D., Song, X. W. (2013). Dye adsorption of mesoporous activated carbons produced from NaOH-pretreated rice husks. *Bioresource technology, 136*, 437-443.

[15] Olèico (2009) European awareness raising campaign for an environmentally sustainable olive mill waste management. available at LIFE07 INF/IT/000438

[16] Ghanbari, R., Anwar, F., Alkharfy, K. M., Gilani, A. H., Saari, N. (2012). Valuable nutrients and functional bioactives in different parts of olive (Olea europaea L.) — a review. *International journal of molecular sciences, 13*(3), 3291-3340.

[17] Daifullah, A. H. M., Rizk, M. A., Aly, H. M., Yakout, S. M., Hassen, M. R. Treatment of some organic pollutants (THMs) using activated carbon derived from local agro-residues.

[18] Puziy, A. M., Poddubnaya, O. I., Martínez-Alonso, A., Castro-Muniz, A., Suárez-García, F., Tascón, J. M. (2007). Oxygen and phosphorus enriched carbons from lignocellulosic material. *Carbon, 45*(10), 1941-1950.

[19] Jayakumar, R., Rajasimman, M., Karthikeyan, C. (2015). Optimization, equilibrium, kinetic, thermodynamic and desorption studies on the sorption of Cu (II) from an aqueous solution using marine green algae: Halimeda gracilis. *Ecotoxicology and environmental safety, 121*, 199-210.

[20] Beveridge, T. J. (1986). Biotechnology and bioengineering symposium No. 16: Biotechnology for the mining, metal-refining, and fossil fuel processing industries.

[21] Ho, Y. S., & McKay, G. (1999). The sorption of lead (II) ions on peat. *Water research, 33*(2), 578-584.

[22] Hall, K. R., Eagleton, L. C., Acrivos, A., Vermeulen, T. (1966). Pore-and solid-diffusion kinetics in fixed-bed adsorption under constant-pattern conditions. *Industrial & engineering chemistry fundamentals, 5*(2), 212-223.

[23] Aksas, H., Boureghda, M.Z., Babaci, H. and Louhab, K. (2015) Kinetics and Thermodynamics of Cr Ions Sorption on Mixed Sorbents prepared from Olive Stone and Date pit from Aqueous Solution. International journal of food and biosystem engineering, 1(1), 1-8.

[24] Weber, W. J., Morris, J. C. (1963). Kinetics of adsorption on carbon from solution. *Journal of the sanitary engineering division, 89*(2), 31-60.

[25] John, A. C., Ibironke, L. O., Adedeji, V., Oladunni, O. (2011). Equilibrium and kinetic studies of the biosorption of heavy metal (cadmium) on Cassia siamea Bark. *American-Eurasian journal of scientific research, 6*(3), 123-130.

[26] Aksu, Z., Akpınar, D. (2000). Modelling of simultaneous biosorption of phenol and nickel (II) onto dried aerobic activated sludge. *Separation and purification technology, 21*(1), 87-99.

[27] Langmuir, I. (1916). The constitution and fundamental properties of solids and liquids. J. Am. *Journal of the American chemical society, 38*(11), 2221-2295.

[28] Uber, F. H. (1985). Die adsorption in losungen. *Zeitschrift for physikalische chemie, 57*, 387-470.

[29] Brady, J. M., Tobin, J. M. (1994). Adsorption of metal ions by Rhizopus arrhizus biomass: Characterization studies. *Enzyme and microbial technology, 16*(8), 671-675.

[30] Şengil, İ. A., Özacar, M. (2009). Competitive biosorption of Pb^{2+}, Cu^{2+} and Zn^{2+} ions from aqueous solutions onto valonia tannin resin. *Journal of hazardous materials, 166*(2), 1488-1494.

[31] Shannon, R. T. (1976). Revised effective ionic radii and systematic studies of interatomic distances in halides

and chalcogenides. *Acta crystallographica section A: Crystal physics, diffraction, theoretical and general crystallography, 32*(5), 751-767.

[32] Gorgievski, M., Božić, D., Stanković, V., Štrbac, N., Šerbula, S. (2013). Kinetics, equilibrium and mechanism of Cu^{2+}, Ni^{2+} and Zn^{2+} ions biosorption using wheat straw. *Ecological engineering, 58*, 113-122.

[33] Liu, C. L., Chang, T. W., Wang, M. K., Huang, C. H. (2006). Transport of cadmium, nickel, and zinc in Taoyuan red soil using one-dimensional convective–dispersive model. *Geoderma, 131*(1), 181-189.

[34] Puls, R. W., Bohn, H. L. (1988). Sorption of cadmium, nickel, and zinc by kaolinite and montmorillonite suspensions. *Soil science society of America journal, 52*(5), 1289-1292.

[35] Sullivan, P. J. (1977). The principle of hard and soft acids and bases as applied to exchangeable cation selectivity in soils. *Soil science, 124*(2), 117-121.

[36] Guo, Z., Fan, J., Zhang, J., Kang, Y., Liu, H., Jiang, L., Zhang, C. (2016). Sorption heavy metal ions by activated carbons with well-developed microporosity and amino groups derived from Phragmites australis by ammonium phosphates activation. *Journal of the Taiwan institute of chemical engineers, 58*, 290-296.

[37] Nguyen, T. A. H., Ngo, H. H., Guo, W. S., Zhang, J., Liang, S., Yue, Q. Y., Nguyen, T. V. (2013). Applicability of agricultural waste and by-products for adsorptive removal of heavy metals from wastewater. *Bioresource technology, 148*, 574-585.

[38] Ozsoy, H. D., Kumbur, H., Saha, B., Van Leeuwen, J. H. (2008). Use of Rhizopus oligosporus produced from food processing wastewater as a biosorbent for Cu (II) ions removal from the aqueous solutions. *Bioresource technology, 99*(11), 4943-4948.

[39] Rao, M. M., Ramesh, A., Rao, G. P. C., Seshaiah, K. (2006). Removal of copper and cadmium from the aqueous solutions by activated carbon derived from Ceiba pentandra hulls. *Journal of hazardous materials, 129*(1), 123-129.

[40] Üçer, A., Uyanik, A., Aygün, Ş. F. (2006). Adsorption of Cu (II), Cd (II), Zn (II), Mn (II) and Fe (III) ions by tannic acid immobilised activated carbon. *Separation and purification technology, 47*(3), 113-118.

[41] Erdoğan, S., Önal, Y., Akmil-Başar, C., Bilmez-Erdemoğlu, S., Sarıcı-Özdemir, Ç., Köseoğlu, E., Icduygu, G. (2005). Optimization of nickel adsorption from aqueous solution by using activated carbon prepared from waste apricot by chemical activation. *Applied surface science, 252*(5), 1324-1331.

[42] Wilson, K., Yang, H., Seo, C. W., Marshall, W. E. (2006). Select metal adsorption by activated carbon made from peanut shells. *Bioresource technology, 97*(18), 2266-2270.

[43] Hasar, H. (2003). Adsorption of nickel (II) from aqueous solution onto activated carbon prepared from almond husk. *Journal of hazardous materials, 97*(1), 49-57.

[44] Demirbaş, E., Kobya, M., Öncel, S., Şencan, S. (2002). Removal of Ni (II) from aqueous solution by adsorption onto hazelnut shell activated carbon: equilibrium studies. *Bioresource technology, 84*(3), 291-293.

[45] Kouakou, U., Ello, A. E. S., Yapo, J. A., Trokourey, A. (2013). Adsorption of iron and zinc on commercial activated carbon. *Journal of environmental chemistry and ecotoxicology, 5*(6), 168-171.

[46] Behnamfard, A., Salarirad, M. M., Vegliò, F. (2014). Removal of Zn (II) ions from aqueous solutions by ethyl xanthate impregnated activated carbons. *Hydrometallurgy, 144*, 39-53.

[47] Galiatsatou, P., Metaxas, M., Kasselouri-Rigopoulou, V. (2002). Adsorption of zinc by activated carbons prepared from solvent extracted olive pulp. *Journal of hazardous materials, 91*(1), 187-203.

[48] Depci, T., Kul, A. R., Önal, Y. (2012). Competitive adsorption of lead and zinc from aqueous solution on activated carbon prepared from Van apple pulp: study in single-and multi-solute systems. *Chemical engineering journal, 200*, 224-236.

[49] Alslaibi, T. M., Abustan, I., Ahmad, M. A., Abu Foul, A. (2014). Preparation of activated carbon from olive stone waste: optimization study on the removal of Cu^{2+}, Cd^{2+}, Ni^{2+}, Pb^{2+}, Fe^{2+}, and Zn^{2+} from aqueous solution using response surface methodology. *Journal of dispersion science and technology, 35*(7), 913-925.

[50] Unuabonah, E. I., Adebowale, K. O., Olu-Owolabi, B. I., Yang, L. Z., Kong, L. (2008). Adsorption of Pb (II) and Cd (II) from aqueous solutions onto sodium tetraborate-modified Kaolinite clay: equilibrium and thermodynamic studies. *Hydrometallurgy, 93*(1), 1-9.

6

Magnetite nanoparticles coated with methoxy polyethylene glycol as an efficient adsorbent of diazinon pesticide from water

Mahboubeh Saeidi[1], Atena Naeimi *[,2], Marzie Komeili[1]
[1]Department of Chemistry, Faculty of Science, Vali-e- Asr University of Rafsanjan, Iran
[2]Faculty of Science, Department of Chemistry, University of Jiroft, Jiroft, Iran

Keywords:
Adsorption
Spectrophotometry
Pesticides
Fe_3O_4

ABSTRACT

Methoxy polyethylene glycol modified magnetite nanoparticles (PEGMNs) were synthesized and characterized by scanning electron microscopy (SEM), vibrating sample magnetometer (VSM), and X-ray diffraction (XRD). The adsorption of diazinon onto PEGMNs was investigated by UV-Vis spectrophotometry at 236 nm, through batch experiments. The effects of adsorbent dosage, solution pH, contact time, solution temperature and water impurities on the adsorption of diazinon onto PEGMNs were investigated. The process of adsorption was increased rapidly in the first contact period of 10 min. The adsorption at equilibrium (qe) was found to increase with increasing pH. The results of diazinon removal at various PEGMNs dosages demonstrated that the optimum dose of PEGMNs was 1mg. The amount of adsorption of diazinon at equilibrium increased with an increasing temperature from 15°C to 45°C that indicateds an endothermic process. Therefore, PEGMNs were used as an efficient absorbent for the removal of diazinon.

1. Introduction

The use of large quantities of pesticides, which includes insecticides in agriculture, is one of the main sources of pollution of surface and ground water [1]. In fact, 17% of the 2.36 billion kg of pesticides used worldwide was insecticides. The Water Frame work Directive (WFD) (Directive, 2000/60/EC – *European Parliament* and Council of the European Union, 2000) established the environmental quality standards(EQS) for pesticides, their relevant metabolites, degradation and reaction products in 0.1 μg/L for individual compounds and 0.5 μg/L for the sum of pesticides in ground water [2]. Conventional technologies have been used to treat all types of organic and toxic waste by adsorption, biological oxidation, chemical oxidation and incineration. In parallel, advances in nanoscale science and engineering suggest that many of the current problems involving water quality could be resolved or greatly diminished by using nonabsorbent, nanocatalysts, and bioactive nanoparticles. In addition to

having high specific surface areas, nanoparticles also have unique adsorption properties due to different distributions of reactive surface sites and disordered surface regions. The mobility of nanomaterials in solution is high and the whole volume can be quickly scanned with small amounts of nanomaterials due to their small size. Magnetic separation has been applied recently in various fields such as analytical biochemistry [3], medical science [4] and biotechnology [5]. From an environmental point of view, on magnetic separation offers advantages due to the easy recovery of the adsorbent without filtration or centrifugation. Several studies have reported magnetic separation using modified magnetite (Fe_3O_4) as an environmentally friendly approach to remove heavy metal ions [7] and organic pollutants [8]. In this work, we attempt to use polyethylene glycol modified magnetic nanoparticles PEGMNs as an adsorbent for the removal of pesticide from aqueous solutions. Diazinon, an organophosphate insecticide, was selected for the present study as it is widely used in pest control, and high

*Corresponding author
E-mail address: at_naimi@yahoo.com, a.naeimi@ujiroft.ac.ir

residual levels had been detected in vegetables contact time, solution temperature, and water impurities were investigated on the adsorption of diazinon.

2. Experimental

2.1. Materials and Methods

Analytical grade diazinon for the experiment was purchased from the Fluka Co. (Germany). A diazinon stock solution of 40 mg/L was prepared in distilled water and kept in a refrigerator at 4 °C until use. All the other chemicals used, were as analytical grade and were purchased from Merck Co. (Germany). The standard solutions and working solutions were prepared by the appropriate dilution of the stock solutions.

2.2. Adsorption experiments and analysis

The adsorption of diazinon on adsorbents such as PEGMNs, Fe_3O_4, Silica-Coated Magnetite Nanoparticles, Iranian natural zeolite, and multiwall carbon nanotubes were investigated by UV-Vis spectrophotometry at 236 nm, through batch experiments. For each adsorption test, 5 mL of diazinon solution was transferred to the beaker, and the solution pH level was adjusted to the desired value; the given amount of adsorbents were added to the solution and the suspensions were subjected to ultrasonic waves to obtain a uniform dispersion. For PEGMNs, Fe_3O_4, and silica-coated magnetite nanoparticles, the mixture was allowed to stand and the adsorbent was precipitated at the bottom of the beaker by a strong magnet and the supernatant was decanted. For Iranian natural zeolite and multiwall carbon nanotubes, the adsorbents were collected by centrifuging and the supernatant was decanted. Then, it was transferred to a cm^{-1} quartz cell and the absorbance at 236 nm was considered for determination of any residual diazinon, using a Cary 100 UV spectrophotometer.

2.3. Synthesis of Magnetite Nanoparticles (Fe$_3$O$_4$, MNP)

A solution of $FeCl_2$ (5.40 g) and $FeCl_3$ (2.00 g) in aqueous hydrochloride acid (2.00 M, 25.00 mL) at room temperature, was sonicated until the salts dissolved completely. Aqueous ammonia (25%, 40.00 mL) was added slowly over 20 min to the mixture under Ar atmosphere at room temperature followed by stirring for about 30 min with a mechanical stirrer. The Fe_3O_4 nanoparticles were separated by an external magnet and washed three times with deionized water and ethanol. The final product was obtained after drying under a vacuum [16-17].

2.4. Synthesis of Silica-Coated Magnetite Nanoparticles (SMNP)

The synthesized Fe_3O_4 was suspended in 35.00 mL of ethanol and 6 mL of deionized water and sonicated for 15 min. 1.50 mL of tetraethyl orthosilicate (TEOS) was slowly added to the mixture and sonicated for 10 min. Then, aqueous ammonia (10%, 1.40 mL) was added slowly over 10 min under mechanical stirrer. The mixture was heated at 40 °C for 12 h. The iron oxide nanoparticles with a thin layer of silica ($Fe_3O_4@SiO_2$, SMNP) were separated by an external magnet and washed three times with ethanol and dried under a vacuum [16-17].

2.5. Synthesis of methoxy polyethylene glycol attached to amino-silane modified magnetic nanoparticles (PEGMNs)

1 eq of methoxy polyethylene glycol (1100 g/mol) was added in the dichloromethane. Then, 1 eq of triethylamine and acryloyl chloride were added to the reaction. After 24h, the mixture of the reaction was filtered to remove the triethylamine hydrochloride. By adding diethylether, Acrylated methoxy polyethylene glycol (AmPEG) was obtained. AmPEG and (3-aminopropyl) triethoxysilane were added to dry DMF. After 48h, 0.5 gr of Magnetite Nanoparticles was added and stirred for another 48h. The final sample was separated by an external magnet and washed three times with DMF and dried under a vacuum.

3. Results and discussion

Initially, Fe_3O_4 NPs were synthesized by a chemical co-precipitation technique of ferric and ferrous ions in an alkali solution and was coated by tetraethyl orthosilicate to obtained the SMNPs. The AmPEG was then allowed to react with an appropriate concentration of 3-aminopropyltrimethoxysilane to give amino-functionalized AmPEG. Then, SMNPs and amino functionlized AmPEG were reacted together to obtained PEGMNs. The size and structure of PEGMNs were evaluated using scanning electron microscopy (SEM). The SEM image (Figure 1) showed uniformity and spherical-like morphology of the nanoparticles with an average diameter from 20-30 nm.

FT-IR spectra of the PEGMNs are shown in Figure 1. The band at around 627–648 cm^{-1} was assigned to the stretching vibrations of the Fe-O bond in these compounds [35]. The peaks positioned at 3424 cm^{-1} and 2922 cm^{-1} in the FT-IR spectrum of the PEGMNs was related to the stretching and bending of the OH and CH bonds, respectively [18]. The SiO stretching bond was observed at about 1000–1110 cm^{-1} [18].

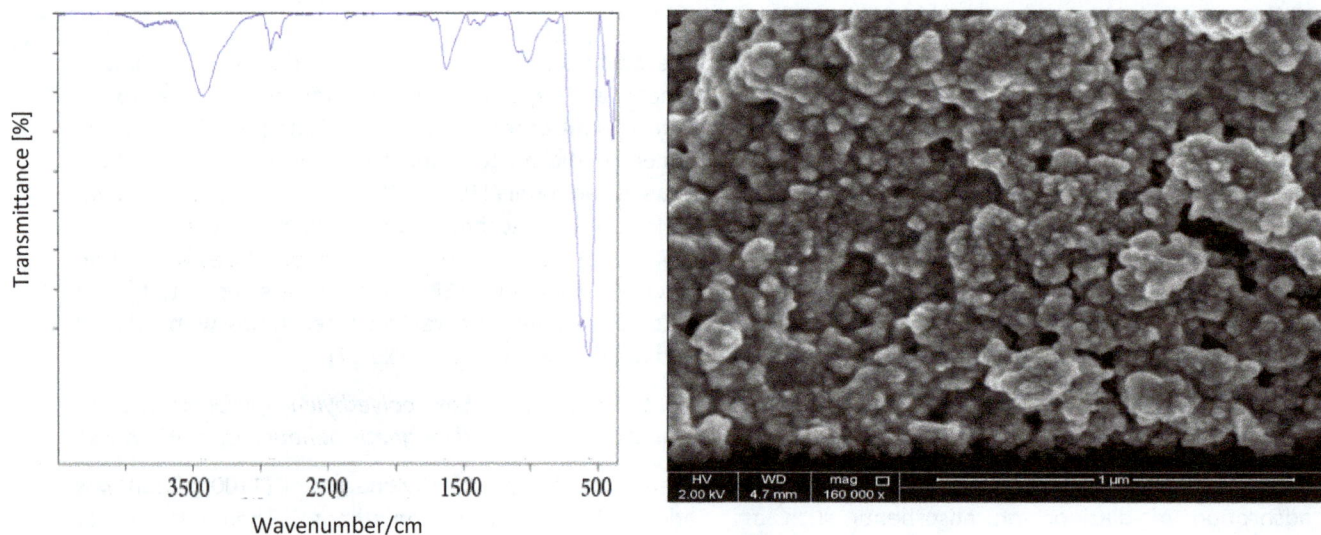

Fig. 1. IR and SEM of PEGMNs

3.1. Effect of adsorbent dosage

The adsorbent dosage is an important parameter because it determines the capacity of the adsorbent for a given diazinon concentration, and it also determines adsorbent–adsorbate equilibrium of the system [2]. Therefore, the effect of adsorbent dosage in the range of 0.1-20 mg on the diazinon adsorption was studied using a solution containing 5 mg/L diazinon. The percentage of the removal of diazinon increased from 55.1% at 0.1 mg to 72.7% at 2.5 mg of adsorbent dosage (Figure 2). The optimum dosage was found to be 1 mg, because after 1 mg, no significantly changes occurred. The improvement of diazinon removal with an increased dose of PEGMNs as an adsorbent can be attributed to the increased adsorbent surface area and the availability of active adsorption sites for a fixed number of diazinon molecules in the solution [12].

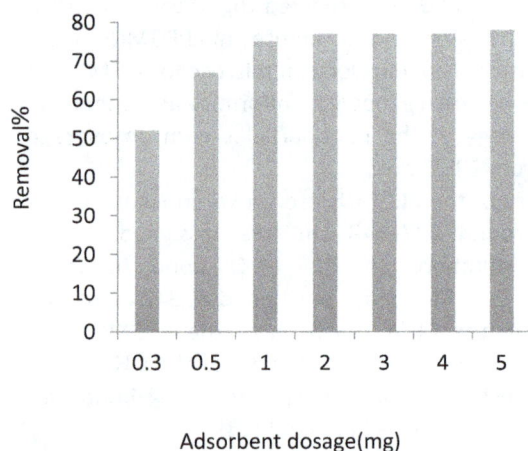

Fig. 2. Effect of adsorbent dosage on removal of diazinon onto (PEGMNs) (diazinon concentration: 5 mg /L PEGMNs dosage: 0.1–20 mg; contact time: 30 min)

3.2. Effect of pH on the diazinon removal

The adjustment of the pH has an important role for the removal of the compounds that can be protonated. In this study, the effect of pH on diazinon adsorption was investigated using a 5 mg/L of initial concentration of diazinon. The pH value varied from 2 to 10 to evaluate the effect of pH on the removal of diazinon onto PEGMNs. As shown in Figure 3, the equilibrium adsorption (q_e) increased by increasing the pH. This behavior suggested that the adsorption was dominated by the van der Waals interaction between the diazinon and adsorbent surface [3]. The existent $-OH$ groups on the surface of PEGMNs were predominantly $-OH_3^+$ in the aqueous medium. So a double layer with negative electric charge could be formed around the PEGMNs. On the other hand, diazinon molecules have positive electric charge in a pH of 2 because of the pK_a=2.6. Thus the van der Waals interaction between the cationic diazinon molecules and negative surface of the nanoparticles could not be improved at this pH level [13].

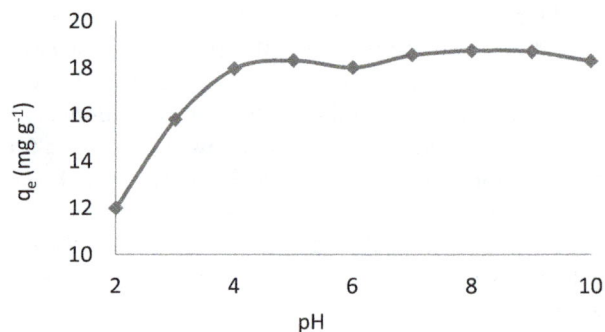

Fig. 3. Effect of solution pH on removal of diazinon onto (PEGMNs). (diazinon concentration: 5 mg/L; solution pH: 2–10; PEGMNs dosage: 1 mg; contact time: 30 min).

3.3. Effect of contact time on the diazinon removal

Another important parameter in the adsorption process is the contact time between adsorbate and adsorbent. The contact time study was performed with initial diazinon concentrations of 5 mg/L, a pH value of 2 and a contact time of 2-60 min. Figure 4 represents the amount of equilibrium adsorption of diazinon onto PEGMNs as a function of contact time. The adsorption process increased rapidly in the first contact period of 10 min, It appeareds that the fast adsorption at the initial stage may be due to the fact that a large number of surface sites were available for adsorption. It was difficult to occupy the remaining vacant surface sites due to the formation of repulsive forces between the diazinon molecules on the solid surface and in the bulk phase [2].

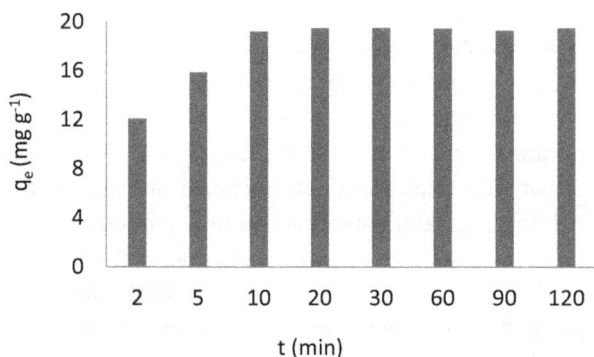

Fig. 4. Effect of contact time on removal of diazinon onto PEGMNs (diazinon concentration: 5 mg /L; solution pH: 7; PEGMNs dosage: 1 mg; contact time: 2-60 min).

3.4 Effect of temperature on the diazinon removal

The effect of temperature on the diazinon adsorption onto PEGMNs was carried out at the range of 15°C to 45°C. The amount of adsorption of diazinon at equilibrium was plotted against the temperature (Figure 5). The amount of the adsorption of diazinon at equilibrium increased with an increasing temperature from 15°C to 45°C indicating an endothermic process. This finding could be due to a tendency for the target molecules to escape from the solid phase to the bulk phase with a decrease in the temperature of the solution [14].

3.5 Effect of water impurities

The effect of several important water impurities including NaCl and ammonia on the adsorption of diazinon onto PEGMNs was investigated. The concentrations of NaCl, and ammonia used were 0–1.5 mg/L and 0–20 mg/L were used respectively [18]. These results showd that the percentage of diazinon adsorption improved in the presence of NaCl (Figure 6(A)). This could be due to the electrostatic interaction that is the main mechanism of diazinon adsorption onto PEGMNs. The increased adsorption of organic molecules with a NaCl concentration of up to 0.1 M

has been reported in other research such as Al-Degs et al. [18-19]. In addition, the diazinon adsorption was enhanced in the presence of ammonia (Figure 6(B)). The ammonia molecules in pH<7 have NH_4^+ and OH^- ions and the existent OH^- ions in the aqueous medium could form a double layer with a negative electric charge around the PEGMNs. Therefore, the electrostatic interaction between the cationic diazinon molecules and the negative surface of nanoparticles was confirmed again.

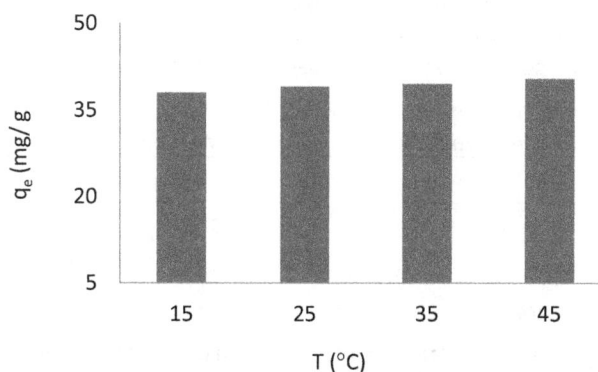

Fig. 5. Effect of temperature on removal of diazinon onto PEGMNs. (diazinon concentration: 5 mg/L; solution pH: 7; PEGMNs dosage: 1 mg; contact time: 10 min, Temperature; 15-45°C).

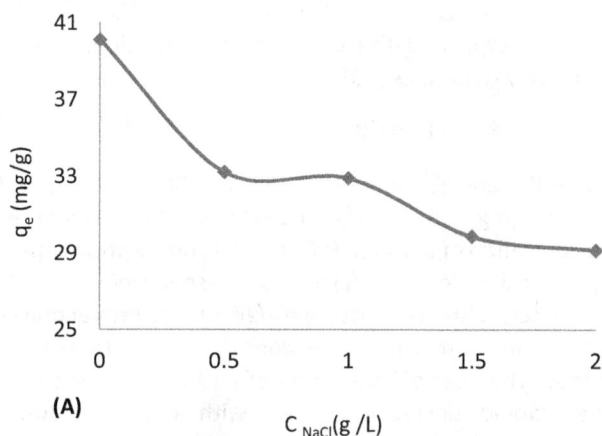

(A)

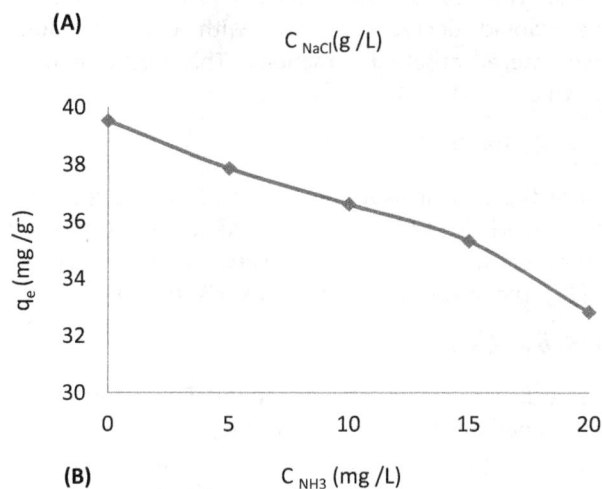

(B)

Fig. 6. The effect of different concentration of NaCl and NH_3 on removal f diazinon on PEGMNs.

Finally, in order to investigate the applicability of the proposed method for the removal of diazinon in water samples, adsorption tests were performed on tap, mineral and well water samples spiked with diazinon (5 mg/L). The experiment was conducted at the optimum condition including 1 mg adsorbent, a pH value of 7, and a contact time of 10 min. The obtained results are presented in Table 1. A low amount of PEGMNs could remove diazinon from an aqueous solution at a relatively short contact time with an acceptable percentage. These findings demonstrate the feasibility of PEGMNs in the removal of diazinon from actual contaminated water samples (Table 1).

Table 1. The applicability of the proposed method for removal of diazinon in water samples: Tap, Mineral, Well water

Entry	C_0 (mg/L)	C_e (mg/L)	%R
Tap	10	3.53	67.79
Mineral Water	10	3.05	70.69
Well	10	3.84	70.75

Diazinon concentration: 5 mg/L; solution pH: 7; PEGMNs dosage: 1 mg; contact time: 10 min, temperature

3.6. Isotherm adsorption

Freundlich and Temkin isotherms described the adsorption behavior of diazinon onto PEGMNs. In fact, the Freundlich equation explained the equilibrium adsorption, which is commonly written as [20]:

$$\ln q_e = \ln k_f + (1/n)\ln C_e \tag{1}$$

where the equilibrium adsorption capacity of the PEGMNs is qe (mg diazinon/g PEGMNs), the equilibrium concentration of diazinon is Ce (mg/L) and sorption capacity and intensity are Kf (mg/L) and 1/n, respectively.

The effects of indirect adsorbate/adsorbate interactions on adsorption isotherms were considered by Temkin and Pyzhev. The heat of adsorption of all the molecules in the layer would decrease linearly with coverage due to adsorbate/adsorbate interactions. This isotherm can be shown as Equation 4:

$$q_e = RT \ln(AC_e)/b \tag{2}$$

where the Temkin isotherm constant is A (L/g), the gas constant is R (8.314 J mol^{-1} K^{-1}), the heat of adsorption is the Temkin constant (J/mol), and the absolute temperature is T (K)[21]. The linear form of the Temkin isotherm is as follows:

$$q_e = B \ln A + B \ln C_e \tag{3}$$

Where B = RT/b [22-23]. The A and B constants can be determined by plotting qe against lnCe.

The result of the curve was fitted with the Freundlich equation (Table 2) instead of the Temkin equation, because the value of R^2 in this isotherm is higher than the other one (R^2=0.9658). Fitting the results with the Freundlich model reflected a heterogeneous surface and the adsorption into

a porous material. Therefore, this system apeared to be porous adsorption.

Table 2. Freundlich equation and temkin equation constants[a]

Freundlich		
n	K_F	R^2
1.25	21.84	0.9920
temkin		
A	B	R^2
1.28	48.67	0.9658

a) diazinon concentration: 5 mg/L solution pH: 7; PEGMNs dosage: 1 mg; contact time: 10 min

To better understand diazinon adsorption on PEGMNs, the thermodynamic of this process was investigated. $\Delta G°$ is negative and $\Delta H°$ and $\Delta S°$ were positive during the diazinon adsorption process (Table 3). The negative value of $\Delta G°$ showeds that the adsorption of diazinon onto PEGMNs was a spontaneous and favorable process [22]. The positive value of $\Delta S°$ indicateds an increase in the state of disorderness in the molecules during the adsorption of diazinon onto adsorbent, which was due to the binding of molecules with the adsorbent surface [23]. The positive value of $\Delta H°$ showeds the endothermic nature of the adsorption process. Typically, physisorption, which is mainly driven by the van der Waals interaction forces, is usually lower than 20 kJ/mol; the electrostatic interaction forces range from 20 to 80 kJ/mol and these kind of interaction forces are, frequently, classified as physisorption. Also, the chemisorption bond strengths can be 80–450 kJ/mol [24]. According to the obtained value for $\Delta H°$ (+ 6.79 kj/mol), it can be concluded that the adsorption process of diazinon onto PEGMNs was based on the van der Waals interactions.

Table 3. The thermodynamic information of diazinon adsorption onto PEGMNs

Solution temperature (K)	K (mL/g)	$\Delta G°$ (kJ/mol)	$\Delta H°$ (kJ/mol)	$\Delta S°$ (J (K mol)$^{-1}$)
288	15.81			
298	17.71	-7.12	+6.79	+46.68
308	18.67			
318	20.91			

In order to evaluate the potential of adsorption of diazinon on PEGMNs, different kind of sorbents were studied. Hence, Fe_3O_4, SMNP, Iranian natural zeolite, and MWCNT which are used in heavy metal removal [24], were checked as adsorbents. According to the obtained results, the time of the proposed method is much shorter (10 min) and the amount of sorbent is significantly lower than other ones

(0.001g), which is indicative of the efficiency of PEGMNs (Table 4).

Table 4. The effect of different sorbents on removal of diazinon[a]

Sorbent	Time (h)	Sorbent dosage (g)	Removal%
Fe₃O₄	1	0.002	23
MWCNTs	8	0.25	1.12
SMNP	1	0.25	39
Iranian natural zeolite	7	0.25	12
PEGMNs	0.16	0.001	83

[a] diazinon concentration: 5 mg/L solution pH: 7

In 2015, F. Chan investigated the effectiveness of chlorine dioxide (CD) to remove phorate and diazinon residues on fresh lettuce and in an aqueous solution. At their optimum condition, 60% of diazinon was remained after 20 min [25]. In 2013, K. S. Ryoo reported fly ash, loess, and activated carbon as adsorbents for the removal of diazinon from water. The equilibrium adsorption times of diazinon by activated carbon and loess were found within 24 h of contact time. Activated carbon showed the best adsorption under the same condition. After 4 hours, approximately 75-85% of the diazinon was removed by the activated carbon. The adsorption data shows that fly ash is not effective for the adsorption of diazinon [24]. The batch removal of diazinon from an aqueous solution by granular-activated carbon was reported by Pirsaheb *et al.* in 2014. After 50 min, the highest removal efficiency of 88% for diazinon was obtained in a 50-min contact time [26]. To our knowledge, the maximum diazinon removed in the minimum contact time was obtained by ASMNPs.

4. Conclusions

The prepared PEGMNs were used for the removal of diazinon from aqueous solutions. For this purpose, the adsorption of diazinon onto PEGMNs was investigated by UV-Vis spectrophotometry at 236 nm, through batch experiments. The effects of adsorbent dosage, solution pH, contact time, and solution temperature were optimized. The results showed that at a natural pH, within 10 min, and the presence of 1mg of PEGMNs, the maximum removal of diazinon was achieved. The isotherm adsorption indicated that this system was fitted with the Freundlich equation but the Temkin equation. Therefore, the application of PEGMNs in the removal of diazinon in environmental remediation could provide lower costs and shorter process times.

Acknowledgements

We are thankful to University of the Jiroft and the Vali-e-Asr University of Rafsanjan Research Council for their support on this work.

References

[1] Salman, J. M., Njoku, V. O., Hameed, B. H. (2011). Adsorption of pesticides from aqueous solution onto banana stalk activated carbon. *Chemical engineering journal*, *174*(1), 41-48.

[2] Omri, A., Wali, A., Benzina, M. (2012). Adsorption of bentazon on activated carbon prepared from Lawsonia inermis wood: equilibrium, kinetic and thermodynamic studies. *Arabian journal of chemistry* doi.org/10.1016/j.arabjc.2012.04.047.

[3] Hameed, B. H., Salman, J. M., Ahmad, A. L. (2009). Adsorption isotherm and kinetic modeling of 2, 4-D pesticide on activated carbon derived from date stones. *Journal of hazardous materials*, *163*(1), 121-126.

[4] Aungpradit, T., Sutthivaiyakit, P., Martens, D., Sutthivaiyakit, S., Kettrup, A. A. F. (2007). Photocatalytic degradation of triazophos in aqueous titanium dioxide suspension: identification of intermediates and degradation pathways. *Journal of hazardous materials*, *146*(1), 204-213.

[5] Mahalakshmi, M., Arabindoo, B., Palanichamy, M., Murugesan, V. (2007). Photocatalytic degradation of carbofuran using semiconductor oxides. *Journal of hazardous materials*, *143*(1), 240-245.

[6] Martín, M. B., Pérez, J. S., Sánchez, J. G., de Oca, L. M., López, J. C., Oller, I., Rodríguez, S. M. (2008). Degradation of alachlor and pyrimethanil by combined photo-Fenton and biological oxidation. *Journal of hazardous materials*, *155*(1), 342-349.

[7] Saritha, P., Aparna, C., Himabindu, V., Anjaneyulu, Y. (2007). Comparison of various advanced oxidation processes for the degradation of 4-chloro-2 nitrophenol. *Journal of hazardous materials*, *149*(3), 609-614.

[8] Murthy, H. R., Manonmani, H. K. (2007). Aerobic degradation of technical hexachlorocyclohexane by a defined microbial consortium. *Journal of hazardous materials*, *149*(1), 18-25.

[9] Ahmad, A. L., Tan, L. S., Shukor, S. A. (2008). Dimethoate and atrazine retention from aqueous solution by nanofiltration membranes. *Journal of hazardous materials*,*151*(1), 71-77.

[10] Maldonado, M. I., Malato, S., Pérez-Estrada, L. A., Gernjak, W., Oller, I., Doménech, X., Peral, J. (2006). Partial degradation of five pesticides and an industrial pollutant by ozonation in a pilot-plant scale reactor. *Journal of hazardous materials*, *138*(2), 363-369.

[11] El-Dib, M. A., Aly, O. A. (1977). Removal of phenylamide pesticides from drinking waters—I. Effect of chemical coagulation and oxidants. *Water research*, *11*(8), 611-616.

[12] Katsumata, H., Kobayashi, T., Kaneco, S., Suzuki, T., Ohta, K. (2011). Degradation of linuron by ultrasound combined with photo-Fenton treatment.*Chemical engineering journal*, *166*(2), 468-473.

[13] Banasiak, L. J., Van der Bruggen, B., Schäfer, A. I. (2011). Sorption of pesticide endosulfan by electrodialysis membranes. *Chemical engineering journal,166*(1), 233-239.

[14] Chang, C. F., Chang, C. Y., Hsu, K. E., Lee, S. C., Höll, W. (2008). Adsorptive removal of the pesticide methomyl using hypercrosslinked polymers. *Journal of hazardous materials, 155*(1), 295-304.

[15] Mishra, P. C., Patel, R. K. (2008). Removal of endosulfan by sal wood charcoal. *Journal of hazardous materials, 152*(2), 730-736.

[16] Rezaeifard, A., Jafarpour, M., Naeimi, A., Haddad, R. (2012). Aqueous heterogeneous oxygenation of hydrocarbons and sulfides catalyzed by recoverable magnetite nanoparticles coated with copper (II) phthalocyanine. *Green chemistry, 14*(12), 3386-3394.

[17] Rezaeifard, A., Jafarpour, M., Farshid, P., Naeimi, A. (2012). Nanomagnet-Supported Partially Brominated Manganese–Porphyrin as a Promising Catalyst for the Selective Heterogeneous Oxidation of Hydrocarbons and Sulfides in Water. *European journal of inorganic chemistry*, 33, 5515-5524.

[18] Moussavi, G., Hosseini, H., Alahabadi, A. (2013). The investigation of diazinon pesticide removal from contaminated water by adsorption onto NH 4 Cl-induced activated carbon. *Chemical engineering journal, 214*, 172-179.

[19] Al-Degs, Y. S., El-Barghouthi, M. I., El-Sheikh, A. H., Walker, G. M. (2008). Effect of solution pH, ionic strength, and temperature on adsorption behavior of reactive dyes on activated carbon. *Dyes and pigments, 77*(1), 16-23.

[20] Ghorbani, F., Younesi, H., Ghasempouri, S. M., Zinatizadeh, A. A., Amini, M., Daneshi, A. (2008). Application of response surface methodology for optimization of cadmium biosorption in an aqueous solution by Saccharomyces cerevisiae. *Chemical engineering journal, 145*(2), 267-275.

[21] Meniai, A. H. (2012). The use of sawdust as by product adsorbent of organic pollutant from wastewater: adsorption of phenol. *Energy procedia, 18*, 905-914.

[22] Bilgili, M. S., Varank, G., Sekman, E., Top, S., Özçimen, D. (2012). Modeling 4-chlorophenol removal from aqueous solutions by granular activated carbon. *Environmental modeling and assessment, 17*(3), 289-300.

[23] Deng, J., Shao, Y., Gao, N., Deng, Y., Tan, C., Zhou, S.,Hu, X. (2012). Multiwalled carbon nanotubes as adsorbents for removal of herbicide diuron from aqueous solution. *Chemical engineering journal, 193*, 339-347.

[24] Ryoo, K. S., Jung, S. Y., Sim, H., Choi, J. H. (2013). Comparative study on adsorptive characteristics of diazinon in water by various adsorbents. *Bulletin of the korean chemical society, 34*(9), 2753-2759.,

[25] Chen, Q., Wang, Y., Chen, F., Zhang, Y., Liao, X. (2014). Chlorine dioxide treatment for the removal of pesticide residues on fresh lettuce and in aqueous solution. *Food control, 40*, 106-112

[26] Pirsaheb, M., Dargahi, A., Hazrati, S., Fazlzadehdavil, M. (2014). Removal of diazinon and 2, 4-dichlorophenoxyacetic acid (2, 4-D) from aqueous solutions by granular-activated carbon. *Desalination and Water treatment, 52*(22-24), 4350-4355.

Response surface methodology and artificial neural network modeling of reactive red 33 decolorization by O_3/UV in a bubble column reactor

Jamshid Behin*, Negin Farhadian

Department of Chemical Engineering, Faculty of Engineering, Razi University, Kermanshah, Iran

ABSTRACT

In this work, response surface methodology (RSM) and artificial neural network (ANN) were used to predict the decolorization efficiency of Reactive Red 33 (RR 33) by applying the O_3/UV process in a bubble column reactor. The effects of four independent variables including time (20-60 min), superficial gas velocity (0.06-0.18 cm/s), initial concentration of dye (50-150 ppm), and pH (3-11) were investigated using a 3-level 4-factor central composite experimental design. This design was utilized to train a feed-forward multilayered perceptron artificial neural network with a back-propagation algorithm. A comparison between the models' results and experimental data gave high correlation coefficients and showed that the two models were able to predict Reactive Red 33 removal by employing the O_3/UV process. Considering the results of the yield of dye removal and the response surface-generated model, the optimum conditions for dye removal were found to be a retention time of 59.87 min, a superficial gas velocity of 0.18 cm/s, an initial concentration of 96.33 ppm, and a pH of 7.99.

Keywords:
Artificial neural network
Bubble column
Ozone/Ultraviolet
Response surface method
Reactive red 33

1. Introduction

Large amounts of chemicals (more than 10000 dyes) are used in the textile industry during the finishing and dying processes [1]. Azo dyes are environmentally hazardous materials due to their toxicity and slow degradation [2-4]. The treatment of azo dyes effluents to meet the stringent environmental regulations is necessary prior to their final discharge into the environment [5-7]. Different conventional methods consisting of various combinations of biological, physical and chemical methods have been used in order to deal with textile wastewater, but these methods are not as efficient as advanced oxidation processes (AOPs) [8-10]. AOPs are chemical methods based on the generation of high reactive hydroxyl radicals (OH˙) that can oxidize the contaminants powerfully and non-selectively. A number of AOPs such as ozonation (O_3), hydrogen peroxide (H_2O_2), O_3/UV, O_3/H_2O_2, O_3/UV/H_2O_2, UV/TiO_2, UV/ZnO and recently O_3/Ultrasound (US) have been well studied [11-15].

The combination of O_3 with UV, which yields hydroxyl, peroxyl, and superoxide radicals, should synergistically accelerate the removal of organic matter from complex wastewater matrices [16]. As illustrated by Beltran [17] and Lucasa et al. [16], the O_3/UV process was capable of oxidizing wastewater faster than O_3 alone, showing a photochemical enhancement oxidation effect. This was principally due to the photolysis of ozone, the enhanced mass transfer of ozone, and the generation of hydroxyl radicals that reacted rapidly with the organic matter in the winery wastewater. Khan et al. [18] showed that the effectiveness of ozonation was enhanced by applying UV. Consequently, the reactant molecules were raised to a higher energy state and reacted more rapidly. Moreover, free radicals for use in the reaction were readily hydrolyzed by water. Another benefit of the combined use of ozone and UV was a substantial reduction in the amount of ozone required as compared to a system using O_3 alone [18]. Several literatures also reported that by combining UV irradiation with O_3, the oxidation power of the systems for

*Corresponding author
E-mail: Behin@razi.ac.ir

organic pollutant degradation could be significantly enhanced [19-24]. Response surface methodology (RSM) has been widely used in process and product improvement. It is efficiently used to examine and optimize the operational variables for experiment designing and model developing [25, 26]. RSM is typically used for mapping a response surface over a particular region of interest, optimizing the responses, and selecting operating conditions to achieve target specifications or consumer requirements [27]. The decolorization of reactive blue 19 dye by Phanerochaete chrysosporium in an aqueous solution was optimized using the Box-Behnken design based on RSM [28]. Based on the central composite design (CCD), the optimization of the UV/TiO$_2$ process in the photo-reactor was carried out using RSM to assess the effects of the main independent parameters on the decolorization efficiency of the azo dye C.I. Basic Red 46 [29]. More recently, artificial neural networks (ANN) are increasingly used as predictive tools in an extensive range of disciplines, such as engineering, due to their ability to employ learning algorithms and discern input–output relationships for complex, nonlinear systems [30-32]. Some literature surveys have shown the application of AANs in water treatment that included the removal of acid orange 7 by activated carbon [33], basic Red 46 degradation using photoelectro-fenton combined with the photocatalytic process [34], and the removal of four different dyes from an aqueous medium by a peroxi-coagulation method using a carbon nanotube (CNT) cathode [35]. Nowadays, RSM and ANN approaches are applied for optimization and process modeling [36-41]. A comparison of the predictive and generalization capabilities, sensitivity analysis, and optimization abilities of ANN and RSM techniques revealed that the ANN model fit the data better and had a higher predictive capability than RSM, even with the limited number of experiments.

Sinha et al. [41] used RSM and ANN modeling of microwave assisted natural dye extraction from pomegranate rind to optimize the effects of processing parameters and to get a good correlation between the input variables and the output parameter. Maran et al. [42] performed a comparative study between ANN and RSM to predict the mass transfer parameters of the osmotic dehydration of papaya. The results showed that the ANN model was more accurate in prediction as compared to the RSM model. The decolorization process of the dye was carried out by bubbling O$_3$ in a bubble column reactor containing the dye solution. The gas flow ensured both the O$_3$ (oxidant) supply and the efficient mixing (high mass transfer of ozone) without the need for mechanical mixing. The experiments were conducted using a batch bubble column to take advantage of the intensive back-mixing that prevailed in the bubble columns. The strong back-mixing reduced the mixing time between the reactants and accelerated the process of decolorization. In addition, the bubble columns were simple

in their design and operated in the absence of mechanical moving parts. A reactor that provided the benefits of high efficiency, low energy input, and easy construction to improve decolorization efficiency was necessary. Decolorization in the bubble column photo-reactor had many advantages such as convenience, economy, safety and high efficiency and as a consequence, it can be considered a good prospect in future applications. The main motivation behind this study was the utilization the RSM and ANN methodologies for predicting the decolorization of Reactive Red 33 (RR 33) by the O$_3$/UV process; the results obtained through RSM were then compared with those obtained through ANN. A number of experiments were carried out based on CCD to collect the output variable (decolorization efficiency) as a function of time, superficial gas velocity, initial concentration of dye, and pH. A feed-forward neural network on back-propagation were developed utilizing the experimental data.

2. Materials and methods

Reactive Red 33 (C$_{27}$H$_{19}$ClN$_7$Na$_3$O$_{11}$S$_3$) was taken from the Boyakhsaz Company, Iran. The experimental setup consisted of a laboratory scale bubble column reactor that was 6.5 cm in diameter and 50 cm in height which was placed inside a photochemical chamber that contained four UV$_{AB}$ lamps (Narva, Germany) of 15 W. Each lamp was placed in a 90° angle to another. The diameter and length of each lamp was 2.5 mm and 45 cm, respectively. Figure 1 shows a schematic of the set-up used for the experimental runs. The reactor was filled with 1 L of aqueous dye solution. An ozone-air mixture was continuously bubbled into the solution throughout a gas distributer that was placed at the bottom of the reactor. Ozone was generated in an ozone generator (Arda, France). The gas flow rate was monitored with a calibrated rotameter incorporated in the ozone generator. The ozone concentration was measured by the iodometry method (KI solution). Liquid samples of 5 mL were withdrawn by a pipette at specific intervals and then analyzed for dye concentration. The dye concentration was determined using a 2100-UV Spectrophotometer (Unico, USA) with a maximum absorption wavelength of 509 nm. The pH of the liquid solution was adjusted using H$_2$SO$_4$ (1 N) or NaOH (1 N). All experiments were carried out at a constant temperature of 25±2°C. The dye decolorization efficiency (Y) was calculated by the following equation:

$$Y\,(\%) = 1 - \frac{C_A}{C_{A_0}} \tag{1}$$

Where C$_A$ is the concentration of dye (ppm) and C$_{A_0}$ is the initial concentration of dye (ppm).

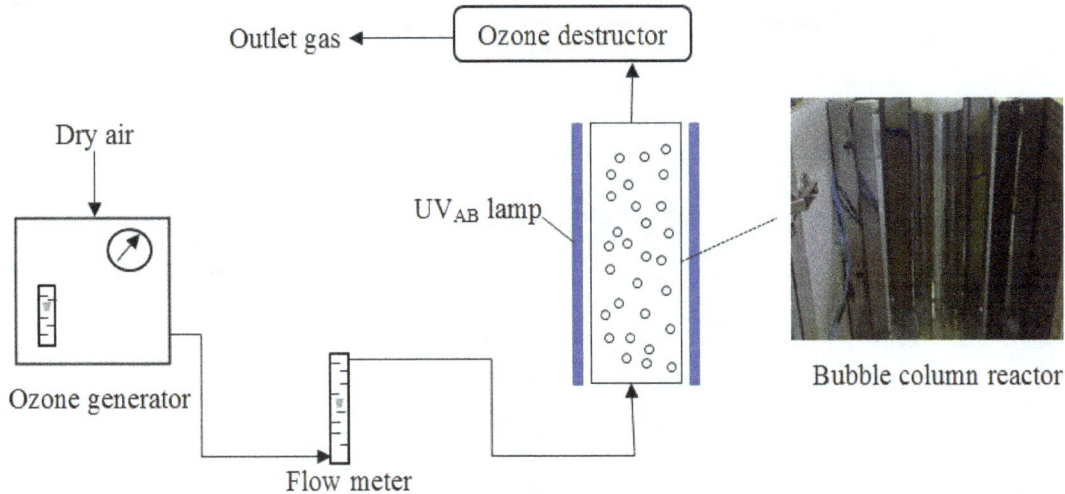

Fig. 1. Experimental set-up and bubble column reactor

2.1. Experimental design

RSM was applied to the experimental data using statistical software, namely Design-expert V7 (trial version). Central composite design (CCD) in RSM was used to develop a response surface quadratic model for describing the dye decolorization process. The ranges and levels of variables investigated in the research including time, superficial gas velocity, initial dye concentration, and pH are given in Table 1. Data from CCD were subjected to the following quadratic equation model to predict the system response and estimate the coefficients by the least-squares regression:

$$Y = \beta_0 + \sum_{i=1}^{k} \beta_i X_i + \sum_{i=1}^{k} \beta_{ii} X_i^2 + \sum_{i=1}^{k-1} \sum_{j=2}^{k} \beta_{ij} X_i X_j + e \qquad (2)$$

Where Y is the predicted decolorization efficiency of RR 33; β_0 is the model intercept coefficient; β_i, β_{ii} and β_{ij} are respectively the linear, quadratic and interaction coefficients; X_i and X_j are the independent variables; and e is the error. The statistical significance of each regression coefficient on the decolorization of RR33 was determined by analysis of variance (ANOVA).

Table 1. Experimental variables and levels

Variable	Symbol	Unit	Low −1	Middle 0	High 1
Time	A	min	20	40	60
Superficial gas velocity	B	cm/s	0.06	0.12	0.18
Initial dye concentration	C	ppm	50	100	150
pH	D	-	3.0	7.0	11.0

2.2. Artificial neural networks

The Back Propagation Algorithm (BPA) was applied to train the neural network. The BPA modified network weights to minimize the MSE between the desired and the actual outputs of the network. A feed forward back propagation neural network with three layers was used. The layers of network contained an input layer, hidden layer and output layer. In the feed forward neural network, information flowed from input to output without feedback [43]. It had one hidden layer with a sigmoid transfer function followed by an output layer with a linear transfer function. Multiple layers of neurons with nonlinear transfer functions allowed the network to learn nonlinear and linear relationships between input and output vectors [44]. It has been reported that multilayer ANN models with only one hidden layer have universal applications [45]. The neural networks toolbox of Matlab 7.12.0 was used. Figure 2 illustrates ANN (4:n:1) for the modeling of the UV/O_3 degradation process in which n is the number of neurons in the hidden layer.

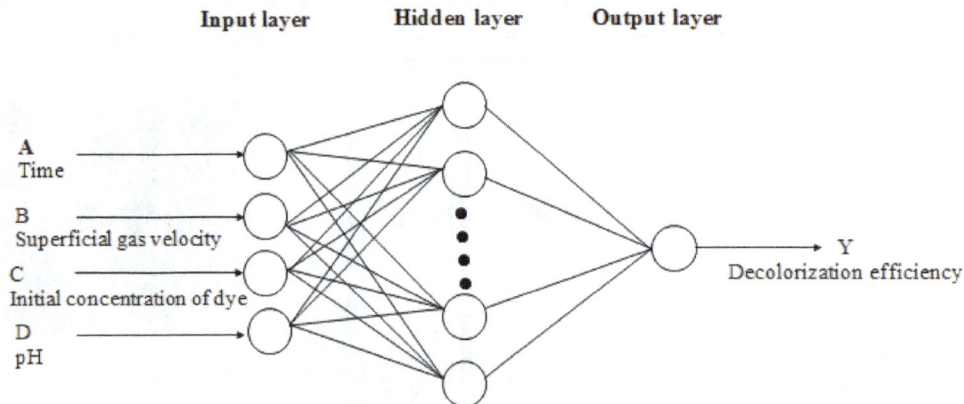

Fig. 2. Conceptual structure of 3 layer ANN model

3. Results and discussion

3.1. RSM modeling

According to the CCD, the experiments were performed in order to determine the optimum combination and study the effect of process variables on the decolorization efficiency of RR33. Table 2 depicts the four-factor, three-level CCD and the observed values for the RR33 decolorization efficiency by the developed quadratic model. The empirical relationships between the response and the four independent variables have been expressed in terms of unit less regression coefficient by the quadratic model and are given as:

$$Y = 90.76 + 13.11\,A - 4.92\,B - 7.76\,C + 3.78\,D + 5.35\,A.\,B + \qquad (3)$$
$$6.63\,A.C - 3.01\,A.\ D - 0.45\,B.C + 3.26\,B.C - 0.29\,C.D - $$
$$5.64\,A^2 - 1.47\,B^2 - 2.20\,C^2 - 0.84\,D^2$$

Without considering the sign of regression model coefficients for variables and interactions, the order of effectiveness of all variables and their binary interactions is as follows (higher model coefficient in absolute values):

$A > C > AC > A^2 > AB > B > D > BC > AD > C^2 > B^2 > D^2 > BC > CD$

Table 3 shows the results of the second-order response surface in the form of analysis of variance (ANOVA); the results indicated that the equation adequately represented the actual relationship between the independent variables and the responses. The positive or negative sign of model coefficient values described the direction of each variable or interaction effect on the response, i.e., positive values variables' increment caused increases in decolorization yields while negative values variables' increment caused a decrease in decolorization yield. Table 3 shows that the two main factors (A, D) and the three binary interaction effects (AB, AC, BC) had positive signs; all other effects (B, C, AD, BC, CD, A^2, B^2, C^2, D^2) had negative signs and a reverse effects on the responses. The ANOVA results (Table 3) for the O_3/UV oxidation system shows the F-value to be 39.93, which implies that the terms in the model have a significant effect on the response. The linear

terms were the four independent variables, which included A: time, B: superficial gas velocity, C: initial concentration of dye and D: pH with the largest effect on the response ($p < 0.0001$). The results suggested that the change of time, superficial gas velocity, initial concentration of dye, and pH had very significant effects on the efficiency of RR33 ($p <0.0001$) when O_3/UV was used in the decolorization of dye. The model's terms with a probability value larger than 0.05 were not significant. The non-significant value of lack of fit (more than 0.05) showed that the quadratic model was valid for the present study. The goodness of fit of the model was examined by the determination of coefficient ($R^2=$ 0.9739), which implied that the sample variation was 97.39% statistically significant and only 0.03% of the total variance could not be explained by the model. The regression model coefficient value, e. g. 13.11, for the retention time (A) was the most significant value in comparison with other variables. Obviously, an increase in the time of the decolorization process can result in a higher decolorization yield. According to the statistical results, the retention time increased the comparable changes in the decolorization yield of RR33 more than the other process variables. The positive sign of the model coefficient for retention time indicated the proportional effect of this variable on the decolorization yield, i.e., increasing retention time will increase the response. The amount of the p-value for the retention time was less than 0.05 for a 95% confident level. In order to analyze the regression equation of the model, three-dimensional (3D) surface and 2D contour plots were obtained by plotting the response (decolorization efficiency) on the Z axis against any two variables while keeping the other variable at middle level. These plots were created to analyze the change in the response surface. The surface and contour plots of the quadratic model are shown in Figure 3 (a-f). The plots were approximately symmetrical in shape; the nonlinear nature of all 3D response surfaces showed considerable interactions between the independent variables and the RR33 decolorization as a response function.

Table 2. Coded central composite design of independent variables and their corresponding experimental values

Run order	A	B	C	D	Decolorization efficiency (Y)
1	1	0	0	0	95.80
2	0	1	0	0	85.70
3	1	1	-1	-1	95.38
4	0	0	0	-1	85.20
5	0	0	0	0	86.40
6	-1	-1	-1	1	94.10
7	-1	-1	-1	-1	91.69
8	1	-1	-1	-1	93.24
9	-1	1	1	1	55.10
10	-1	1	-1	-1	57.80
11	1	-1	1	-1	94.20
12	0	0	1	0	80.88
13	-1	-1	1	-1	62.42
14	1	-1	1	1	93.80
15	0	0	0	0	92.60
16	0	-1	0	0	91.34
17	0	0	0	0	92.40
18	1	-1	-1	1	94.20
19	0	0	0	0	93.10
20	-1	1	-1	1	84.00
21	0	0	0	0	91.20
22	1	1	1	1	92.90
23	0	0	0	1	93.10
24	1	1	-1	1	98.20
25	-1	0	0	0	72.90
26	0	0	0	0	93.50
27	1	1	1	-1	90.29
28	0	0	-1	0	94.70
29	-1	-1	1	1	63.50
30	-1	1	1	-1	30.60

Table 3. Analysis of variance (ANOVA) for RR 33 decolorization efficiency (%)

Parameters	Statistics				
	Sum of squres	Degree of freedom	Mean square	F-value	P-value
Model	6962.11	14	497.29	39.93	< 0.0001
A: time	3091.60	1	3091.60	248.23	< 0.0001
B: superficial gas velocity	435.32	1	435.32	34.95	< 0.0001
C: initial concentration of dye	1082.99	1	1082.99	86.95	< 0.0001
D: pH	257.49	1	257.49	20.67	0.0004
AB	457.32	1	457.32	36.72	< 0.0001
AC	704.11	1	704.11	56.53	< 0.0001
AD	145.20	1	145.20	11.66	0.0038
BC	3.22	1	3.22	0.26	0.6184
BD	169.52	1	169.52	13.61	0.0022
CD	1.32	1	1.32	0.11	0.7490
A^2	82.37	1	82.37	6.61	0.0213
B^2	5.59	1	5.59	0.45	0.5132
C^2	12.52	1	12.52	1.01	0.3319
D^2	1.82	1	1.82	0.15	0.7075
Residuals	186.82	15	12.45		
Lack of fit	152.15	10	15.21	2.19	0.1996
Pure error	34.67	5	6.93		
Total	7148.93	29			

R^2: 0.9739
Adjusted R^2: 0.9495
Adeqate Precision: 27.004
C.V. % : 4.17
Pred R-Squared: 0.8289

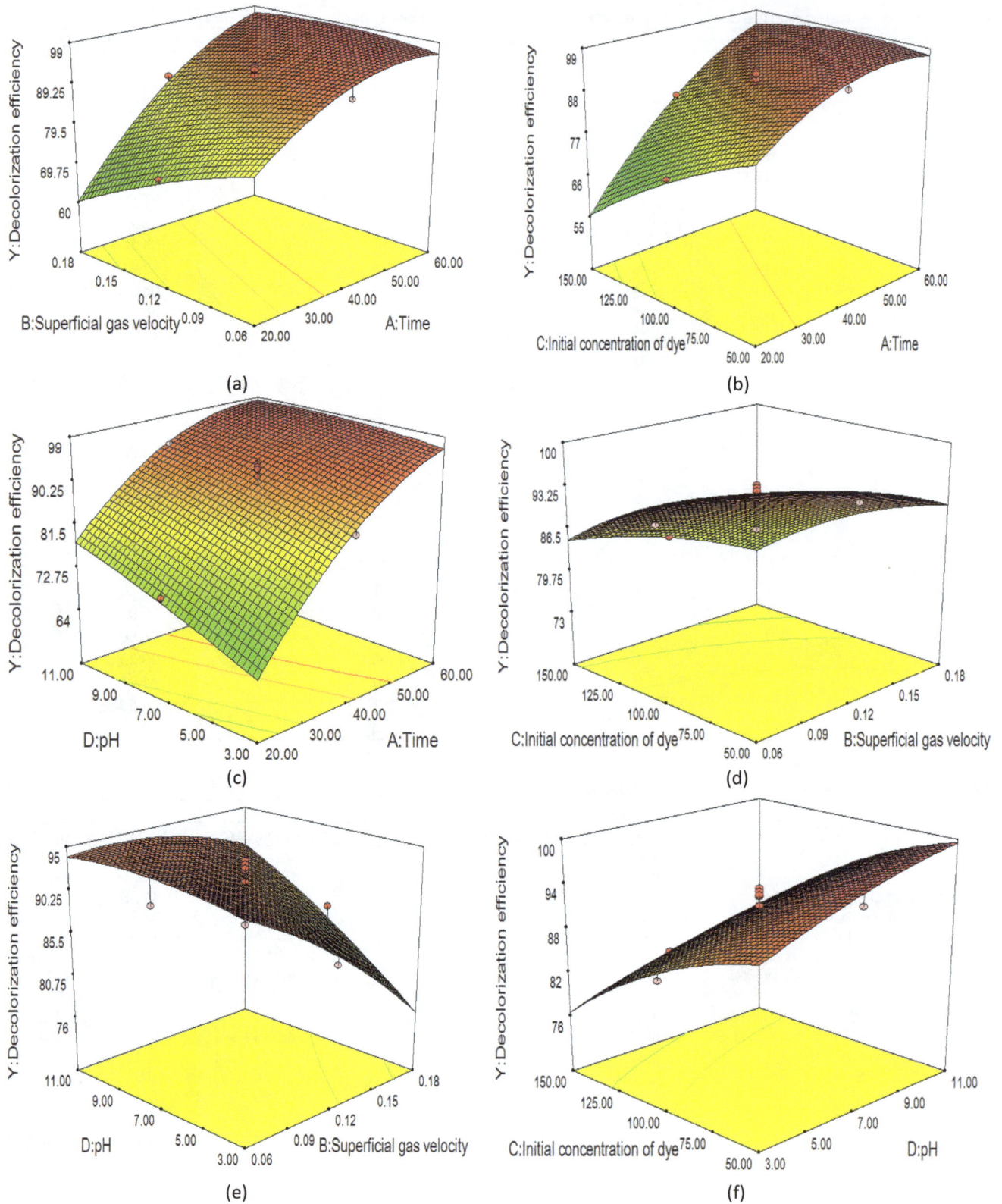

Fig. 3. Response surface plots showing the effect of independent variables on decolorization of RR 33

The AOP with UV irradiation and ozone was initiated by the photolysis of the ozone. The photo-decomposition of the ozone led to the formation of H_2O_2 in the following reaction:

$$O_3 + H_2O + hv \rightarrow H_2O_2 + O_2 \qquad (4)$$

Consequently, the generation of two hydroxyl radicals occurred as a result of the following reaction:

$$H_2O_2 + hv \rightarrow 2OH^\circ \qquad (5)$$

This system contained three components to produce OH radicals and/or oxidize the pollutant for subsequent reactions: UV irradiation, ozone and hydrogen peroxide [46]. All degradation mechanisms should be taken into consideration: OH radical attack as predominant, direct ozone attack, direct photolysis of organics by UV irradiation, and direct oxidation by H_2O_2 [47].

Figure 3 (c, e and f) represents the influence of pH on the color removal efficiency of ozone. There was an increase in the removal efficiency with an increase in the pH of the dye solution [48-50]. Higher color removal efficiency at a pH of 11 can best be explained by the fact that in a highly alkaline medium, the ozone dissociated to the hydroperoxide anions. The HO·radical was especially important in the decolorization process because of its high oxidation potential of 2.8 eV [51, 52]. The free radicals cleaved the conjugated bonds of the dye, resulting in decolorization. The pH tolerance was quite important because the reactive azo dyes bind to cotton fibers by the addition or substitution mechanisms under alkaline conditions and high temperatures [53, 54]. An increase in time enhanced the mass transfer, which resulted in increased ozone content in the liquid phase and an enhanced degradation rate constant [55]. Higher decolorization was achieved at low concentrations of RR33, as is shown in Figure 2 (b, d and f). It can be found that as the dye concentration increased, the decolorization rate constant decreased. Several studies have reported similar observations. The dye had a UV-screening effect and hence a significant quantity of UV light may be absorbed by the high concentration of dye molecules which reduced decolorization efficiency. Moreover, there are more dyes and reaction intermediates that competed with the OH radicals in the high initial concentration [3, 56].From Figure 3(a, d, and e), it can be seen that the decolorization yield increased with the increase of superficial gas velocity. The increase in flow rate corresponded to a larger net surface area for the mass transfer of the ozone from the gas phase to the aqueous phase, and hence increased the volumetric mass transfer coefficient of the ozone. The concentration of the hydroxyl radicals increased with an increase in the ozone concentration. This resulted in a higher dye removal rate [57]. The optimum condition for the removal of RR33 was determined in order to obtain the maximum decolorization efficiency. In order to obtain maximum desirability, the decolorization efficiency was maximized while the independent variables were within range. The optimum condition was found to be a time of 59.87 min, a superficial gas velocity of 0.18 cm/s, an initial concentration of 96.33 ppm, and a pH of 7.99, respectively, with an overall desirability value of 1. The decolorization efficiency of the dye under these optimum conditions was found to be 98.19 and the experimental value was 99.10; the deviation of the experimental and theoretical results was found to be 0.92%. This indicated the suitability of the developed model.

3.2. ANN modeling

ANN methodology was performed to provide a nonlinear mapping between the input variables (time, superficial gas velocity, initial dye concentration and pH) and the output variable (decolorization efficiency) for the runs reported in Table 4. In order to increase the convergence speed and minimize the errors, the experimental data were normalized at {0 1} using a min-max formula:

$$x_n = (x_i - x_{min})/x_{max} - x_{min} \qquad (6)$$

Where x_n, x_i, x_{min} and x_{max} are normalized, real, minimum, and maximum value, respectively. The deviations used for selecting the best ANN architecture were the mean square errors (MSE) and the absolute fraction of variance (R^2) which can be defined as follows [58, 59]:

$$MSE = \frac{1}{N}\sum_{i=1}^{N}(t_i - y_i)^2 \qquad (7)$$

$$R^2 = \frac{\sum_{i=1}^{N}(t_i - t_m^2 - \sum_{i=1}^{N}(t_i - y_i)^2)}{\sum_{i=1}^{N}(t_i - y_i)^2} \qquad (8)$$

Where N is the number of data points, t is the target (experimental) data, and y is the predicted value. The back propagation algorithm was applied for the network training as the most suitable algorithm. In order to determine the optimum number of neurons in the hidden layer, a series of topologies was examined. The different number of neurons in the range of 1-15 was tested in the hidden layer. According to Table 4, the network with 6 neurons in hidden layer had the best results of MSE (2.89×10^{-5}) and R for both the training and testing data. As a result, in this study a three layered feed forward back propagation neural network (4:6:1) was used for the modeling of the decolorization process. The actual and predicted values by RSM and ANN are presented in Figure 4. The values of R^2 for the ANN and RSM models were found to be 0.9739 and 0.9989, respectively. The ANN model was able to capture the nonlinearities of the experimental data better than the RSM model with a combined regression coefficient of 0.998 for RR33 decolorization efficiency.

Table 4. Detail results of the various investigated neural networks structure

Model Structure	Training		Testing	
	MSE	R	MSE	R
4:1:1	6.31×10^{-3}	0.9718	1.45×10^{-2}	0.9567
4:2:1	3.47×10^{-3}	0.9757	2.54×10^{-3}	0.9341
4:3:1	1.07×10^{-4}	0.9976	4.33×10^{-2}	0.9743
4:4:1	5.06×10^{-4}	0.9959	4.85×10^{-3}	0.9711
4:5:1	1.00×10^{-4}	0.9990	1.34×10^{-3}	0.9384
4:6:1	2.89×10^{-5}	0.9996	7.46×10^{-3}	0.9989
4:7:1	1.78×10^{-3}	0.9851	3.85×10^{-3}	0.9875
4:8:1	5.28×10^{-4}	0.9960	3.09×10^{-3}	0.9640
4:9:1	1.88×10^{-4}	0.9986	2.36×10^{-2}	0.9789
4:10:1	3.79×10^{-4}	0.9969	5.63×10^{-3}	0.9290
4:11:1	2.89×10^{-3}	0.9920	1.34×10^{-2}	0.9726
4:12:1	2.28×10^{-4}	0.9955	7.62×10^{-2}	0.8257
4-13:1	9.69×10^{-4}	0.9885	8.41×10^{-2}	0.9720
4:14:1	3.28×10^{-4}	0.9947	2.24×10^{-2}	0.9573
4:15:1	7.28×10^{-4}	0.9957	5.57×10^{-2}	0.9533

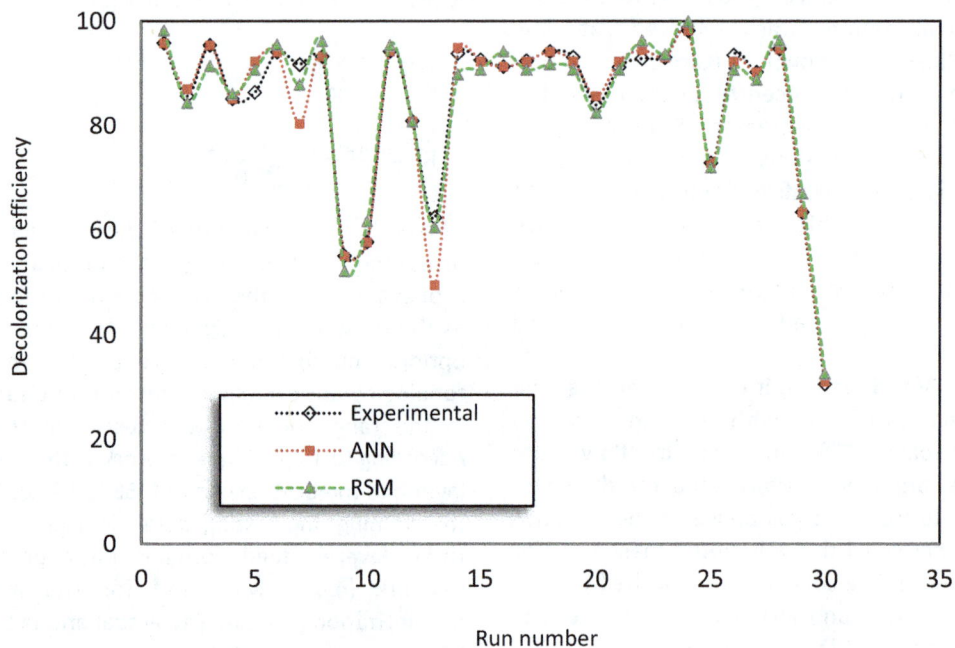

Fig. 4. Plot of the experimental and theoretical results of the RSM and ANN models

3.3. Decolorization kinetics

Several investigators [60-62] have reported that the decolorization process mainly follows first-order kinetics. The kinetics experiments were conducted under optimized reaction conditions for applied AOP. In the experiments, the disappearance of dye was described as first-order reaction kinetics with regard to the dye concentration. The corresponding first-order correlation is shown in Figure 5 which is the typical plot of linear regression (ln C_o/C_t) verses time for color removal. It can be observed that the correlation between in C_o/C_t and the irradiation time was linear. This was a typical pseudo first-order reaction plot. The kinetic expression can be presented as follows:

$$\ln \frac{C_t}{C_0} = -k.t \tag{9}$$

Where C_t is the dye concentration at instant t (ppm), C_o is the initial dye concentration (ppm), k is the pseudo first-order rate constant (min^{-1}), and t is the reaction time (min). The correlation coefficient that can explain the fitting extent of the function equation and the experimental data is presented by R^2. In this case, the value of R^2 is greater than 0.9, which confirms the accuracy of the assumed kinetics for the O_3/UV decolorization reactions of RR33.

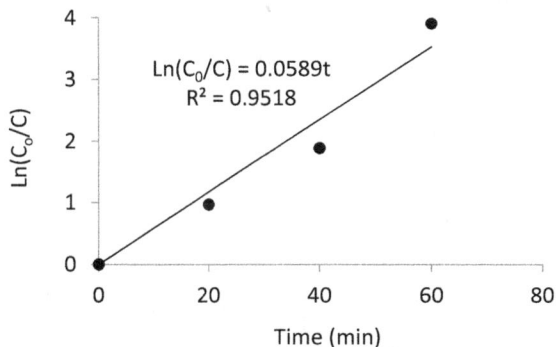

Fig. 5. First-order reaction kinetics for the decolorization of RR33 by O_3/UV process (dye concentration, 100 ppm, pH, 7.99, superficial gas velocity, 0.18 cm/s)

The advantages of ozone over other oxidants are as follows: the fact that the degradable products of ozonation are generally non-toxic; its final products are CO_2 and H_2O; and the residual O_3 in the system changes in a few minutes to O_2 [63]. A comparison between different oxidants was also carried out by Atchariyawuta et al. [64] and they found that O_3 generally produced nontoxic products which were finally converted to CO_2 and H_2O if the conditions were extreme enough. Finally, as an example, the products of a decolorization break-down resulting from a direct dye using ozonation were subjected to toxicity and biodegradability tests. It was found that the oxidation products were non-toxic to algae and had a high tendency for biodegradation [51].

4. Conclusions

Both RSM and ANN techniques were applied for the modeling of the degradation of a colored solution of Reactive Red 33 by the UV/O_3 process in a bubble column reactor. The effects of time, superficial gas velocity, initial concentration of dye, and pH on the decolorization efficiency of RR33 were investigated. The overall efficiency of the UV/O_3 process was enhanced by operating at a basic pH. The efficiency of the AOPs gradually decreased with an increase in the initial concentration of the dye. It can be seen that the decolorization yield increased with an increase in time and superficial gas velocity. The ANN model was found to be capable of better predicting the decolorization efficiency of RR33 within the range it trained than the RSM model. The results of the ANN model indicated that it was much more robust and accurate in estimating the values of dependent variables when compared with the RSM model.

References

[1] Essadki, A. H., Bennajah, M., Gourich, B., Vial, C., Azzi, M., Delmas, H. (2008). Electrocoagulation /electroflotation in an external-loop airlift reactor—application to the decolorization of textile dye wastewater: a case study. *Chemical engineering and processing: Process intensification, 47*(8), 1211-1223.

[2] Daneshvar, N., Khataee, A. R., Ghadim, A. A., Rasoulifard, M. H. (2007). Decolorization of CI Acid Yellow 23 solution by electrocoagulation process: Investigation of operational parameters and evaluation of specific electrical energy consumption (SEEC). *Journal of hazardous materials, 148*(3), 566-572.

[3] Wu, C. H., Chang, C. L. (2006). Decolorization of Reactive Red 2 by advanced oxidation processes: Comparative studies of homogeneous and heterogeneous systems. *Journal of hazardous materials, 128*(2), 265-272.

[4] Anjaneyulu, Y., Chary, N. S., Raj, D. S. S. (2005). Decolourization of industrial effluents–available methods and emerging technologies–a review. *Reviews in environmental science and biotechnology, 4*(4), 245-273.

[5] Saratale, R. G., Saratale, G. D., Chang, J. S., Govindwar, S. P. (2011). Bacterial decolorization and degradation of azo dyes: A review. *Journal of the Taiwan institute of chemical engineers, 42*(1), 138-157.

[6] Wu, C. H., Kuo, C. Y., Chang, C. L. (2007). Decolorization of azo dyes using catalytic ozonation. *Reaction kinetics and catalysis letters, 91*(1), 161-168.

[7] Golder, A. K., Hridaya, N., Samanta, A. N., Ray, S. (2005). Electrocoagulation of methylene blue and eosin yellowish using mild steel electrodes. *Journal of hazardous materials, 127*(1), 134-140.

[8] Do, J. S., Chen, M. L. (1994). Decolourization of dye-containing solutions by electrocoagulation. *Journal of applied electrochemistry, 24*(8), 785-790.

[9] Jiang, J. Q., Graham, N. J. D. (1996). Enhanced coagulation using Al/Fe (III) coagulants: effect of coagulant chemistry on the removal of colour-causing NOM. *Environmental technology, 17*(9), 937-950.

[10] Slokar, Y. M., Le Marechal, A. M. (1998). Methods of decoloration of textile wastewaters. *Dyes and pigments, 37*(4), 335-356.

[11] Li, G., Zhao, X. S., Ray, M. B. (2007). Advanced oxidation of orange II using TiO$_2$ supported on porous adsorbents: The role of pH, H$_2$O$_2$ and O$_3$. *Separation and purification technology, 55*(1), 91-97.

[12] He, Z., Lin, L., Song, S., Xia, M., Xu, L., Ying, H., Chen, J. (2008). Mineralization of CI reactive blue 19 by ozonation combined with sonolysis: Performance optimization and degradation mechanism. *Separation and purification technology, 62*(2), 376-381.

[13] Sanches, S., Crespo, M. T. B., Pereira, V. J. (2010). Drinking water treatment of priority pesticides using

low pressure UV photolysis and advanced oxidation processes. *Water research, 44*(6), 1809-1818.

[14] Gogate, P. R., Pandit, A. B. (2004). A review of imperative technologies for wastewater treatment I: oxidation technologies at ambient conditions. *Advances in environmental research, 8*(3), 501-551.

[15] Mohajerani, M., Mehrvar, M., Ein-Mozaffari, F. (2012). Using an external-loop airlift sonophotoreactor to enhance the biodegradability of aqueous sulfadiazine solution. *Separation and purification technology, 90*, 173-181.

[16] Lucas, M. S., Peres, J. A., Puma, G. L. (2010). Treatment of winery wastewater by ozone-based advanced oxidation processes (O_3, O_3/UV and O_3/UV/H_2O_2) in a pilot-scale bubble column reactor and process economics. *Separation and purification technology, 72*(3), 235-241.

[17] Beltran, F. J. (2003). *Ozone reaction kinetics for water and wastewater systems*. CRC Press.

[18] Khan, H., Ahmad, N., Yasar, A., Shahid, R. (2010). Advanced oxidative decolorization of Red Cl-5B: Effects of dye concentration, process optimization and reaction kinetics. *Polish journal of environmental studies., 19*(1), 83-92.

[19] Beltrán, F. J., Encinar, J., González, J. F. (1997). Industrial wastewater advanced oxidation. Part 2. Ozone combined with hydrogen peroxide or UV radiation. *Water research, 31*(10), 2415-2428.

[20] Alnaizy, R., Akgerman, A. (2000). Advanced oxidation of phenolic compounds. *Advances in environmental research, 4*(3), 233-244.

[21] Gimeno, O., Carbajo, M., Beltrán, F. J., Rivas, F. J. (2005). Phenol and substituted phenols AOPs remediation. *Journal of hazardous materials, 119*(1), 99-108.

[22] Esplugas, S., Gimenez, J., Contreras, S., Pascual, E., Rodriguez, M. (2002). Comparison of different advanced oxidation processes for phenol degradation, *Water research, 36*(4), 1034-1042.

[23] Shu, H. Y., Chang, M. C. (2005). Decolorization effects of six azo dyes by O_3, UV/O_3 and UV/H_2O_2 processes. *Dyes and pigments, 65*(1), 25-31.

[24] Peternel, I., Koprivanac, N., Kusic, H. (2006). UV-based processes for reactive azo dye mineralization. *Water research, 40*(3), 525-532.

[25] Maran, J. P., Sivakumar, V., Sridhar, R., Immanuel, V. P. (2013). Development of model for mechanical properties of tapioca starch based edible films. *Industrial crops and products, 42*, 159-168.

[26] Maran, J. P., Sivakumar, V., Sridhar, R., Thirugnanasambandham, K. (2013). Development of model for barrier and optical properties of tapioca starch based edible films. *Carbohydrate polymers, 92*(2), 1335-1347.

[27] Maran, J. P., Manikandan, S., Nivetha, C. V., Dinesh, R. (2013). Ultrasound assisted extraction of bioactive compounds from Nephelium lappaceum L. fruit peel using central composite face centered response surface design. *Arabian journal of chemistry* http://dx.doi.org/10/1016/j.arabjc.2013.002.007

[28] Hemmat, J., MazaheriAssadi, M. (2013). Optimization of reactive blue 19 biodegradation by Phanerochaetechrysosporium. *International journal of environmental research, 7*(4), 957-962.

[29] Berkani, M., Bouhelassa, M., Bouchareb, M. K. (2015). Implementation of a venturi photocatalytic reactor: Optimization of photodecolorization of an industrial azo dye. *Arabian Journal of Chemistry, http://dx.doi.org/10.1016/j.arabjc.2015.07.004.*

[30] Zobel, C. W., Cook, D. F. (2011). Evaluation of neural network variable influence measures for process control. *Engineering applications of artificial intelligence, 24*(5), 803-812.

[31] Alavala, C. R. (2007). Logic and Neural Networks: Basic concepts and applications. *New Age. New Age Publications.*

[32] Pilkington, J. L., Preston, C., Gomes, R. L. (2014). Comparison of response surface methodology (RSM) and artificial neural networks (ANN) towards efficient extraction of artemisinin from Artemisia annua. *Industrial crops and products, 58*, 15-24.

[33] Aber, S., Daneshvar, N., Soroureddin, S. M., Chabok, A., Asadpour-Zeynali, K. (2007). Study of acid orange 7 removal from aqueous solutions by powdered activated carbon and modeling of experimental results by artificial neural network. *Desalination, 211*(1), 87-95.

[34] Zarei, M., Khataee, A. R., Ordikhani-Seyedlar, R., Fathinia, M. (2010). Photoelectro-Fenton combined with photocatalytic process for degradation of an azo dye using supported TiO_2 nanoparticles and carbon nanotube cathode: neural network modeling. *Electrochimica Acta, 55*(24), 7259-7265.

[35] Zarei, M., Niaei, A., Salari, D., Khataee, A. R. (2010). Removal of four dyes from aqueous medium by the peroxi-coagulation method using carbon nanotube–PTFE cathode and neural network modeling. *Journal of electroanalytical chemistry, 639*(1), 167-174.

[36] Cheok, C. Y., Chin, N. L., Yusof, Y. A., Talib, R. A., Law, C. L. (2012). Optimization of total phenolic content extracted from Garcinia mangostana Linn. Hull using response surface methodology versus artificial neural network. *Industrial crops and products, 40*, 247-253.

[37] Khajeh, M., Kaykhaii, M., Sharafi, A. (2013). Application of PSO-artificial neural network and response surface methodology for removal of methylene blue using silver nanoparticles from water samples. *Journal of industrial and engineering chemistry, 19*(5), 1624-1630.

[38] Ebrahimzadeh, H., Tavassoli, N., Sadeghi, O., Amini, M. M. (2012). Optimization of solid-phase extraction using artificial neural networks and response surface methodology in combination with experimental design for determination of gold by atomic absorption spectrometry in industrial wastewater samples. *Talanta*, *97*, 211-217.

[39] Khayet, M., Cojocaru, C. (2013). Artificial neural network model for desalination by sweeping gas membrane distillation. *Desalination*, *308*, 102-110.

[40] Lakshminarayanan, A. K., Balasubramanian, V. (2009). Comparison of RSM with ANN in predicting tensile strength of friction stir welded AA7039 aluminium alloy joints. *Transactions of nonferrous metals society of China*, *19*(1), 9-18.

[41] Sinha, K., Saha, P. D., Datta, S. (2012). Response surface optimization and artificial neural network modeling of microwave assisted natural dye extraction from pomegranate rind. *Industrial crops and products*, *37*(1), 408-414.

[42] Maran, J. P., Sivakumar, V., Thirugnanasambandham, K., Sridhar, R. (2013). Artificial neural network and response surface methodology modeling in mass transfer parameters predictions during osmotic dehydration of Carica papaya L. *Alexandria engineering journal*, *52*(3), 507-516.

[43] Yang, S. H., Chung, P. W. H., Brooks, B. W. (1999). Multi-stage modelling of a semi-batch polymerization reactor using artificial neural networks. *Chemical engineering research and design*, *77*(8), 779-783.

[44] Bassam, A., Ortega-Toledo, D., Hernandez, J. A., Gonzalez-Rodriguez, J. G., Uruchurtu, J. (2009). Artificial neural network for the evaluation of CO2 corrosion in a pipeline steel. *Journal of solid state electrochemistry*, *13*(5), 773-780.

[45] Baughman, D. R., Liu, Y. A. (2014). *Neural networks in bioprocessing and chemical engineering*. Academic press.

[46] Kusic, H., Koprivanac, N., Bozic, A. L. (2006). Minimization of organic pollutant content in aqueous solution by means of AOPs: UV-and ozone-based technologies. *Chemical engineering journal*, *123*(3), 127-137.

[47] Peternel, I., Koprivanac, N., Kusic, H. (2006). UV-based processes for reactive azo dye mineralization. *Water research*, *40*(3), 525-532.

[48] Sevimli, M. F., Sarikaya, H. Z. (2002). Ozone treatment of textile effluents and dyes: effect of applied ozone dose, pH and dye concentration. *Journal of chemical technology and biotechnology*, *77*(7), 842-850.

[49] Azbar, N., Yonar, T., Kestioglu, K. (2004). Comparison of various advanced oxidation processes and chemical treatment methods for COD and color removal from a polyester and acetate fiber dyeing effluent. *Chemosphere*, *55*(1), 35-43.

[50] Sevimli, M. F., Sarikaya, H. Z. (2005). Effect of some operational parameters on the decolorization of textile effluents and dye solutions by ozonation. *Environmental technology*, *26*(2), 135-144.

[51] Konsowa, A. H. (2003). Decolorization of wastewater containing direct dye by ozonation in a batch bubble column reactor. *Desalination*, *158*(1), 233-240.

[52] Selcuk, H. (2005). Decolorization and detoxification of textile wastewater by ozonation and coagulation processes. *Dyes and pigments*, *64*(3), 217-222.

[53] Aksu, Z. (2003). Reactive dye bioaccumulation by Saccharomyces cerevisiae. *Process biochemistry*, *38*(10), 1437-1444.

[54] Asad, S., Amoozegar, M. A., Pourbabaee, A., Sarbolouki, M. N., Dastgheib, S. M. M. (2007). Decolorization of textile azo dyes by newly isolated halophilic and halotolerant bacteria. *Bioresource technology*, *98*(11), 2082-2088.

[55] Wu, J., Wang, T. (2001). Ozonation of aqueous azo dye in a semi-batch reactor. *Water research*, *35*(4), 1093-1099.

[56] Wang, K. H., Hsieh, Y. H., Wu, C. H., Chang, C. Y. (2000). The pH and anion effects on the heterogeneous photocatalytic degradation of o-methylbenzoic acid in TiO$_2$ aqueous suspension. *Chemosphere*, *40*(4), 389-394.

[57] Zhang, H., Duan, L., Zhang, D. (2006). Decolorization of methyl orange by ozonation in combination with ultrasonic irradiation. *Journal of hazardous materials*, *138*(1), 53-59.

[58] Yetilmezsoy, K., Demirel, S. (2008). Artificial neural network (ANN) approach for modeling of Pb (II) adsorption from aqueous solution by Antep pistachio (Pistacia Vera L.) shells. *Journal of hazardous materials*, *153*(3), 1288-1300.

[59] Bingöl, D., Hercan, M., Elevli, S., Kılıç, E. (2012). Comparison of the results of response surface methodology and artificial neural network for the biosorption of lead using black cumin. *Bioresource technology*, *112*, 111-115.

[60] Bali, U., Çatalkaya, E., Şengül, F. (2004). Photodegradation of reactive black 5, direct red 28 and direct yellow 12 using UV, UV/H$_2$O$_2$ and UV/H$_2$O$_2$/Fe^{2+}: a comparative study. *Journal of hazardous materials*, *114*(1), 159-166.

[61] Wu, J., Wang, T. (2001). Ozonation of aqueous azo dye in a semi-batch reactor. *Water research*, *35*(4), 1093-1099.

[62] Neamtu, M., Yediler, A., Siminiceanu, I., Kettrup, A. (2003). Oxidation of commercial reactive azo dye aqueous solutions by the photo-Fenton and Fenton-like processes. *Journal of photochemistry and photobiology A: Chemistry*, *161*(1), 87-93.

[63] Guendy, H. R. (2007). Ozone treatment of textile wastewater relevant to toxic effect elimination in

marine environment. *Egyptian journal of aquatic research*, 33 (1), 98-115.

[64] Atchariyawut, S., Phattaranawik, J., Leiknes, T., Jiraratananon, R. (2009). Application of ozonation membrane contacting system for dye wastewater treatment. *Separation and purification technology*, *66*(1), 153-158.

8

Synthesis of nanocomposite based on Semnan natural zeolite for photocatalytic degradation of tetracycline under visible light

Farzaneh Saadati[1], Narjes Keramati[1*], Mohsen Mehdipour Ghazi[2]
[1]*Faculty of Nanotechnology, Semnan University, Semnan, Iran*
[2]*Faculty of Chemical, Petroleum and Gas Engineering, Semnan University, Semnan, Iran*

Keywords:
Photocatalyst
P25
Semnan Natural Zeolite
Tetracycline
Visible light

ABSTRACT

This study investigated the photocatalytic behaviors for the nanocomposite of TiO_2 P25 and Semnan natural zeolite in the decomposition of tetracycline under visible light in an aqueous solution. The structural features of the composite were investigated by a series of complementary techniques that included X-ray diffractometer (XRD), Fourier transform infrared spectroscopy (FTIR), scanning electron microscopy (SEM), surface area (BET) measurement, and ultraviolet-visible diffuse reflectance spectroscopy (DRS). The surface area measurement disclosed an enhancement of surface area by ~2 times for the synthesized TiO_2/Semnan natural zeolite than that of commercial TiO_2 P25. The as-prepared photocatalyst (TiO_2/Semnan natural zeolite) showed pH dependence and more than 87% of the tetracycline could be degraded from the solution under visible irradiation within 90 min at a pH of 6. This excellent catalytic ability was mainly attributed to the hybrid effect of the photocatalyst and adsorbent. The results provided new insight into the performance of active photocatalysts on the treatment of pharmaceutical wastewater. In addition, the immobilization of TiO_2 onto Semnan natural zeolite permitted easier separation of the adsorbent from the treated water.

1. Introduction

Pharmaceutical antibiotics are used extensively worldwide in human therapy and the farming industry [1, 2]. Among all the antibiotics, tetracycline (TC) is one of the most widely used in aquaculture and veterinary medicines. Exposure to antibiotic residues and their transformed products could cause a variety of adverse effects, including acute and chronic toxicity and antibiotic resistance [3]. This resistance can be transmitted to humans through complex biochemical transference processes [1]. Hence, the removal or degradation of pharmaceutical pollutants is an important research subject. Considerable efforts have been devoted to developing a suitable purification method that can easily destroy these bio-recalcitrant organic contaminants [4]. Due to their antibacterial nature, antibiotic residues cannot be effectively eliminated by traditional biological methods. Several studies have shown that the biodegradation of certain antibiotics is low during wastewater treatment [4, 5]. Most conventional wastewater treatment plants, which employ processes such as activated carbon adsorption, membrane filtration, chemical coagulation, ion exchange on synthetic adsorbent resins, etc., are inefficient in the removal of TC antibiotics. Also, they generate wastes during the treatment of contaminated water which requires additional steps and cost [1]. Nowadays, it is well confirmed that advanced oxidation processes (AOPs) are the most recommended technologies for the removal of different pollutants from water [5]. Heterogeneous photocatalysis oxidation is the most popular AOP method and has been widely used for the removal of water pollutants because of its advantages, especially its capability to be employed under ambient conditions [4, 6-10]. In past decades, titanium dioxide semiconductor (TiO_2) has attracted great interest due to its photocatalytic properties. It is inexpensive, non-toxic, stable, and potentially reusable in water [4, 11]. Zhu et al. [12] used nano sized TiO_2 (P25) as a photocatalyst and studied its ability

*Corresponding author
E-mail: narjeskeramati@semnan.ac.ir

for TC degradation. More than 95% of the TC was removed within 40 min (40 ppm of TC, 1 g/L of TiO_2) under UV irradiation. Recent publications indicate that TC is completely oxidized by photocatalysis in short time periods [1, 4]. The TC degradation efficiency changes by different degrees with respect to different operating conditions. However, the practical applications of TiO_2 as an adsorbent or a photocatalyst in aqueous solutions are limited because of the recovery problems of fine TiO_2 particles [13]. Recently, attempts have been made to immobilize the TiO_2 particles on different supports, such as activated carbon [14], clay [15] and zeolites [16], to improve their separation from bulk water [17]. Zeolites are three dimensional alumino silicate minerals with a porous structure that have valuable physicochemical properties, such as cation exchange, molecular sieving, catalysis and adsorption [16, 17]. Natural zeolites are cheap, abundant and easily available [17]. It is a promising support for TiO_2 because they have regular pores and channels that can confine substrate molecules, thereby enhancing adsorption and photocatalysis. TiO_2 supported on zeolite combines the adsorptive capability of zeolite with the photocatalysis behavior of TiO_2, thus resulting in a synergistic and enhanced degradation efficiency. A zeolite/TiO_2 composite adsorbent that is easily recoverable via TiO_2 photocatalytic oxidation will lead to a cost-effective and environmentally friendly water treatment process, thus avoiding the production of waste or need for a high cost regeneration process. A simple method for combining TiO_2 with adsorbents involves the mechanical mixing of TiO_2 powders into adsorbents powders. Some researchers have utilized this method using zeolites as the adsorbent and applied the resulting photocatalyst for the purification of water [18]. In this work, the particles of an Iranian natural zeolite were used to obtain TiO_2/natural zeolite as a heterogeneous catalyst for the degradation of a TC pharmaceutical capsule in an aqueous solution under visible light via a simple method. TiO_2 P25 was used as the base photocatalyst because of its good photocatalytic activity. Semnan natural zeolite (SNZ) was the adsorbent of choice due to its good adsorption property and availability at an affordable cost. A pharmaceutical capsule was used because of its similarity to an actual waste water sample.

2. Materials and methods

2.1. Materials and chemicals

Titanium dioxide P25 (80% anatase and 20% rutile) was supplied by Evonik industries. Semnan natural zeolite tuffs were obtained from the Semnan region in the north-east of Iran. The TC ($C_{20}H_{24}N_2O_8$) pharmaceutical 250 mg capsule was obtained from the Iran Daru Company. Distilled water was used throughout the experiments.

2.2. Preparation of TiO_2/SNZ nanocomposite

To prepare the mixed photocatalyst, SNZ and TiO_2 P25 were mixed mechanically. In a typical preparation of the supported

photocatalyst, TiO_2 P25 (1 g) was mixed with 5 g of SNZ in powder form and added to 60.0 mL of deionized water at room temperature. The mixture was stirred and ultra-sonicated for 60 min. The solid was collected and washed by deionized water and then dried under 100 °C for one night in an oven. Finally, the solid was calcined at 450 °C for 8 h. Figure 1 shows the commercial TiO_2 P25 and as-prepared TiO_2 supported on SNZ.

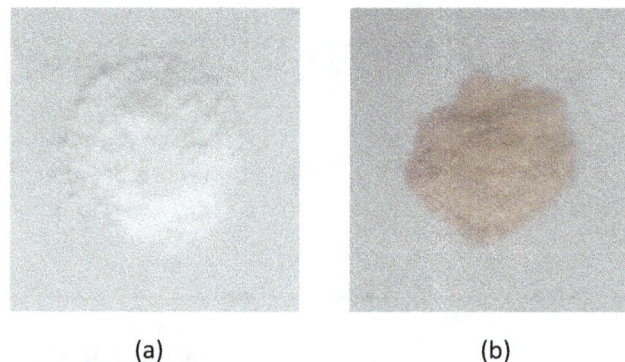

(a)   (b)

Fig. 1. Photo of a) Commercial TiO_2 P25, b) as-prepared TiO_2/SNZ nanocomposite

2.3. Analysis and characterization methods

The structure of the commercial TiO_2 P25, natural zeolite, and TiO_2/SNZ was examined by using a diffractometer Bruker, D8 (with X-ray tube anode and Cu K_α wavelength: 1.5406 Å). The surface morphology of the samples was studied using a KYKY-EM-3200 scanning electron microscope (SEM). The BET surface areas of the samples were determined from the isotherm data of the nitrogen adsorption data in the relative pressure (P/P_0) range of 0.05–0.30 obtained at 77.35 K using a Quanta chrome Autosorb-1 analyzer. The UV–Vis spectra were obtained on a Shimadzu UV-1650PC spectrophotometer. Fourier transform-infrared spectroscopy (FTIR) in the range of 400-4000 cm^{-1} was recorded (Shimadzo FT-IR 8400S).

2.4. Photocatalytic degradation experiments

A photocatalytic reactor with two lamps of 60 Watt as the visible irradiation source was used to degrade the TC as a pollutant (Figure 2). To prepare the TC solution, the contents of a 250 mg capsule was dissolved in water and shaken for 30 min; then it was filtered in a 100 mL volumetric flask and diluted with water. The photoreactor was filled with 300 ml of 8 mg/L pollutant and 0.8 g/L of photocatalyst at an irradiation time of 90 min. The whole reactor was cooled with a water-cooled jacket on its outside, and the temperature was kept at 25°C. In order to set the adsorption/desorption equilibrium of the pollutant on a heterogeneous catalysts surface, the reactor was kept under dark conditions for 30 min. An aliquot suspension was withdrawn and centrifuged at 5000 rpm to remove the solid particles. The UV–Vis absorption spectra of the TC solution was scanned and showed an optimum band centered at 360 nm, which can be used as the absorbance wavelength for the establishment of the TC standard curve by the principle of the Lambert–Beer law. The concentrations of different TC samples could be measured from the standard curve. The degradation extent was calculated according to the following formula based on the absorbance of the solution (at λ_{max} = 360 nm) before and after irradiation.

$$TC\ degradation\ \% = [(C_0-C_t)/C_0]\ *100 = [(A_0-A_t)/A_0]*100 \qquad (1)$$

Where C_0 and C_t are the initial and final concentration of TC at time t, respectively. A_0 and A_t are respectively the initial and final absorbance of the samples. The absorbance values were used to calculate the percent of degradation.

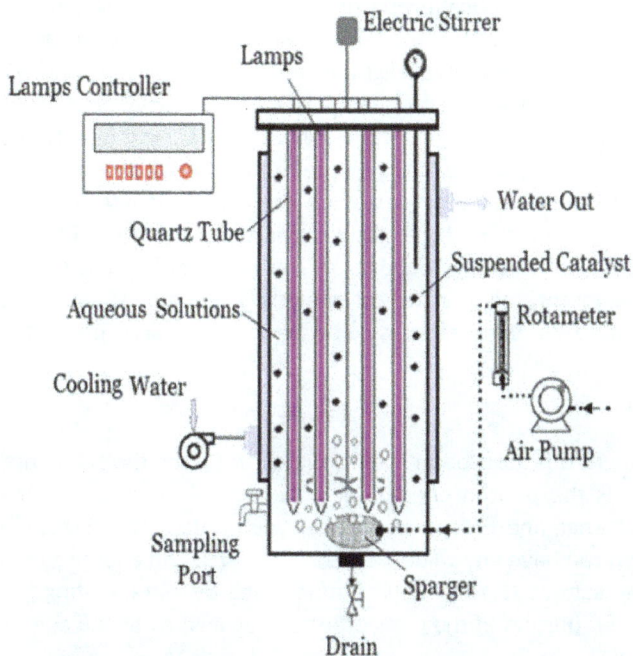

Fig. 2. Schematic diagram of the home-made photo reactor

3. Results and discussion

3.1. TiO₂/SNZ nanocomposite characteristics

3.1.1. XRD analysis

In order to confirm the structure and crystallinity of the TiO₂/SNZ catalyst, an XRD study was carried out. The XRD patterns of natural zeolite and TiO₂/SNZ nanocomposite were recorded and are shown in Figure 3 for comparative purposes, the diffraction pattern of commercial TiO₂ P25 was also included. The presented results revealed that the SNZ particles used were mostly Clinoptilolite (CP) (2θ=11; 22.5; 30) [6, 19]. There was no significant shift of the zeolite in the TiO₂/SNZ peaks by comparison with those of pure zeolite; however, the intensity for the SNZ peaks was lower when compared to the natural zeolite particles. This suggested that the TiO₂ precursor interacted with the support, decreasing the crystallinity of the SNZ. This was in good agreement with other research [17]. It can be seen from Figure 3 that the main peak positions of natural zeolite were remarkably unchanged, indicating that the structure of natural zeolite had good thermal stability. This also revealed that most of the TiO₂ was distributed on the surface of natural zeolite while a part was encapsulated in the cavities.

A similar XRD pattern was obtained for TiO₂ loaded on natural zeolite as reported previously [20-22].

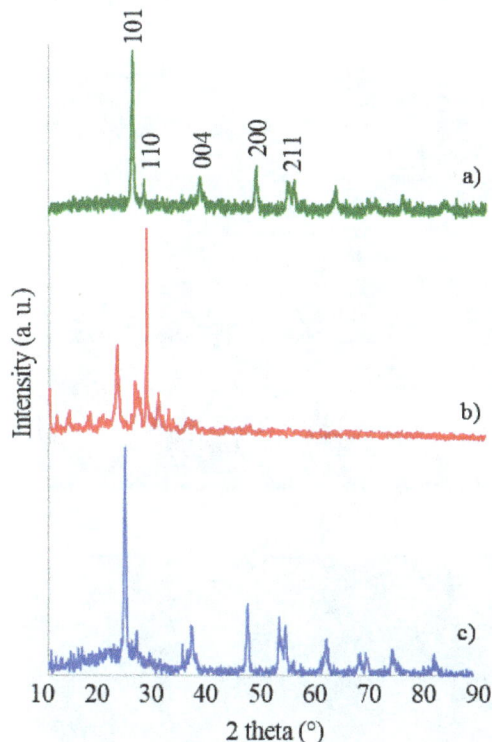

Fig. 3. XRD Spectra of a) Commercial TiO₂ p25, b) SNZ particles c) as-prepared TiO₂/SNZ nanocomposite

The characteristic diffraction peaks of TiO₂ P25 can be observed at 2ϑ=25.4 corresponding to the (101) plane of anatase, indicating that TiO₂ has been impregnated within the sample. A small portion of rutile was also observed in the TiO₂ sample as the appearance of the peak at 2ϑ=27.8 corresponding to (110) plane of rutile, which was overlapped with those of the Semnan natural zeolite [23]. No peaks of impurities were detected, indicating the high phase purity of the as-prepared sample. The average crystallite size of the commercial TiO₂ P25 and TiO₂/SNZ was calculated using the Scherrer equation to be about 21.6 and 27 nm, respectively.

3.1.2. SEM surface morphology

The morphology of the SNZ and TiO₂ supported on SNZ was investigated using SEM, as illustrated in Figure 4. It shows the SEM images of the SNZ surface before and after coating with the TiO₂ P25. The SEM results were also in agreement with XRD as there was no major change in the structure of the SNZ. Also, the surface grain of SNZ was much smoother than that of the TiO₂/SNZ. The TiO₂ spherical nanoparticles (size 30 nm) were present as clusters attached to the zeolite matrix of the modified SNZ. Collectively, these images supported the conclusion that TiO₂ was well dispersed on the external surface of SNZ. It could also be seen that the TiO₂ coating was relatively uniform and there were no apparent sites of uncoated zeolite. Figure 4 also shows pore openings and cavities resulting in an increased specific surface area of the composite, thus providing higher adsorption sites for contaminant adsorption. These findings were consistent with other research [11, 17, 24].

(a)

(b)

Fig. 4. SEM images of (a) SNZ particles, (b) as-prepared TiO_2/SNZ nanocomposite

3.1.3. Specific surface area

The surface area (S_{BET}) of the samples was determined from the N_2 adsorption-desorption isotherms using the Brunner-Emmet-Teller method. The BET specific surface area of the TiO_2 and synthesized composite were measured to be 50 and 93 m^2/g, respectively. The increase in the surface area after modification resulted from the distribution of TiO_2 throughout the natural zeolite matrix. As the pore size of the zeolite was smaller than the size of the TiO_2 particles supported by the zeolite, the TiO_2 particles were coated on the zeolite surface and not on the inside of the internal pores of the zeolite. In other words, the pore diameter for the zeolite was considered to be too small for TiO_2, and therefore during the preparation of zeolite/TiO_2, the sorption of TiO_2 on natural zeolite was limited to the sites on the external surface. Surface area measurement disclosed an enhancement of surface area by ~2

times for the synthesized TiO_2/SNZ than that of the commercial TiO_2 P25, which rendered SNZ as an ideal support for TiO_2. The high surface area enabled TiO_2/SNZ to simplify the adsorption of TC (model pollutant) on the catalyst and improved the photocatalytic activity of the catalyst with respect to TC degradation.

3.1.4. UV-vis DRS

The UV–vis diffuse reflectance spectroscopy (DRS) was used to analyze the optical properties of catalyst and is depicted in Figure 5. Clearly, the TiO_2/SNZ catalyst was able to well absorb both the UV and visible light. It can be seen that the TiO_2/SNZ sample, unlike the commercial TiO_2 P25, showed absorbance in the visible range which suggested its potential to be activated by visible light. The determination of the band gap from the UV-vis spectra was an alternative method to study the modification of the electronic property of the synthesized species. The energy band structure was a key factor affecting the photocatalytic activity of catalysts [25]. The band gap of the composite was estimated by Tauc's equation using the absorption data:

$$\alpha = \alpha_0 (h\nu - E_g)^n / h\nu$$

where α is absorption coefficient, α_0 and h are the constants, $h\nu$ is the photon energy, E_g is the optical band gap of the material, and n depends on the type of electronic transition and can have any value between 0.5 and 3. The energy gap of the sample (E_g) has been distinguished by extrapolating the linear portion of the plots of $(\alpha h\nu)^{0.5}$ against $h\nu$ to the energy axis.

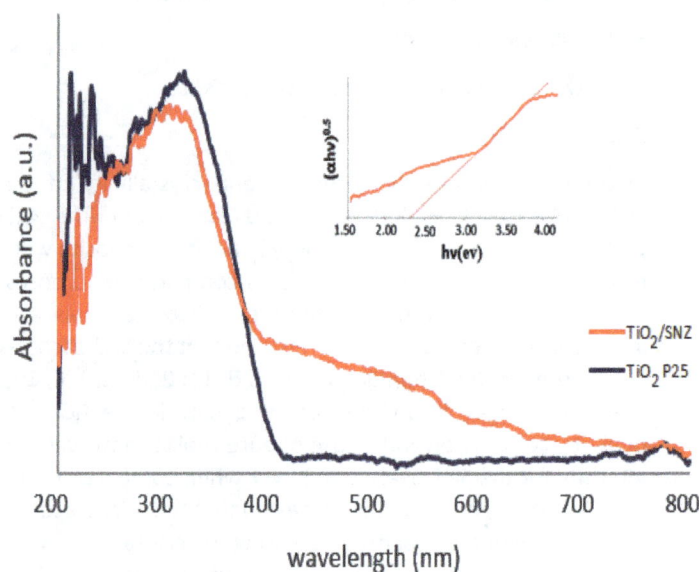

Fig. 5. UV–Vis diffuse reflectance spectra of commercial TiO_2 P25 and TiO_2/SNZ sample (inset: estimated band gap of TiO_2/SNZ)

The valence band (VB) and conduction band (CB) potentials of the semiconductor at the point of zero charge can be calculated by the following formula:

$$E_{VB}=X-E^c+0.5E_g$$

where E_{VB} or E_{CB} are the VB or CB edge potentials of semiconductor, respectively, X is the absolute electronegativity of its constituent atoms, E^c is the energy of free electrons on the hydrogen scale (ca. 4.5 eV), and E_g is the band gap of semiconductor. The CB position can be calculated by $E_{CB}=E_{VB}-E_g$ [26]. Since zeolites generally do not absorb any light in UV–vis regions, they are one of the most suitable supports for photocatalysis. The results revealed a clear trend in decreasing the band gap of commercial TiO_2 P25 from 3.2 to 2.4 eV for TiO_2 supported on SNZ. This suggested that the particle size of TiO_2 supported on zeolite was larger than that of commercial TiO_2 used in the present study (quantum size effect) [27, 28]. Furthermore, zeolite delocalizes the band-gap excited electrons of TiO_2, thereby minimizing electron/hole recombination.

3.1.5. FTIR spectroscopy results

Figure 6 presents the FTIR spectrum of TiO_2/SNZ in comparison with SNZ to demonstrate the possible bonds of TiO_2 with the natural zeolite. The strongest absorption peak at 1047 cm^{-1} was assigned to the framework stretching vibration band of Si (Al)-O in natural zeolite, its position was unchanged at a 450 ^0C treatment temperature, indicating that the zeolite structure was not destroyed at 450 °C . Also, the adsorption bands of TiO_2/SNZ were mostly similar to those of SNZ. This indicated that SNZ was highly suitable to act as a TiO_2 P25 immobilizer, as the structure was not affected by the high annealing temperatures used. The infrared analysis confirmed that the TiO_2 P25 particles combined with the active sites of SNZ and Ti–O–Al and Ti–O– Si which make up the load of TiO_2 P25. A new absorption band was found in the spectra of TiO_2/SNZ comparable with the SNZ. The band covered a range from 945 to 900 cm^{-1} corresponding to the stretching vibration of Ti–O–Si and Ti–O–Al, which was similar to the results of Li et al. [29].

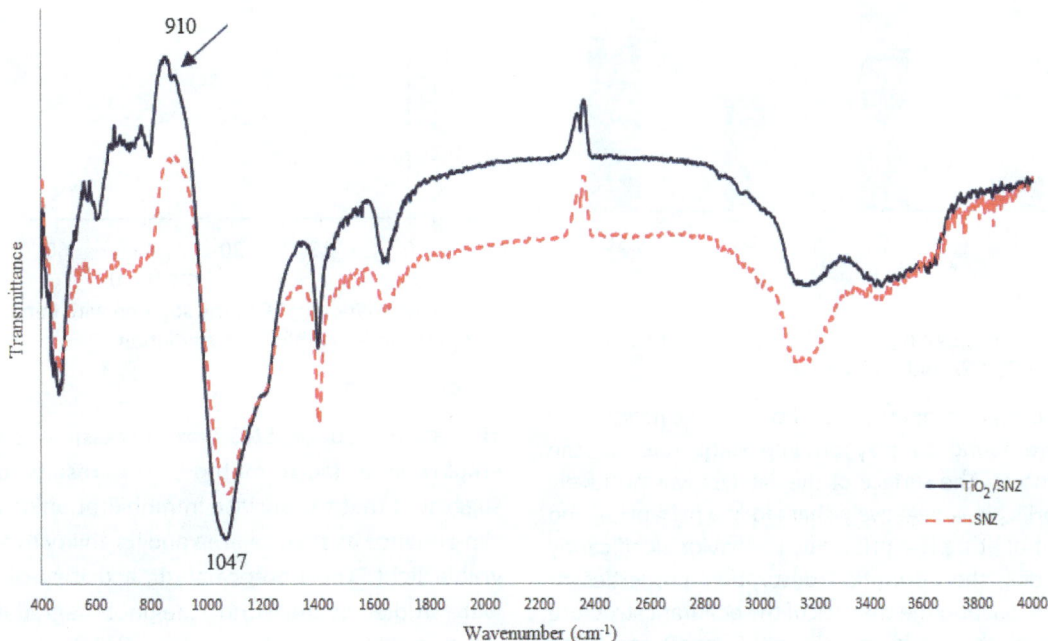

Fig. 6. Comparison of the FTIR spectra of pure SNZ and TiO_2/SNZ nanocomposite

3.2. Photocatalytic degradation of TC

The influence of pH on the photocatalytic degradation extent of TC by supported TiO_2 onto SNZ (TiO_2/SNZ) was evaluated by varying the initial pH of the TC solution and keeping all other experimental conditions constant. In order to compare the degradation efficiency, some reaction was carried out with initial acidic pH and alkaline pH in the range of 4-10. The effect of pH on the photocatalytic reaction was generally attributed to the surface charge of the catalyst and its relation with the ionic form of the organic compound (anionic or cationic). Electrostatic attraction or repulsion between the surface of the

catalyst and the organic molecule took place and consequently enhanced or inhibited, respectively, the photodegradation rate. The photocatalytic degradation efficiencies of TC under visible light by TiO_2/SNZ were obtained in the order of pH 6 > pH 10 > pH 8 > pH 4. Further, 87% of TC was photodegraded at a pH of 6 after 90 min, while only 10% of TC was removed at a pH of 4 under the same experimental condition. The point of zero charge (pzc) of TiO_2/SNZ, i.e., the point when the surface charge density is zero, was found to be of 7 and is shown in Figure. 7. This value corresponded to the pH at which the straight line ($pH_{initial} = pH_{final}$) crossed the sigmoid curve passing through the experimental points [30].

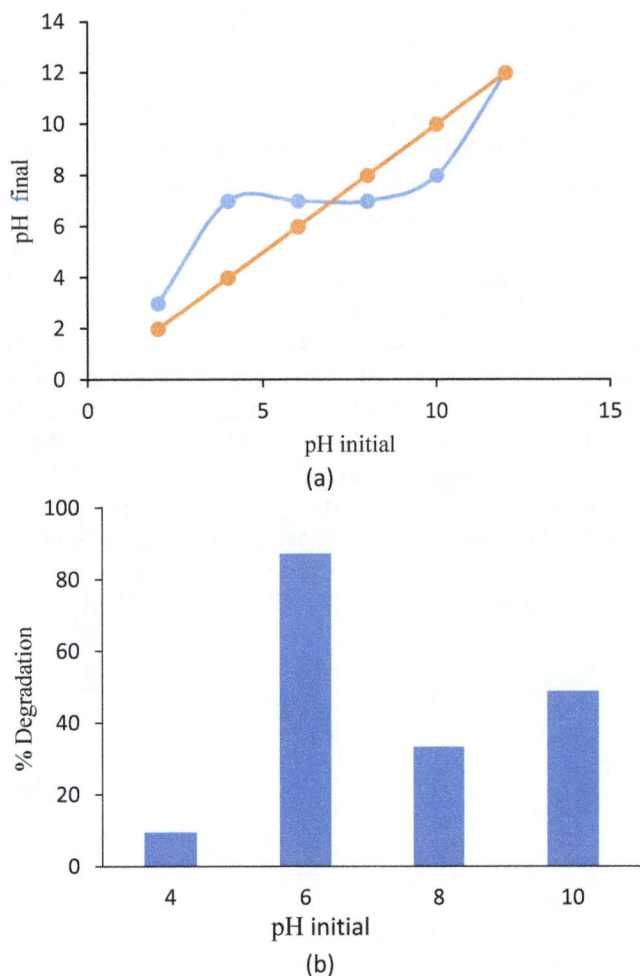

Fig. 7. (a) pH_{PZC} of TiO_2/SNZ, (b) Effect of pH on TC photocatalytic degradation by TiO_2/SNZ under visible light

The electric charge properties of both the catalyst and substrate were found to play an important role on the adsorption process. The surface of the catalyst was positively charged at a pH < pH_{pzc}, negatively charged at a pH > pH_{pzc}, and remained neutral at a pH = pH_{pzc}. Such behavior significantly affected not only the adsorption–desorption properties of catalyst surface, but also the changes of the pollutant structure at various pH values. In addition, the pKa1, pKa2 and pKa3 values of the TC molecule are 3.3, 7.68, and 9.7, respectively, which are related to equilibriums ($TCH_3^+ \rightarrow TCH_2$, $TCH_2 \rightarrow TCH^-$, and $TCH^- \rightarrow TC_2^-$). The adsorption mode and the concentration of OH were two key factors influencing the photocatalytic degradation efficiencies of TC at different pH values [4]. The pH_{PZC} of the synthesized catalyst was found to be about 7. The adsorption of TC on the catalyst might be inhibited by the enhanced electrostatic repulsion between H_3TC^+ and the positively charged catalyst at a pH of 4 as well as between HTC^-/TC_2^- and a negatively charged catalyst at a pH of 8 and 10, respectively. Also, in a pH =6, the TCH^- and TC_2^- species were dominant and the surface of the nano composite particles had a positive charge. Thus, the TC molecule seemed to be attracted to the positively charged TiO_2/SNZ surface. Furthermore, the decrease of the reaction rate at acid pH can

be ascribed to the lower hydroxylation of the catalyst's surface due to the presence of small amounts of OH^- ions. To assure of the removal of TC by photocatalytic degradation, several control experiments were investigated and are shown in Figure 8. It clearly indicated that the TC could hardly be degraded without any photocatalyst (photolysis) while the TiO_2/SNZ nanocomposite showed a much higher activity than the TiO_2 (alone) with visible light exposure. Also, the removal of TC by adsorption on the SNZ was not comparable by photocatalytic degradation of TC on the TiO_2/SNZ. After irradiation for 90 min, only 25% TC removal over the SNZ was observed, while the removal ratio on the TiO_2/SNZ reached 87%. The degradation of TC by commercial TiO_2 P25 under visible light did not occur, due to the inability of the absorption of visible light.

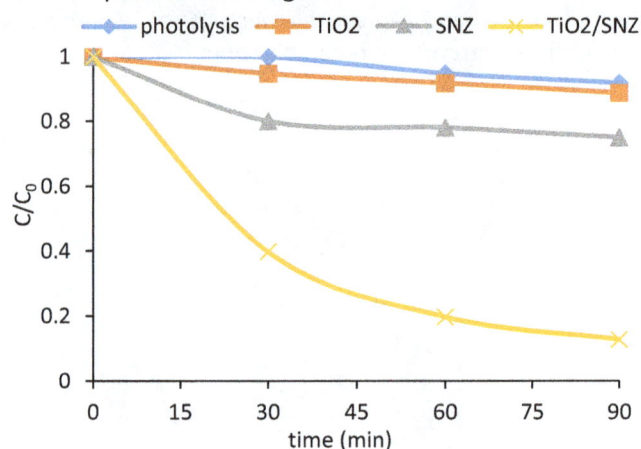

Fig. 8. Degradation of TC in the solution with initial concentration of 8 mg/L under visible light irradiation

4. Conclusions

TiO_2 supported on SNZ was successfully synthesized by employing a facile method. The results of this study suggested that the surface immobilization of TiO_2 onto SNZ can enhance its surface area and its ability to activate under visible light. The photocatalytic activities of the catalysts were studied by measuring the photodegradation of the TC solution. The catalysts showed pH dependence and more than 87% of the TC could be removed from the solution within 90 min at a pH of 6. In addition, the immobilization of TiO_2 onto zeolite particles permitted for easier separation of the adsorbent (compared to finer nanoparticles of TiO_2) from the treated water. Diffuse reflectance spectroscopy revealed a clear trend of decreasing the band gap to 2.4 eV when TiO_2 was supported on SNZ. The specific surface area for the composited material reached 93m^2/g, which was two times larger than the unmodified TiO_2. With these improvements, the TiO_2/SNZ was able to absorb more pollutant because of a higher surface area and enhanced visible light absorption rather than commercial TiO_2 P25. The finding of this study could benefit future applications in water and wastewater treatment.

References

[1] Daghrir, R., Drogui, P. (2013). Tetracycline antibiotics in the environment: a review. *Environmental chemistry letters*, *11*(3), 209-227.

[2] Liu, H., Yang, Y., Kang, J., Fan, M., Qu, J. (2012). Removal of tetracycline from water by Fe–Mn binary oxide. *Journal of environmental sciences*, 24(2), 242–247.

[3] Kümmerer, K. (2009). The presence of pharmaceuticals in the environment due to human use–present knowledge and future challenges. *Journal of environmental management*, *90*(8), 2354-2366.

[4] Saadati, F., Keramati, N., Ghazi, M. M. (2016). Influence of parameters on the photocatalytic degradation of tetracycline in wastewater: A review. *Critical reviews in environmental science and technology*, *46*(8), 757-782.

[5] Yuan, F., Hu, C., Hu, X., Wei, D., Chen, Y., Qu, J. (2011). Photodegradation and toxicity changes of antibiotics in UV and UV/H_2O_2 process. *Journal of hazardous materials*, *185*(2), 1256-1263.

[6] Nezamzadeh-Ejhieh, A., Shirzadi, A. (2014). Enhancement of the photocatalytic activity of ferrous oxide by doping onto the nano-clinoptilolite particles towards photodegradation of tetracycline. *Chemosphere*, *107*, 136-144.

[7] Yue, L., Wang, S., Shan, G., Wu, W., Qiang, L., Zhu, L. (2015). Novel MWNTs–Bi_2WO_6 composites with enhanced simulated solar photoactivity toward adsorbed and free tetracycline in water. *Applied catalysis B: Environmental*, *176*, 11-19.

[8] Choina, J., Duwensee, H., Flechsig, G. U., Kosslick, H., Morawski, A. W., Tuan, V. A., Schulz, A. (2010). Removal of hazardous pharmaceutical from water by photocatalytic treatment. *Central european journal of chemistry*, *8*(6), 1288-1297.

[9] Zhang, Y. P., Jia, C. G., Peng, R., Ma, F., Ou, G. N. (2014). Heterogeneous photo-assisted Fenton catalytic removal of tetracycline using Fe-Ce pillared bentonite. *Journal of central south university*, *21*, 310-316.

[10] Yu, X., Lu, Z., Si, N., Zhou, W., Chen, T., Gao, X., Yan, C. (2014). Preparation of rare earth metal ion/TiO_2 Hal-conducting polymers by ions imprinting technique and its photodegradation property on tetracycline. *Applied clay science*, *99*, 125-130.

[11] Chang, C. T., Wang, J. J., Ouyang, T., Zhang, Q., Jing, Y. H. (2015). Photocatalytic degradation of acetaminophen in aqueous solutions by TiO_2/ZSM-5 zeolite with low energy irradiation. *Materials science and engineering: B*, *196*, 53-60.

[12] Zhu, X. D., Wang, Y. J., Sun, R. J., Zhou, D. M. (2013). Photocatalytic degradation of tetracycline in aqueous solution by nanosized TiO_2. *Chemosphere*, *92*(8), 925-932.

[13] Abbasi, A., Ghanbari, D., Salavati-Niasari, M., Hamadanian, M. (2016). Photo-degradation of methylene blue: photocatalyst and magnetic investigation of Fe_2O_3–TiO_2 nanoparticles and nanocomposites. *Journal of materials science: Materials in electronics*, *27*(5), 4800-4809.

[14] Shi, J. W., Zheng, J. T., Ji, X. J. (2010). TiO_2-SiO_2/activated carbon fibers photocatalyst: preparation, characterization, and photocatalytic activity. *Environmental engineering science*, *27*(11), 923-930.

[15] Paul, B., Martens, W. N., Frost, R. L. (2012). Immobilised anatase on clay mineral particles as a photocatalyst for herbicides degradation. *Applied clay science*, *57*, 49-54.

[16] Wang, H., Yang, B., Zhang, W. J. (2010). Photocatalytic degradation of methyl orange on Y zeolite supported TiO_2. In *Advanced Materials Research* (Vol. 129, pp. 733-737). Trans Tech Publications.

[17] Liu, S., Lim, M., Amal, R. (2014). TiO_2-coated natural zeolite: rapid humic acid adsorption and effective photocatalytic regeneration. *Chemical engineering science*, *105*, 46-52.

[18] Durgakumari, V., Subrahmanyam, M., Rao, K. S., Ratnamala, A., Noorjahan, M., Tanaka, K. (2002). An easy and efficient use of TiO_2 supported HZSM-5 and TiO_2+HZSM-5 zeolite combinate in the photodegradation of aqueous phenol and p-chlorophenol. *Applied catalysis A: General*, *234*(1), 155-165.

[19] Treacy, M. M. J., Higgins, J. B. (2001). Collection of simulated XRD powder patterns for zeolites. Published on behalf of the Structure Commission of the 'International Zeolite Association'. *Powder patterns*, *203, 204*.

[20] Huang, M., Xu, C., Wu, Z., Huang, Y., Lin, J., Wu, J. (2008). Photocatalytic discolorization of methyl orange solution by Pt modified TiO_2 loaded on natural zeolite. *Dyes and pigments*, *77*(2), 327-334.

[21] Alshameri, Aref, Chunjie Yan, and Xinrong Lei. "Enhancement of phosphate removal from water by TiO_2/Yemeni natural zeolite: preparation, characterization and thermodynamic." *Microporous and mesoporous materials* 196 (2014): 145-157.

[22] Kanakaraju, D., Kockler, J., Motti, C. A., Glass, B. D., Oelgemöller, M. (2015). Titanium dioxide/zeolite integrated photocatalytic adsorbents for the degradation of amoxicillin. *Applied catalysis B: Environmental*, *166*, 45-55.

[23] Ogura, M., Kawazu, Y., Takahashi, H., Okubo, T. (2003). Aluminosilicate species in the hydrogel phase formed during the aging process for the crystallization of FAU zeolite. *Chemistry of materials*, *15*(13), 2661-2667.

[24] Zhao, C., Deng, H., Li, Y., Liu, Z. (2010). Photodegradation of oxytetracycline in aqueous by 5A and 13X loaded with TiO_2 under UV irradiation. *Journal of hazardous materials*, *176*(1), 884-892.

[25] Zhong, S., Zhang, F., Yu, B., Zhao, P., Jia, L., Zhang, S. (2016). Synthesis of PVP-Bi2WO6 photocatalyst and

degradation of tetracycline hydrochloride under visible light. *Journal of materials science: Materials in electronics*, 27(3), 3011-3020.

[26] Ghorai, T. K., Biswas, N. (2013). Photodegradation of rhodamine 6G in aqueous solution via $SrCrO_4$ and TiO_2 nano-sphere mixed oxides. *Journal of materials research and technology*, 2(1), 10-17.

[27] Anpo, M., Takeuchi, M. (2003). The design and development of highly reactive titanium oxide photocatalysts operating under visible light irradiation. *Journal of catalysis*, 216(1), 505-516.

[28] Ohno, T., Tagawa, S., Itoh, H., Suzuki, H., Matsuda, T. (2009). Size effect of TiO_2–SiO_2 nano-hybrid particle. *Materials chemistry and physics*, 113(1), 119-123.

[29] Li, F., Jiang, Y., Yu, L., Yang, Z., Hou, T., Sun, S. (2005). Surface effect of natural zeolite (clinoptilolite) on the photocatalytic activity of TiO_2. *Applied surface science*, 252(5), 1410-1416.

[30] Keramati, N., Nasernejad, B., Fallah, N. (2014). Synthesis of N-TiO_2: Stability and visible light activity for aqueous styrene degradation. *Journal of dispersion science and technology*, 35(10), 1476-1482.

9

Comparison and analysis of two natural adsorbents of Sorghum and Ziziphus nummularia pyrene for the removal of erythrosine dye from aquatic environments

Nayereh Yahyaei[a], Javad Mousavi[b], Mehdi Parvini[b,*], Pedram Mohebbi[a]

[a] Department of Chemical Engineering, Shahrood Branch, Islamic Azad University, Shahrood, Iran
[b] Department of Chemical Engineering, Gas and Petroleum, Semnan University, Semnan, Iran

ABSTRACT

One pollutant which seriously threatens water resources is dye. Therefore, finding a suitable method to separate the dye in water resources is very important. An adsorption process that uses low cost adsorbents is considered as an efficient strategy for this purpose. In this study, Erythrosine dye removal from an aquatic environment using natural absorbents, namely Sorghum and Ziziphus nummularia pyrene, was reviewed. The effects of different parameters such as pH, contact time, initial density, and the adsorbent amount in the batch system were investigated. The results indicated that increased temperature has no significant effect on the removal of Erythrosine dye, and the highest adsorption was achieved in the first 30 min of adsorbent- dye contact time. Also, most of the adsorption occurred at pH values of 4-8. Moreover, the highest amount of dye removal was observed in a concentration of 20 mg/L for the Ziziphus nummularia pyrene adsorbent and 5 mg/L of the Sorghum adsorbent. Also, the Langmuir and Freundlich equations were used to analyze the adsorption process, where both the Sorghum and Ziziphus nummularia pyrene adsorbents showed a better agreement with the Langmuir isotherm.

Keywords:
Adsorption
Exhaust emissions
EGR
Injection timing
Methanol biofuel

1. Introduction

Nowadays, increasing importance has been attached to the issue of preserving water resources. The consequences of rapid population growth including the excessive use of limited water resources and the associated water pollution due to a variety of agricultural, biological and industrial activities have added to the problem of water shortages in recent years. Environmental pollution from dyes and the contamination of water supplies have proved to be a considerable challenge for the modern world. The discharge of colored wastewater from textile, paper, plastics, cosmetics and food industries into the waterways can be considered as the first detectable sign of contamination. Since light and heat do not affect the stability of most dyes and impedes biological decomposition, it is difficult to remove them from water; therefore, in most countries, the removal of textile dyes from industrial wastewater proves to be a major problem for environmental management even though environmental laws are enforced [1]. There are many different methods for reducing the toxicity of dyes including coagulation, oxidation, membrane separation, electrochemical methods, and aerobic and non-aerobic microbial decomposition. However, these methods have some limitations. Among the different methods, adsorption is more economical and also removes high amounts of the dye. The most commonly used adsorbent is active carbon. However, application of active carbon has some limitations such as its high cost. Therefore, adsorption via agricultural products can be an economical and practical method for the removal of most pollutants such as dyes, heavy metals, fennel, and gasses [2]. Erythrosine is one of the most well-known dyes generally

*Corresponding author
E-mail address: m.parvini@ semnan.ac.ir

employed in cosmetics, food, medicine, and textile industries. Erythrosine is a water-soluble dye from the Zanten category. It is a highly poisonous dye which can cause many different kinds of health problems such as allergies, triode illness, cancer and DNA behavioral abnormalities. A large number of qualitative studies have already been carried out in regard to Erythrosine dye removal, and these studies have attracted attention among different researchers [3-4]. In the present study, two natural adsorbents, Sorghum and Ziziphus nummularia pyrene, were applied to remove Erythrosine dye from the effluent.   They are low-priced and available from agricultural wastes. In addition, the abilities of these adsorbents have been analyzed and compared. The main issues considered in the present paper are as follows:

1. Analysis of contact time, pH, initial density of the dye, temperature and heat adsorption in the removal of Erythrosine dye.
2. Determination of the best isotherm and Kinetic model for the dye adsorption under study.

2. Materials and methods

2.1. Sorbents

In this study, Sorghum and Ziziphus nummularia pyrene procured from agriculture wastes were used as adsorbents for the removal of Erythrosine dye. Sorghum and Ziziphus nummularia pyrene were obtained from a farm in the city of Garmsar and from agriculture wastes in the city of Behbahan, respectively. Initially, the obtained wastes were extensively washed with tap water to remove the initial impurities, sprayed with distilled water, and dried at room temperature to a constant weight. Then they were ground into small fine particles by a home type grinder and particles with sizes of $35 < d \leq 60\mu m$ were obtained; these particles were used directly for the adsorption experiments without any pre-treatment.

2.2. Chemicals

Erythrosine (empirical formula $C_{20}H_8I_4O_5$; molecular weight= 835.89) is a commercial Nitrozo dye containing -NO-groups. The mentioned dye was purchased from Merk (Germany). The test solutions containing the required dye concentration were prepared by diluting 100ppm of stock solutions of dye which was obtained by dissolving a weighed quantity of 0.05 g of Erythrosine dye in 0.5 L of distilled water. The concentration range of the prepared dye solutions was between 5 and 100 mg/L.

2.3. Sorption studies

Sorption studies were conducted in a routine manner via the batch technique. A number of Pyrex flasks containing a definite volume (100 ml in each case) of Erythrosine dye solution with the desired concentration and pH were placed in a thermostatic rotary shaker. In the batch experiments, 0.3 g of sorghum and 0.5 g of Ziziphus

nummularia pyrene were treated with 100 ml of dye solution. The flasks were agitated at a constant shaking rate of 300 rpm for 30 min to ensure that equilibrium was reached. The dye solutions were separated from the sorbents by filtering. Uptake values were determined as the difference between the initial dye concentration and the one in the supernatant. The effects of the experimental parameters on the adsorption capacity of the adsorbent in the experiments were investigated.

Equation (1) was used to calculate the removal efficiency of the dye from the solution. To calculate the adsorption capacity or the amount of dye adsorbed per unit weight of adsorbent, Equation (2) is commonly used

$$R = \frac{C_0 - C_e}{C_0} \times 100 \tag{1}$$

$$q_e = \frac{(C_0 - C_e)V}{m} \tag{2}$$

which C_0 and C_e are initial concentration and dye concentration at time t, in terms of milligrams per liter V is volume of solution in terms of Liter, and M is the mass of sorbent in term of gram [5]. All the experiments were carried out in duplicates and the average values were used for further calculation. For the calculation of average value, the percent relative standard deviation for samples was calculated.

2.4. Analysis of Erythrosine

The concentration of unadsorbed Erythrosine dye in the adsorption medium was measured colorimetrically using a spectrophotometer (UNICO2100). The absorbance of the green dye was at 527 nm, where the corresponding maximum adsorption peak existed.

3. Results and discussion

3.1. Analysis of contact time

The known density of the dye material (100ppm) has been exposed to certain amounts of adsorbent (0.3 g of sorghum and 0.5 g of Ziziphus nummularia) with the adjusted pH of solution, and then the resulting mixture was stirred with a magnetic stirrer. The percentages of dye adsorption at different contact times of 5 to 90 minutes have been investigated. The results show that an increase in contact time from 5 to 20 minutes quickly increases the amount of adsorption, which can be ascribed to the increase in the active sites on the adsorbent. This incremental trend continues for 30 minutes and then reaches a constant level. Therefore, the time of 30 minutes has been considered as the optimal time for both adsorbents, as shown in Figure 1.

3.2. Analysis of the initial concentration

Solutions of different initial Erythrosine concentrations (5, 10, 20, 25,50ppm) were shaken for 30 min at ambient temperature with an adjusted solution pH to investigate the effect of the concentration on the removal of Erythrosine. Obviously, the total amount of adsorbed Erythrosine always increased with an increase in the initial Erythrosine concentration. Figure 2 shows the effect of pollutant concentration on dye removal using Erythrosine. As it can be observed from this figure, the optimal concentration of the pollutant for the adsorption of sorghum and Ziziphus nummularia were found to be 5ppm and 20ppm, respectively. At a lower concentration, the exposed surface of adsorbent to the dyes was higher while this exposure was reduced as the concentration was increased. Therefore, removal efficiency decreased with an increase in initial concentration. The results indicated that the maximum amounts of adsorption were 93.7% and 92% for Ziziphus nummularia and sorghum, which was reduced to 91% and 75.6%, respectively, when initial concentration was increased.

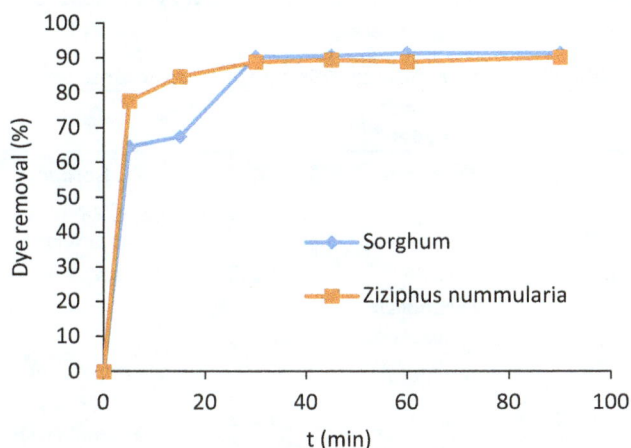

Fig. 1. Comparison of optimization of contact time of two adsorbents

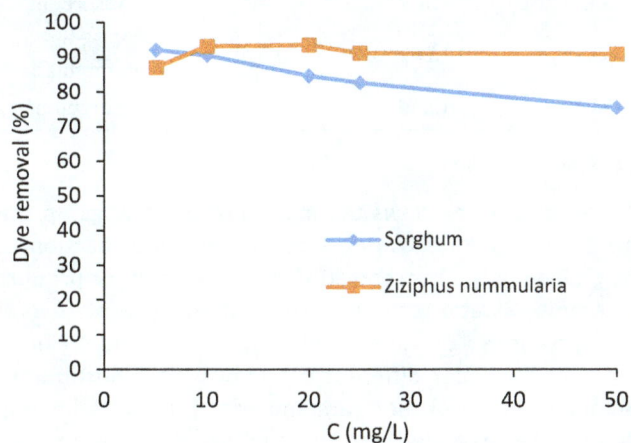

Fig. 2. Comparison of pollutant density optimization for two adsorbent

3.3. Effect of adsorbent dose

Different weighed amounts of Sorghum (0.1-1g) and Ziziphus nummularia (0.025-1) adsorbents were added to

the dye solution of 5 mg/L and 20 mg/L, respectively. Subsequently, they were stirred for the optimal time achieved in the previous stage (30 min) using a magnetic stirrer. After sifting the mixture, the remaining pollutant concentration in the solution was measured by a spectrophotometer device; resultantly, the optimal gram of sorghum and Ziziphus nummularia adsorption were found to be 0.3 gr and 0.1gr, respectively. The results are shown in Figure 3.

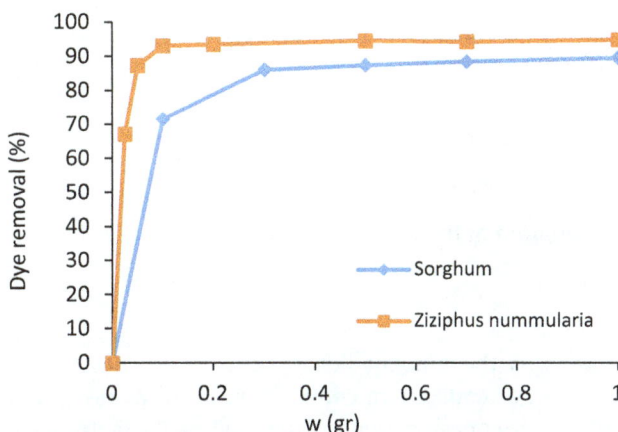

Fig. 3. Optimal comparison of adsorbent dose

3.4. Effect of pH

The pH of aqueous solution is an important controlling parameter in the adsorption process. Variations in pH result in the variations in the adsorption load, degree of homogeneity of adsorbent material and homogeneity of active absorbent groups [6]. Studies on the effect of Ziziphus nummularia adsorbent on the Erythrosine dye removal from aqueous environments have been carried out in the pH range of 2-10. Sodium hydroxide and hydrochloric acid were used to adjust the pH. The effect of pH on the Erythrosine dye adsorption process is shown in Figure 4. According to this figure, the best percentage of dye removal was achieved in the neutral range. Furthermore, a pH = 2 was not considered due to the color change and wavelength variation.

3.5. Adsorption isotherm

Adsorption is a process of mass transformation where the different compounds may compete to reach a balance. The initial effective forces in the adsorption between the adsorbent and the adsorbate are electrostatic attraction and electrostatic repulsion, which can be physical or chemical. As recorded in the literature, different equations have been formulated to explain the balance between the adsorbate and the adsorbent, among which are Langmuir, Freundlich and others [7].

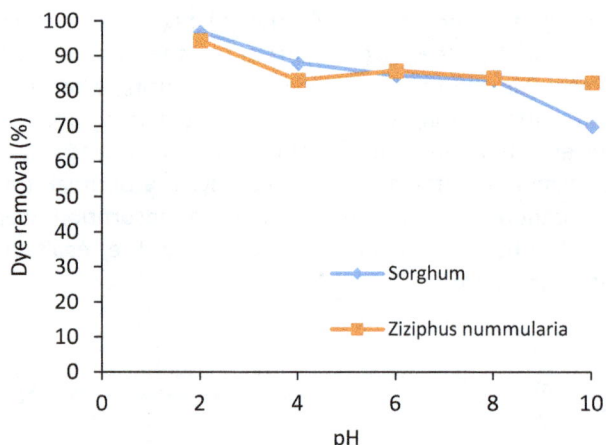

Fig. 4. Optimization of pH comparison

3.5.1. Freundlich isotherm

The equation of the Freundlich isotherm [8-9] is as follows:

$$Lnq_e = \frac{1}{n} LnC_e + LnK_F \tag{3}$$

where q_e is the amount adsorbed per unit mass of the adsorbent at equilibrium condition in terms of mg/g, C_e is equilibrium concentration of the pollutants in terms of mg/L, K_F is adsorption capacity at unit concentration and $1/n$ is intensity of adsorption. $1/n$ signifies the type of isotherm that if $1/n = 0$, then it is irreversible; $0 < 1/n < 1$ is desirable; and $1/n > 1$ is undesirable.

3.5.2. Langmuir isotherm

The Langmuir isotherm [10] is:

$$\frac{C_e}{q_e} = \frac{1}{Q_o b} + \frac{C_e}{Q_o} \tag{4}$$

where q_e and C_e are defined as before, q_0 is the amount of sorbent required for single-layer capacity per unit mass of the adsorbent in terms of mg/g and the constant b is related to the binding energy in terms of mg/l. The basic characteristic of a Langmuir isotherm is a non-dimensional constant called the equilibrium parameter (R_L) [11], which can be defined through the following equation:

$$R_L = \frac{1}{(1 + bC_0)} \tag{5}$$

In the above equation, C_o is the initial concentration of dye in terms of mg/L; R_L indicates the type of isotherm according to which if $R_L = 0$, it is irreversible; $0 < R_L < 1$ is desirable; $R_L = 1$ is linear; and $R_L > 1$ is undesirable [12-13].

3.5.3. Temkin isotherm

The Temkin model [14-15] is another adsorption isotherm model constructed to describe the adsorption phenomenon. The corresponding equation can be expressed as follows:

$$q_e = (\frac{RT}{b})LnA + (\frac{RT}{b})LnC_e \tag{6}$$

$$\frac{RT}{b} = B \tag{7}$$

$$q_e = B_T LnA_T + B_T LnC_e \tag{8}$$

In the above equation, A is in terms of L/gr and B is in terms of J/mol. They are both Temkin's constants which are determined by plotting q_e versus LnC_e. The parameters of the Langmuir, Freundlich and Temkin isotherms and the regression correlation coefficients (R^2) are listed in Table 1. According to Table 1, the adsorption was better fitted by Langmuir ($R^2 = 0.998–0.999$) when compared to Freundlich ($R^2 = 0.953–0.941$) and Temkin ($R^2 = 0.911–0.977$) models. Thus, adsorption happened in a single layer or on a fixed number of adsorption sites in the surface of the adsorbent. Also, all locations had the same adsorption energy and were based on the assumption that the adsorbent structure was homogeneous.

Table 1. Isotherm coefficients of Erythrosine Dye adsorption

Langmuir model			
b (L/mg)	Q_0 (mg/gr)	R^2	Adsorbent
0.2329	101	0.998	Ziziphus nummularia
0.1551	17.857	0.999	Sorghum
Freundlich model			
n	K_f (mg/g)	R^2	Adsorbent
1.5723	3.4452	0.953	Ziziphus nummularia
1.675	2.846	0.941	Sorghum
Temkin model			
B	A	R^2	Adsorbent
43.52	1.642	0.9107	Ziziphus nummularia
3.748	1.546	0.9774	Sorghum

3.6. Kinetic studies

The term 'kinetic' shows the motion or the change. In the present study, kinetics refers to the rate of a reaction as the change in the concentration of the reactant or product over time. Kinetic tests are usually performed at several concentrations of adsorbent. By conducting kinetic experiments under different conditions, the contributory factors in the adsorption rate and adsorption rate-limiting step can be identified. For the analysis of the adsorption kinetics, the first and second order kinetics models can be used.

3.6.1. First order kinetic model

To express the adsorption rate of the dissolved substances from the aqueous solution, the first equation is employed. The equation can be stated as follows:

$$\frac{dq}{dt} = K_1(q_e - q_t) \tag{9}$$

By integrating both sides of the above equation for t = 0 to t = t and q = 0 to q = q_e as boundary conditions, the following linear form is obtained [16]:

$$Ln(q_e - q_t) = Lnq_e - K_1 t \tag{10}$$

This equation was applied to describe the adsorption kinetics of different systems.

3.6.2. Second order kinetic model

Another equation for kinetic analysis of adsorption is the quadratic equation, which can be expressed as follows:

$$\frac{dq}{dt} = K_2(q_e - q)^2 \tag{11}$$

K_2 : Quadratic equation rate constant of absorbent (g mol^{-1} min^{-1}). By integrating the above equation for t = 0 to t = t and q = 0 to q = q_i as boundary conditions, the following equation can be obtained [17]:

$$\frac{t}{q_t} = \frac{1}{K_2 q_e^2} + \frac{t}{q_e} \tag{12}$$

By plotting t/q_t versus t, a straight line is obtained, from which the slope and the intercept are $1/q_e$ and $1/K_2 q_e^2$, respectively. Consequently, the equilibrium adsorption capacity, q_e, and the adsorption rate constant can be determined.

3.6.3. Elovich model

The Elovich model is shown as follows [18-19]:

$$q_t = \frac{1}{b} Ln(ab) + \frac{1}{b} Lnt \tag{13}$$

where a is the initial adsorption rate in terms of mg.g^{-1}.min^{-1}; b is related to the extent of surface coverage and activation energy for chemisorption(g/mg); and q_t is the amount of the adsorbed dye at time t in terms of mg/g. By plotting q_t versus Ln (t), the values of a and b are obtained.

3.6.4. Richie Kinetic Model

The Richie Kinetic Model is expressed as follows:

$$\frac{1}{q_t} = \frac{1}{K_r q_e t} + \frac{1}{q_e} \tag{14}$$

In the above equation, K_r is the rate constant in terms of min^{-1} whereas q_e and q_t are defined as before. By plotting

$1/q_t$ versus $1/t$, the values of q_e and K_r are obtained [20]. The corresponding kinetic parameters obtained from the above models are listed in Table 2. However, the correlation coefficient (R^2) obtained from the second-order adsorption kinetic model was higher than that from the other kinetic models, suggesting that the pseudo-second-order equation was more appropriate to simulate the experimental kinetic data. The pseudo second-order model was based on the assumption that the adsorption rate was linearly related to the square of the number of unoccupied sites and the adsorption rate between adsorbent and adsorbate appeared to be controlled by the chemical adsorption.

Table 2. Calculation of Kinetic parameters for the first and second order reaction

First order Kinetic model		
R^2	K_1 (min^{-1})	q_e (mg/g)
0.4887	0.00005	65.313
0.6184	13.818×10^{-5}	16.428
Second order Kinetic model		
R^2	K_2 (mg.g^{-1}min^{-1})	q_e (mg/g)
0.9999	0.05192	35.714
1	0.601	1.535
Richie Kinetic model		
R^2	K_r (min^{-1})	q_e (mg/g)
0.9783	1.146	36.496
0.9957	1.009	1.529
Elovich Kinetic model		
R^2	b (g/mg)	a (mg.g^{-1}.min^{-1})
0.9025	0.4455	503.048
0.9278	12.048	1.722

4. Conclusions

To investigate the effect of natural adsorbents of Ziziphus nummularia pyrene and sorghum on the amount of Erythrosine dye removal, several important and influential factors including contact time, dose of adsorbent, dye initial concentration, temperature, and pH were analyzed. Through pH analysis, it was observed that the dye removal for both types of adsorption does not depend on pH; thus, all the tests were performed in neutral pH. With an increase in the adsorbent dosage and also an increase in contact time, the adsorption amount increased. This can be attributed to the fact that as the amount of adsorption dose increased, larger surface was available for removal and more sites could be placed in the environment for adsorption. The analysis of the initial concentration effects indicated that with the increase in the density of the pollutant, the amount of dye removal for the sorghum adsorption decreased while this amount increased for Ziziphus nummularia pyrene. This indicated the different capabilities of natural adsorptions for dye removal. The data obtained from the tests were analyzed using the Langmuir and Fernedlich models. According to the results, the data were best fitted by the Langmuir model.

References

[1] McKay, G. (1982). Adsorption of dyestuffs from aqueous solutions with activated carbon I: Equilibrium and batch contact-time studies. *Journal of chemical technology and biotechnology, 32*(7-12), 759-772.

[2] Salleh, M. A. M., Mahmoud, D. K., Karim, W. A. W. A., Idris, A. (2011). Cationic and anionic dye adsorption by agricultural solid wastes: A comprehensive review. *Desalination, 280*(1), 1-13.

[3] Gupta, V. K., Mittal, A., Kurup, L., Mittal, J. (2006). Adsorption of a hazardous dye, erythrosine, over hen feathers. *Journal of colloid and interface science, 304*(1), 52-57.

[4] Al-Degs, Y. S., Abu-El-Halawa, R., Abu-Alrub, S. S. (2012). Analyzing adsorption data of erythrosine dye using principal component analysis. *Chemical engineering journal, 191*, 185-194.

[5] Bauer, C., Jacques, P., Kalt, A. (2001). Photooxidation of an azo dye induced by visible light incident on the surface of TiO_2. *Journal of photochemistry and photobiology A: Chemistry, 140*(1), 87-92.

[6] Langmuir, I. (1916). The constitution and fundamental properties of solids and liquids. *Journal of the American chemical society, 38*(11), 2221-2295.

[7] H. Freundlish, (1906). Over the Adsorption in Solution, *Journal of physical chemistry, 57*, 385-470.

[8] Zawani, Z., Chuah, A. L., Choong, T. S. Y. (2009). Equilibrium, kinetics and thermodynamic studies: adsorption of Remazol Black 5 on the palm kernel shell activated carbon. *European journal of scientific research, 37*(1), 67-76.

[9] Li, Y. H., Di, Z., Ding, J., Wu, D., Luan, Z., Zhu, Y. (2005). Adsorption thermodynamic, kinetic and desorption studies of Pb^{2+} on carbon nanotubes. *Water research, 39*(4), 605-609.

[10] Tan, I. A. W., Hameed, B. H., Ahmad, A. L. (2007). Equilibrium and kinetic studies on basic dye adsorption by oil palm fibre activated carbon. *Chemical engineering journal, 127*(1), 111-119.

[11] Bulut, E., Özacar, M., Şengil, İ. A. (2008). Equilibrium and kinetic data and process design for adsorption of Congo Red onto bentonite. *Journal of hazardous materials, 154*(1), 613-622.

[12] Alzaydien, A. S., Manasreh, W. (2009). Equilibrium, kinetic and thermodynamic studies on the adsorption of phenol onto activated phosphate rock. *International journal of physical sciences, 4*(4), 172-181.

[13] Venkateswaran, V., Priya, V. T. (2012). Adsorption kinetics and thermodynamics of removal of basic dyes by stishovite clay-TiO_2 nanocomposite. *Journal of applied technology in environmental sanitation, 2*(1), 7–16

[14] Allen, S. J., Gan, Q., Matthews, R., Johnson, P. A. (2003). Comparison of optimised isotherm models for basic dye adsorption by kudzu. *Bioresource technology, 88*(2), 143-152.

[15] Hamdaoui, O., Naffrechoux, E. (2007). Modeling of adsorption isotherms of phenol and chlorophenols onto granular activated carbon: Part I. Two-parameter models and equations allowing determination of thermodynamic parameters. *Journal of hazardous materials, 147*(1), 381-394.

[16] Ho, Y. S., McKay, G. (1998). A comparison of chemisorption kinetic models applied to pollutant removal on various sorbents. *Process safety and environmental protection, 76*(4), 332-340.

[17] Li, L., Luo, C., Li, X., Duan, H., Wang, X. (2014). Preparation of magnetic ionic liquid/chitosan/grapheneoxide composite and application for water treatment. *International journal of biological macromolecules, 66*, 172-178.

[18] Özacar, M., Şengil, İ. A. (2004). Application of kinetic models to the sorption of disperse dyes onto alunite. *Colloids and Surfaces A: Physicochemical and engineering aspects, 242*(1), 105-113.

[19] Örnek, A., Özacar, M., Şengil, İ. A. (2007). Adsorption of lead onto formaldehyde or sulphuric acid treated acorn waste: equilibrium and kinetic studies. *Biochemical engineering journal, 37*(2), 192-200.

[20] Ho, Y. S. (2006). Review of second-order models for adsorption systems. *Journal of hazardous materials, 136*(3), 681-689.

10

Removal of copper (II) from aqueous solutions by adsorption onto granular activated carbon in the presence of competitor ions

Saeed Almohammadi, Masoomeh Mirzaei[*]

Department of Chemical Engineering, Mahshahr Branch, Islamic Azad University, Mahshahr, Iran

Keywords:
Activated carbon
Adsorption
Competitor ion
Copper

ABSTRACT

In this work, the removal of copper from an aqueous solution by granular activated carbon (GAC) in the presence of competitor ions was studied. A batch adsorption was carried out and different parameters such as pH, contact time, initial copper concentration and competitor ions concentration were changed to determine the optimum conditions for adsorption. The optimum pH required for maximum adsorption was found to be 4.5 for copper. Equilibrium was evaluated at 144 h at room temperature. The removal efficiency of Cu(II) was 71.12% at this time. The kinetics of copper adsorption on activated carbon followed the pseudo second-order model. The experimental equilibrium sorption data were tested using the Langmuir, Freundlich, Temkin and Dubinin–Radushkevich (D-R) equations and the Langmuir model was found to be well fitted for copper adsorption onto GAC. The maximum adsorption capacity of the adsorbent for Cu(II) was calculated from the Langmuir isotherm and found to be 7.03 mg/g. Subsequently, the removal of copper by granular activated carbon in the presence of Ag^{1+} and Mn^{2+} as competitor ions was investigated. The removal efficiency of Cu(II) ions without the presence of the competitor ions was 46% at 6 h, while the removal efficiency of Cu(II) ions in the presence of competitor ions, Ag^{1+} and Mn^{2+}, was 34.76% and 31.73%, respectively.

1. Introduction

The contamination of water by toxic heavy metals is a worldwide environmental problem. For this reason, the discharge of industrial wastewaters contaminated with heavy metals into waterways or sewage systems is stringently regulated to reduce the environmental impact [1]. Heavy metals are toxic pollutants that are released into the environment from many industries such as petrochemical, leather, paint, metallurgical, battery and car radiator manufacturing, metal plating, and textile industries as well as from agricultural sources and mining operations [1-8]. Metals such as lead, cadmium, copper, arsenic, nickel, chromium, zinc and mercury have been recognized as hazardous heavy metals [9]. These metals tend to accumulate in ecological systems causing serious soil and water pollution which can also be harmful to humans, animals, and plants, even in low concentrations [10]. Unlike biodegradable organic matter, metal ions are not eliminated from natural aqueous ecosystems by natural processes. As a result, scientists strive to develop novel methods to remove metal ions from water [11]. The presence of hazardous metals in water streams and marine water cause a significant health threat to the aquatic community, most commonly its damage to the gills of the fish [9]. Consequently, many countries have introduced stricter legislation to control water pollution. In addition, various regulatory bodies have set maximum prescribed limits for the discharge of toxic heavy metals into aquatic systems. However, metal ions are still being added to the water stream at a much higher concentration than the prescribed limits via industrial activities, thus leading to

*Corresponding author
E-mail address: mirzaei_fateme@yahoo.com

health hazards and environmental degradation [9]. The World Health Organization recommends a maximum acceptable concentration of Cu(II) in drinking water of less than 1.5mg/L[12]. Copper exists mainly as a divalent cation in aqueous solutions [13]. Copper is an essential metal for human metabolism because it acts as a cofactor for several cellular enzymes such as catalase (amineoxidase), cytochrome oxidase, dopamine b-hydroxylase, and peroxidase (galactose oxidase) [14]. In addition to being carcinogenic, high concentrations of Cu can be harmful or fatal to humans as well. Thus, it has been classified as one of the priority pollutants by the US EPA [15]. An excessive intake of copper by humans leads to nausea, vomiting, severe mucosal irritation, widespread capillary damage, hepatic and renal damage, headache, diarrhea, and central nervous problems followed by depression and gastrointestinal irritation [16-19]. Several methods have been proposed for the removal of copper ions from water and wastewaters including chemical precipitation, ion exchange, membrane filtration, electrolysis and adsorption [20, 21]. Each method has its limitations because of their cost, complexity, and efficiency as well as the secondary wastes they produce. For example, the electrolysis processes often have higher operational costs and chemical precipitation may generate secondary wastes [22]. In the case of membrane filtration, which is often guided by the pores in the membrane structure, it is inefficient in meeting the requirement of low metal concentration; another disadvantage is its high cost [21]. Adsorption is a highly effective and economical method to remove heavy metal ions from aqueous solutions [23] because of its high efficiency, easy handling, and the availability of different adsorbents [11]. Up to now, various materials like silica gels, activated alumina, oxides and hydroxides of metals, zeolite, clay minerals, sawdust, fly ash, kaolinite, bentonite, peanut hulls, polymeric materials, and activated carbons, etc. have been used as adsorbents [2, 16, 24-27]. Activated carbon is the most widely employed adsorbent used to remove various classes of compounds from contaminated streams [28]. It is mainly composed of carbonaceous material with various microporous structures. Its industrial usage can be found in the treatment process for flue gas, volatile solvents, etc. In the treatment of wastewater, it is used for purification, decolorization, and the removal of toxic organics and heavy metal ions [29]. Due to its porosity, large surface area, high adsorption capacity and surface reactivity[30], metals such as Cu(II) [12], Hg(II) [31], Pb(II) [32], Cd(II) [33], Ni(II) [34], Zn(II) [35], Cr(VI) [36], Co(II) [3], Mn(II) [37], and As(III) [38] can be removed using activated carbon. Commercial activated carbons are prepared from a variety of carbonaceous raw materials [39]. The qualities and characteristics of activated carbons depend on the physical and chemical properties of the starting materials and the activation methods used. Presently, lignite, peat, wood,

and coconut shell are the main common precursors for commercial activated carbons [39, 40]. The purpose of this paper was to study the removal of copper from aqueous solutions by granular activated carbon in the presence of competitor ions, and then compare it with the removal of copper without competitor ions. The effect of the solution pH, contact time and initial adsorbate concentration on the removal of Cu(II) was studied. The thermodynamic parameters for the adsorption of Cu(II) were also computed and discussed. The kinetics and factors controlling the adsorption process were also studied. Finally, the removal of copper ions in the presence of Ag(I) and Mn(II) ions were investigated.

2. Materials and methods

2.1. Materials and Batch adsorption experiments

The granular activated carbon obtained from the Merk Company was used as the adsorbent. The granular activated carbon (GAC) was used directly without any treatment. The mean diameter of the GAC particles was 1.5 mm. In Table 1, the physical characteristics of the granular activated carbon are presented. Stock solutions of Cu^{2+}, Ag^{1+} and Mn^{2+} were prepared by dissolving given amounts of $CuSO_4.5H_2O$, $AgNO_3$ and $MnCl_2.4H_2O$, respectively, in distilled water. The solutions of different concentrations used in various experiments were obtained by diluting the stock solutions. All reagents used were of analytical reagent grade and purchased from Merck. Several experiments were done to identify the optimum values of pH and contact time. The experiment to remove copper ions from the aqueous solution was performed under optimal conditions in the presence of competitor ions. The pH values were adjusted at 2.5, 3.5, and 4.5 using a 1 M HCl solution; the initial pH of the copper solution was 4.5. The solution pH was adjusted using a 1 M HCl solution before adding the adsorbent to the batch. At the optimum pH, other experiments were done. Samples were collected at 0.25, 0.5, 1, 2, 4, 6, 12, 24, 48, 72, 96, 120, 144 and 168 h to determine the optimum value of time. Isotherm studies were carried out at 100, 250, 500, 750 and 1000 mg/L as initial Cu^{2+} concentration at optimum pH and equilibrium time. The experiments were carried out using a 100 mL conical flask filled with 50 mL of test solution. All adsorption experiments were carried out at room temperature (about 30 ^{0}C). The contents of the flasks were filtrated through filter paper. After adding 25%wt ammonia solution to the filtrate solution as the indicator, the copper residual concentration was measured at 615 nm using a UV/vis spectrophotometer (JENWAY 7315). A previously established linear Beer−Lambert relationship was used for the concentration determination. For each test run, the average of three replicates was reported. The equilibrium adsorption capacity (q_e, mg/g) and percentage removal (Re %) for copper were determined using

$$q_e = \frac{(C_0 - C_e)V}{m} \qquad (1)$$

$$Re\% = \frac{C_0 - C_e}{C_0} \times 100 \qquad (2)$$

where C_0 is the initial copper concentration (mg/L), Ce is the concentrations of copper at the equilibrium time (mg/L), m is the mass of adsorbent (g), and V is the volume of copper solution (L).

Table 1. Physical properties of adsorbent of granular activated carbon (GAC)

Provider	TFC Company of Korea
BET surface area (m2/g)	1261
Average pore diameter (mm)	1.5
Total pore volume (cm3/g)	0.610
Total volume of micropores (cm3/g)	0.473

2.2. Effect of pH

One of the most important parameters controlling the uptake of metal ions from an aqueous solution is pH. The pH determines the surface charge of the adsorbent and the degree of ionization and speciation of the adsorbate. The pH of the solution controls the electrostatic interactions between the adsorbent and the adsorbate. It is known that the percent removal of heavy metal ions generally increases with pH. At low pH, the cations compete with the H^+ ions in the solution for the sorption sites and therefore adsorption declines. In contrast, as the pH increases, the competition between proton and metal cation decreases which means that there are more negative groups available for the binding of metal ions and this results in greater metal uptake. On the other hand, at a higher pH, metal cations start to form hydroxide complexes or precipitate as their hydroxides, which decrease the adsorption of metal ions. In aqueous solutions, metal cations hydrolyze according to the generalized expression for divalent metals as presented in Eq. (3):

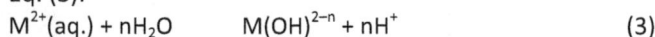

$$M^{2+}(aq.) + nH_2O \longrightarrow M(OH)^{2-n} + nH^+ \qquad (3)$$

The distribution of various hydroxyl-metal complexes depends on the pH of the solution and the corresponding stability constants. Hydroxyl-metal complexes are known to adsorb with a higher affinity than the completely hydrated metals [10].

2.3. Adsorption kinetic studies

To evaluate the adsorption kinetics to Cu^{2+} ions, two different kinetic models were applied to the experimental data: the pseudo first-order model and pseudo second-order model. The pseudo first-order kinetic model is as follows:

$$\frac{dq_t}{dt} = k_1(q_e - q_t) \qquad (3)$$

where k_1 (1/min) is the rate constant of the pseudo first-order sorption and q_t (mg/g) denotes the amount of sorption at time t (min).

After definite integration by the application of the conditions $q_t = 0$ at $t = 0$ and $q_t = q_t$ at $t = t$, Eq. (3) becomes

$$\log(q_e - q_t) = \log q_e - \frac{k_1}{2.303}t \qquad (4)$$

The values of k_1 and qe can be obtained from the slopes and intercepts of $\log(q_e-q_t)$ against the t plots [27,30].
The pseudo second-order equation can be written as

$$\frac{dq_t}{dt} = k_2(q_e - q_t)^2 \qquad (5)$$

where k_2 (g/mg min) is the rate constant. The integration of Eq. (5) and the application of the conditions $q_t = 0$ at $t = 0$ and $q_t = q_t$ at $t = t$ give

$$\frac{1}{(q_e - q_t)} = \frac{1}{q_t} + k_2 t \qquad (6)$$

The following equation can be obtained by rearranging Eq. (6) into a linear form

$$\frac{t}{q_t} = \frac{1}{k_2 q_e^2} + \frac{1}{q_e}t \qquad (7)$$

The values of q_e and k_2 can be gained from the slopes and the intercepts of the t/q_t versus t plots [27, 33, 41].

2.4. Adsorption isotherms

Adsorption isotherms are an important part of this study, as they can provide information about adsorption capacity as well as the surface properties and affinity of the adsorbent, thus providing a better understanding of how an adsorption system can be improved. In order to investigate the adsorption isotherm, four equilibrium models were analyzed. These included the Langmuir, Freundlich, Temkin and Dubinin–Radushkevich isotherms.

2.4.1. Langmuir isotherm

The Langmuir isotherm assumes that adsorption occurs at specific homogeneous sites within the adsorbent without any interactions between the adsorbed substances. The Langmuir isotherm model is given by Eq. (8):

$$q_e = \frac{q_m K_L C_e}{1 + K_L C_e} \qquad (8)$$

The above equation can be rearranged to the following linear form:

$$\frac{C_e}{q_e} = \frac{1}{K_L q_m} + \frac{C_e}{q_m} \qquad (9)$$

where qm (mg/g) is the maximum adsorption capacity and KL (L/mg) is the Langmuir constant which is related to the adsorption energy [10,26,42].
The shape of the isotherm may be considered to predict if an adsorption system is favorable or unfavorable. The

essential characteristic of a Langmuir isotherm can be expressed in terms of a dimensionless separation factor or an equilibrium parameter R, which is defined by the following equation:

$$R_L = \frac{1}{1 + K_L C_e} \tag{10}$$

According to the value of R, the isotherm shape may be interpreted as follows:
- R > 1: Unfavorable adsorption.
- R = 1: Linear adsorption.
- 0 < R < 1: Favorable adsorption.
- R = 0: Irreversible adsorption [43,44].

2.4.2. Freundlich isotherm

The Freundlich model is empirical in nature and assumes that the uptake of ions occurs on a heterogeneous surface. The Freundlich isotherm model is given by Eq. (11):

$$q_e = K_F C_e^{\frac{1}{n}} \tag{11}$$

Eq. (11) can be linearized in logarithmic form and Freundlich constants can be determined.

$$\text{Log } q_e = \text{Log } K_F + \frac{1}{n} \text{Log } C_e \tag{12}$$

where K_F is the intercept and n is the slope, which are the Freundlich constants representing the adsorption capacity and the adsorption intensity, respectively. In general, the greater the value of K_F, the greater the heterogeneity and the larger the value of n (n > 1) and the more spontaneous the adsorption process is [10, 42].

2.4.3. Temkin isotherm

Temkin considered the effects of some indirect adsorbent–adsorbate interaction on adsorption isotherms and suggested that the heat of adsorption of all the molecules in the layer would linearly decrease with coverage because of these interactions. The Temkin isotherm has been applied in the following form:

$$q_e = B \text{ Ln } K_T + B \text{ Ln } C_e \tag{13}$$

where, B = RT/b, T (K) is the absolute temperature; R is the universal gas constant (8.314 J/mol K); K_T (L/mg) is the equilibrium binding constant that corresponds to the maximum binding energy; and b is the variation of adsorption energy (J/mol). The Temkin constants can be derived from the plot of qe versus ln Ce [26, 45].

2.4.4. Dubinin–Radushkevich isotherm

The Dubinin–Radushkevich (D-R) isotherm does not assume a homogeneous surface or constant adsorption potential. The linear form of the D–R isotherm model can be represented by the following equation:

$$\text{Ln } q_e = \text{Ln } q_m - \beta \varepsilon^2 \tag{14}$$

where q_m (mg/g) is the maximum adsorption capacity, β (mol^2/J^2) is a constant related to the mean free energy of adsorption per mole of the metal, and ε (J/mol) is the Polanyi potential, which is described in Eq. (15):

$$\varepsilon = RT \text{ Ln}\left(1 + \frac{1}{C_e}\right) \tag{15}$$

where R (J/mol K) is the gas constant and T (K) is the absolute temperature. By plotting ln qe versus ε^2, it is possible to generate the value of q_m (mg/g) from the intercept and the value of β from the slope. The mean free energy E (kJ/mol) describes free energy change when one mole of ion is transferred from the solution to the surface of the sorbent. The mean free energy E (kJ/mol) can be calculated from the following equation:

$$E = \frac{1}{\sqrt{2\beta}} \tag{16}$$

In the D–R parameter, the mean free energy E (kJ/mol) gives an idea about the type of adsorption mechanism as to whether it is a chemical ion-exchange or physical adsorption. A magnitude of E between 8 and 16 kJ/mol corresponds to a chemical ion-exchange process. If E is less than 8 kJ/mol, the adsorption is physical in nature [10, 46].

3. Results and discussion

3.1. pH

The solution pH plays a crucial role in affecting Cu adsorption. Figure 1 shows the results of the pH dependence of the adsorption. An increase of Cu adsorption was observed when the pH increased from 2.5 to 4.5. The reason for this increase has been explained in section 2.2. The maximum amount of Cu adsorption (46.08%) was observed at a pH = 4.5 which was the same as the initial pH of the copper solution. When NaOH solution was added to the copper solution in order to increase the pH to more than 4.5, Cu (OH)$_2$ precipitation was formed. Therefore, at a pH > 4.5, the adsorption of Cu ions was not only the result of adsorption.

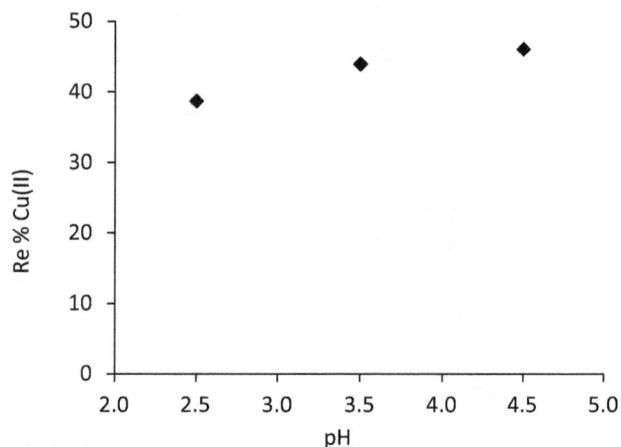

Fig. 1. Effect of pH on the removal of copper by GAC (initial Cu(II) concentration 1000 mg/L; adsorbent dose 100 g/L; contact time 6 h; temperature 30 °C).

3.2. Adsorption affected by contacting time

Figure 2 shows the extent of Cu^{2+} adsorption affected by contacting time. As shown, the uptake of Cu^{2+} by GAC was very rapid within the first 6 h. After 6 h, the uptake of Cu progressively decreased with time. Equilibrium was established at 144 h. It is interesting to note that there were two periods observed in the kinetic data: a fast adsorption and a progressive adsorption achieving equilibrium thereafter. In the fast adsorption stage, around 46% Cu removal occurred. The practical implication of the fast adsorption phenomenon can facilitate the design of the treatment process with energy saving expenditure by shortening the contacting time. The fast increase of Cu^{2+} adsorption rate at the beginning of the process was possibly due to a high availability of active surface sites on GAC surfaces. Since the readily available sites were mostly occupied, the subsequent slow adsorption is normally considered as being influenced by diffusion into the interior pore spaces of GAC.

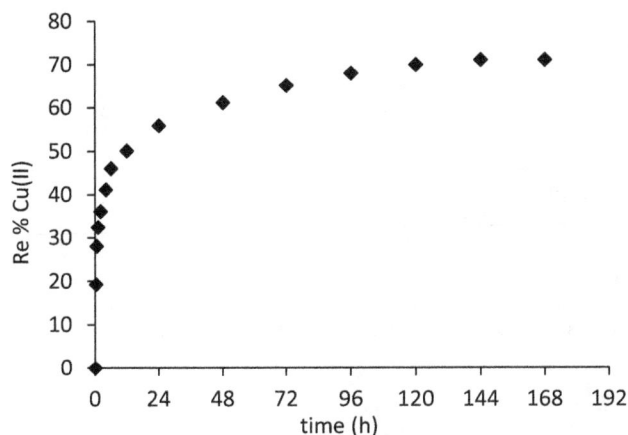

Fig. 2. Effect of contact time on the removal of copper by GAC (initial Cu(II) concentration 1000 mg/L; adsorbent dose 100 g/L; pH 4.5; temperature 30 °C).

3.2.1 Kinetic

The results showed that the correlation coefficient for the pseudo second-order kinetic model was more than 0.99. Therefore, the pseudo second-order kinetic model was selected to be the best-fit model and only the results from the pseudo second-order kinetics studies are presented (Table 2 and Figure 3). For the pseudo second-order kinetic model, the experimental $q_{e(exp)}$ corresponded to the calculated $q_{e(cal)}$ value. Furthermore, the linear regression value (R^2) was higher than 0.99 which indicated that the kinetic of sorption can be well described by the pseudo second-order equation.

3.3. Effect of initial copper concentration

The experimental results for the removal of copper ions on the GAC adsorbent in the Cu^{2+} concentration range of 100–1000 mg/L at the optimum initial pH is shown in Figure 4.

The results showed that in the removal of copper by GAC, the sorption capacity q_e initially rose sharply. This indicated that specific adsorption sites were available for adsorbent sites of the metal ions. At higher initial metal concentrations, the adsorbent became saturated and no more sites were available for further sorption. The maximum sorption capacity of copper ($q_{m,exp}$) by activated carbon was 7.11 mg/g.

3.3.1. Adsorption isotherms

The constants of Langmuir, Freundlich, Temkin and Dubinin–Raduschkevich (D–R) isotherm were obtained from the linear plots of C_e/q_e versus C_e, log q_e versus log C_e, q_e versus ln C_e and ln q_e versus ε^2, respectively (Fig. 5 and Table 3). The correlation coefficients, values of R^2, were regarded as a measure of the goodness-of-fit for experimental data (by the isotherms models). According to these values, the sorption of the studied copper by activated carbon fitted better to the Langmuir model than the Freundlich, Temkin and D–R isotherm models. It can be seen from Table 4 that the experimentally obtained values of $q_{m,exp}$ were comparable to the maximum sorption obtained from the Langmuir adsorption isotherm (q_m). This suggested the monolayer coverage of copper ions onto GAC. The maximum adsorption capacity of the adsorbent for Cu(II) was calculated from the Langmuir isotherm and found to be 7.03 mg/g. The Langmuir parameter R_L indicated the feasibility of the adsorption process: the adsorption process was unfavorable if $R_L > 1$; linear when $R_L = 1$; favorable in the range $0 < R_L < 1$; or irreversible ($R_L = 0$). Based on the results, R_L values lie between 0 and 1 for the studied copper, suggesting that the Langmuir model was favorable for adsorbents GAC (Table 4). Furthermore, higher R_L values at lower copper concentrations show that the adsorption was more favorable at lower copper concentrations. The results (Table 3) show that the mean free energy E (kJ/mol) values were lower than 8 kJ/mol in the study of the removal of copper by GAC. This indicated that the sorption process may be physical in nature.

3.4. The effect of competitor ions

In this experiment, the removal of Cu(II) ions from an aqueous solution in the presence of Ag^{1+} and once again in the presence of Mn^{2+} ions was performed by activated carbon. The experiment was done under the following conditions: an initial Cu(II) concentration of 1000 mg/L; a contact time of 6 h; an adsorbent dose of 100 g/L; a pH of 4.5; and a temperature of 30 °C. In this experiment, the initial concentration of Cu^{2+} was constant and the competitor ion concentration was changed from 100 to 1000 mg/L. The results are shown in Table 5. In this table, CI is the symbol of the competitor ions. Figure 6 shows the effect of the competitor ions on the copper removal efficiency. As seen in the Figure 6, both silver and manganese ions decreased the removal of Cu(II) from aqueous solutions by GAC; by increasing the value of competitor ions, the copper removal rate was reduced. These competitor ions occupied many activated sites of the adsorbent instead of the Cu^{2+}. The behaviors of the two selected competitor ions were not the same. It can be seen that the Mn^{2+} ions caused more trouble to remove the Cu^{2+} by the activated carbon than the Ag^{1+}. This may occur because the Mn^{2+} ionic radius was smaller than that of the Ag^{1+}. So the smaller Mn^{2+} ions penetrated to the pores of the activated carbon more than the Ag^{1+} ions and caused more occupation of the activated sites. Therefore, under the same conditions, the Cu^{2+} removal percent was reduced from 46% to 31.7% in the presence of the Mn^{2+}

ion, but decreased from 46% to 34.8% with the Ag^{2+} ion presence.

Table 2. Models rate constants for copper sorption kinetics by GAC

Pseudo-first-order rate model			
$q_{e,exp}$ (mg/g)	$q_{e,cal}$ (mg/g)	k_1 ($\times 104$ min^{-1})	R^2
7.11	4	4.8	0.956
Pseudo-second-order rate model			
$q_{e,exp}$ (mg/g)	$q_{e,cal}$ (mg/g)	k_2 ($\times 104$ min^{-1})	R^2
7.11	7.11	7.1	0.998

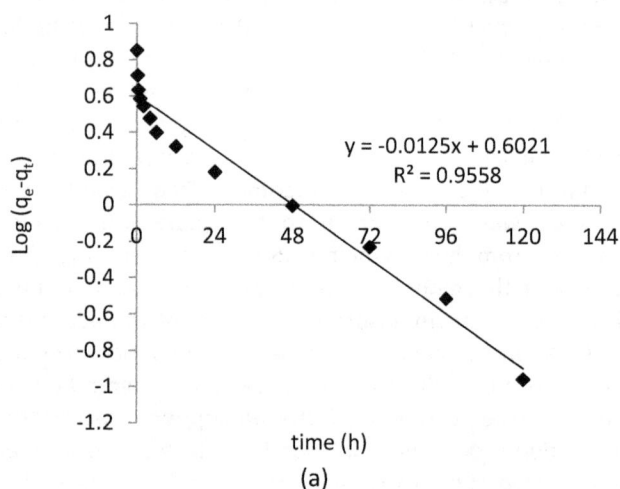

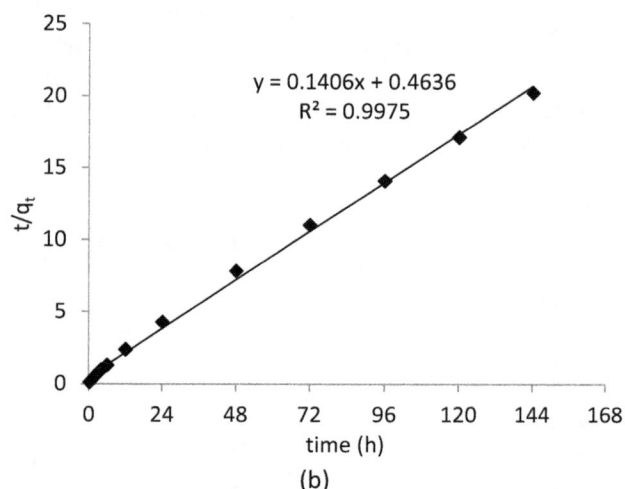

Fig. 3. Pseudo first-order (a) and pseudo second-order (b) for Cu(II) adsorption onto GAC (initial Cu(II) concentration 1000 mg/L; adsorbent dose 100 g/L; pH 4.5; temperature 30 °C).

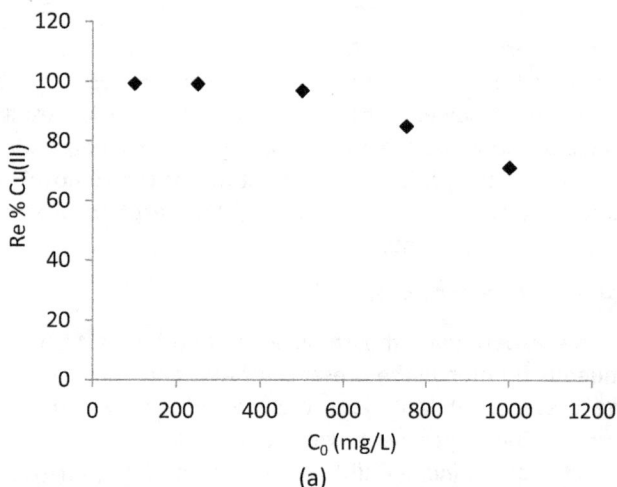

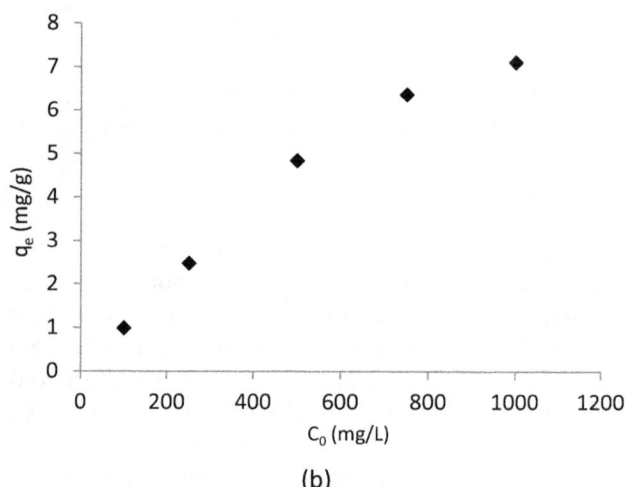

Fig. 4. The percentage of (a) removal and (b)adsorption capacity of various concentrations of copper (100–1000 mg/L) onto GAC under equilibrium condition (contact time 144 h; adsorbent dose 100 g/L; pH 4.5; temperature 30 °C).

(a)

(b)

(c)

(d)

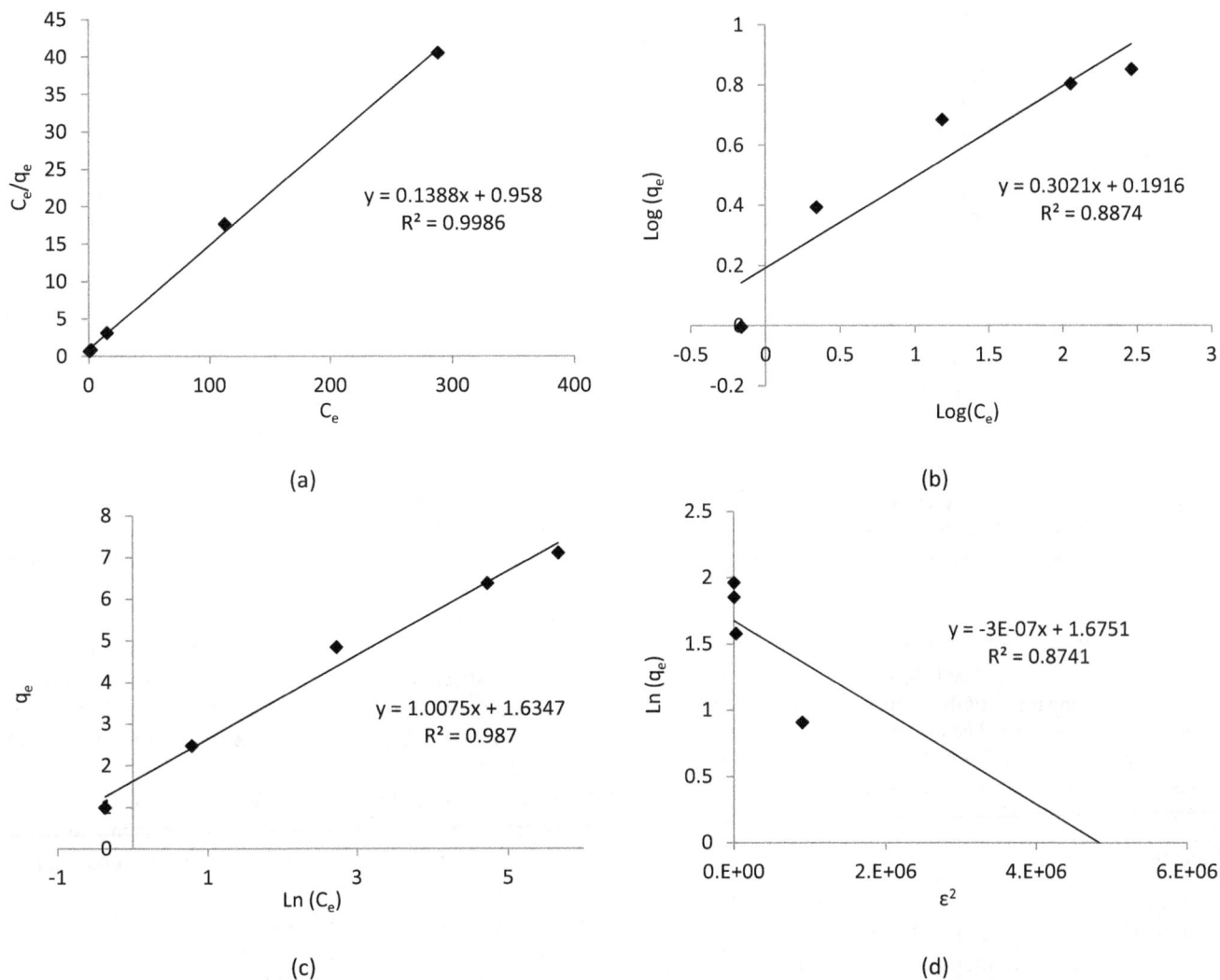

Fig. 5. Langmuir (a), Freundlich (b), Temkin (c) and Dubinin–Raduschkevich (d) isotherms for Cu(II) adsorption onto GAC (contact time 144 h; adsorbent dose 100 g/L; pH 4.5; temperature 30 °C).

Table 3. Constants of models for copper sorption by GAC at pH=4.5

Langmuir			Freundlich		
q_m (mg/L)	K_L (L/mg)	R^2	K_F (mg[1-1/n]L1/n/g)	n	R^2
7.20	0.14	0.998	1.57	3.31	0.881

Temkin			Dubinin–Radushkevich		
K_T (L/mg)	B	R^2	q_m (mg/L)	E (KJ/mol)	R^2
5.07	1.01	0.987	5.34	1.29	0.945

Table 4. RL values at different concentrations of copper ions

C0 (mg/L)	100	250	500	750	1000
RL	0.065	0.027	0.014	0.009	0.007

Table 5. Copper removal efficiency in the presence of competitor ions Ag^{1+} and Mn^{2+}

$C_0(Cu^{2+})$ [mg/L]	$C_0(Ag^{1+})$ [mg/L]	$C_0(Mn^{2+})$ [mg/L]	Re% Cu^{2+} without the presence of Cl	Re% Cu^{2+} in the presence of Ag^{1+}	Re% Cu^{2+} in the presence of Mn^{2+}
1000	100	100	46	45	43.1
1000	250	250	46	42.1	37
1000	500	500	46	37.9	32.6
1000	750	750	46	35	32
1000	1000	1000	46	34.8	31.7

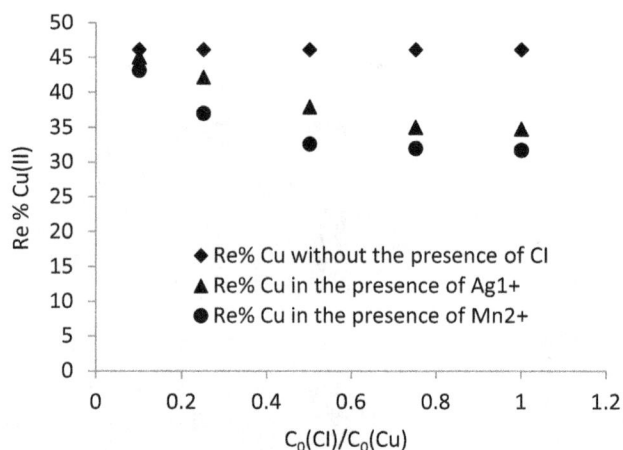

Fig. 6. The comparison of the copper removal efficiency without the presence of competitor ions and the presence of competitor ions Ag^{1+} and Mn^{2+} (initial Cu(II) concentration 1000 mg/L; contact time 6 h; adsorbent dose 100 g/L; pH 4.5; temperature 30 °C)

4. Conclusions

The results of the present study demonstrated the potential of granular activated carbon (GAC) for the removal of copper from aqueous solutions in the presence of competitor ions. The effects of pH value, contact time, initial concentration on adsorption were investigated. The optimum pH required for maximum adsorption was found to be 4.5 for copper which was also the pH of the solution. The results from the kinetic experiments showed that the adsorption rate of Cu(II) ions onto the GAC was slow. The adsorption process can be well described by the pseudo second-order kinetic model. The equilibrium data correlated well with the Langmuir isotherm model. The presence of the competitor ions Ag^{1+} and Mn^{2+} reduced the amount of copper removed by GAC.

Acknowledgement

This paper was prepared from an MSc thesis conducted in the Department of Chemical Engineering, Mahshahr Branch, Islamic Azad University, Mahshahr, Iran

References

[1] Butter, T. J., Evison, L. M., Hancock, I. C., Holland, F. S., Matis, K. A., Philipson, A., Zouboulis, A. I. (1998). The removal and recovery of cadmium from dilute aqueous solutions by biosorption and electrolysis at laboratory scale. *Water research*, *32*(2), 400-406.

[2] Jiang, T., Liu, W., Mao, Y., Zhang, L., Cheng, J., Gong, M., Zhao, Q. (2015). Adsorption behavior of copper ions from aqueous solution onto graphene oxide–CdS composite. *Chemical engineering journal*, *259*, 603-610.

[3] Kobya, M., Demirbas, E., Senturk, E., Ince, M. (2005). Adsorption of heavy metal ions from aqueous solutions by activated carbon prepared from apricot stone. *Bioresource technology*, *96*(13), 1518-1521.

[4] Nadaroglu, H., Kalkan, E., Demir, N. (2010). Removal of copper from aqueous solution using red mud. *Desalination*, *251*(1), 90-95.

[5] Li, Q., Zhai, J., Zhang, W., Wang, M., Zhou, J. (2007). Kinetic studies of adsorption of Pb (II), Cr (III) and Cu (II) from aqueous solution by sawdust and modified peanut husk. *Journal of hazardous materials*, *141*(1), 163-167.

[6] Amarasinghe, B. M. W. P. K., Williams, R. A. (2007). Tea waste as a low cost adsorbent for the removal of Cu and Pb from wastewater. *Chemical engineering journal*, *132*(1), 299-309.

[7] Sarioglu, M., Atay, Ü. A., Cebeci, Y. (2005). Removal of copper from aqueous solutions by phosphate rock. *Desalination*, *181*(1), 303-311.

[8] Hou, J., Wen, Z., Jiang, Z., Qiao, X. (2014). Study on combustion and emissions of a turbocharged compression ignition engine fueled with dimethyl ether and biodiesel blends. Journal of the energy institute, 87(2), 102-113.

[9] Ahmaruzzaman, M. (2011). Industrial wastes as low-cost potential adsorbents for the treatment of wastewater laden with heavy metals. *Advances in colloid and interface science*, *166*(1), 36-59.

[10] Runtti, H., Tuomikoski, S., Kangas, T., Lassi, U., Kuokkanen, T., Rämö, J. (2014). Chemically activated carbon residue from biomass gasification as a sorbent for iron (II), copper (II) and nickel (II) ions. *Journal of water process engineering*, *4*, 12-24.

[11] Godino-Salido, M. L., Santiago-Medina, A., Arranz-Mascarós, P., López-Garzón, R., Gutiérrez-Valero, M. D., Melguizo, M., López-Garzón, F. J. (2014). Novel active carbon/crown ether derivative hybrid material for the selective removal of Cu (II) ions: The crucial role of the surface chemical functions. *Chemical engineering science*, *114*, 94-104.

[12] Larous, S., Meniai, A. H. (2012). Removal of copper (II) from aqueous solution by agricultural by-products-sawdust. *Energy procedia*, *18*, 915-923.

[13] Hu, X. J., Liu, Y. G., Wang, H., Chen, A. W., Zeng, G. M., Liu, S. M., Zhou, L. (2013). Removal of Cu (II) ions from aqueous solution using sulfonated magnetic graphene oxide composite. *Separation and purification technology*, *108*, 189-195.

[14] Zacaroni, L. M., Magriotis, Z. M., das Graças Cardoso, M., Santiago, W. D., Mendonça, J. G., Vieira, S. S., Nelson, D. L. (2015). Natural clay and commercial activated charcoal: Properties and application for the removal of copper from cachaça. *Food control*, *47*, 536-544.

[15] Wen, Y., Ma, J., Chen, J., Shen, C., Li, H., Liu, W. (2015). Carbonaceous sulfur-containing chitosan–Fe (III): a novel adsorbent for efficient removal of copper (II) from water. *Chemical engineering journal*, *259*, 372-380.

[16] Kalavathy, M. H., Karthikeyan, T., Rajgopal, S., Miranda, L. R. (2005). Kinetic and isotherm studies of Cu (II) adsorption onto H_3PO_4-activated rubber wood sawdust. *Journal of colloid and interface science*, *292*(2), 354-362.

[17] Cho, H., Oh, D., Kim, K. (2005). A study on removal characteristics of heavy metals from aqueous solution by fly ash. *Journal of hazardous materials*, *127*(1), 187-195.

[18] SenthilKumar, P., Ramalingam, S., Sathyaselvabala, V., Kirupha, S. D., Sivanesan, S. (2011). Removal of copper (II) ions from aqueous solution by adsorption using cashew nut shell. *Desalination*, *266*(1), 63-71.

[19] Ajmal, M., Khan, A. H., Ahmad, S., Ahmad, A. (1998). Role of sawdust in the removal of copper (II) from industrial wastes. *Water research*, *32*(10), 3085-3091.

[20] Benaissa, H., Elouchdi, M. A. (2007). Removal of copper ions from aqueous solutions by dried sunflower leaves. *Chemical engineering and processing: Process intensification*, *46*(7), 614-622.

[21] Igberase, E., Osifo, P., Ofomaja, A. (2014). The adsorption of copper (II) ions by polyaniline graft chitosan beads from aqueous solution: equilibrium, kinetic and desorption studies. *Journal of environmental chemical engineering*, *2*(1), 362-369.

[22] Zhou, Y. T., Nie, H. L., Branford-White, C., He, Z. Y., Zhu, L. M. (2009). Removal of Cu $^{2+}$ from aqueous solution by chitosan-coated magnetic nanoparticles modified with α-ketoglutaric acid. *Journal of colloid and interface science*, *330*(1), 29-37.

[23] Wen, Y., Ma, J., Chen, J., Shen, C., Li, H., Liu, W. (2015). Carbonaceous sulfur-containing chitosan–Fe (III): a novel adsorbent for efficient removal of copper (II) from water. *Chemical engineering journal*, *259*, 372-380.

[24] Machida, M., Yamazaki, R., Aikawa, M., Tatsumoto, H. (2005). Role of minerals in carbonaceous adsorbents for removal of Pb (II) ions from aqueous solution. *Separation and purification technology*, *46*(1), 88-94.

[25] Salam, M. A., Al-Zhrani, G., Kosa, S. A. (2012). Simultaneous removal of copper (II), lead (II), zinc (II) and cadmium (II) from aqueous solutions by multi-walled carbon nanotubes. *Comptes rendus chimie*, *15*(5), 398-408.

[26] Mi, X., Huang, G., Xie, W., Wang, W., Liu, Y., Gao, J. (2012). Preparation of graphene oxide aerogel and its adsorption for Cu $^{2+}$ ions. *Carbon*, *50*(13), 4856-4864.

[27] Olgun, A., Atar, N. (2011). Removal of copper and cobalt from aqueous solution onto waste containing boron impurity. *Chemical engineering journal*, *167*(1), 140-147.

[28] Moussavi, G., Alahabadi, A., Yaghmaeian, K., Eskandari, M. (2013). Preparation, characterization and adsorption potential of the NH 4 Cl-induced activated carbon for the removal of amoxicillin antibiotic from water. *Chemical engineering journal*, *217*, 119-128.

[29] Kim, J. W., Sohn, M. H., Kim, D. S., Sohn, S. M., Kwon, Y. S. (2001). Production of granular activated carbon from waste walnut shell and its adsorption characteristics for Cu $^{2+}$ ion. *Journal of hazardous materials*, *85*(3), 301-315.

[30] Kurniawan, T. A., Chan, G. Y., Lo, W. H., Babel, S. (2006). Comparisons of low-cost adsorbents for treating wastewaters laden with heavy metals. *Science of the total environment*, *366*(2), 409-426.

[31] Rao, M. M., Reddy, D. K., Venkateswarlu, P., Seshaiah, K. (2009). Removal of mercury from aqueous solutions using activated carbon prepared from agricultural by product/waste. *Journal of environmental management*, *90*(1), 634-643.

[32] Ayyappan, R., Sophia, A. C., Swaminathan, K., Sandhya, S. (2005). Removal of Pb (II) from aqueous solution using carbon derived from agricultural wastes. *Process biochemistry*, *40*(3), 1293-1299.

[33] Rao, M. M., Ramesh, A., Rao, G. P. C., Seshaiah, K. (2006). Removal of copper and cadmium from the aqueous solutions by activated carbon derived from Ceiba pentandra hulls. *Journal of hazardous materials*, *129*(1), 123-129.

[34] Hasar, H. (2003). Adsorption of nickel (II) from aqueous solution onto activated carbon prepared from almond husk. *Journal of hazardous materials*, *97*(1), 49-57.

[35] Ramos, R. L., Jacome, L. B., Barron, J. M., Rubio, L. F., Coronado, R. G. (2002). Adsorption of zinc (II) from an aqueous solution onto activated carbon. *Journal of hazardous materials*, *90*(1), 27-38.

[36] Selvi, K., Pattabhi, S., Kadirvelu, K. (2001). Removal of Cr (VI) from aqueous solution by adsorption onto

activated carbon. *Bioresource technology, 80*(1), 87-89.

[37] Omri, A., Benzina, M. (2012). Removal of manganese (II) ions from aqueous solutions by adsorption on activated carbon derived a new precursor: Ziziphus spina-christi seeds. *Alexandria engineering journal, 51*(4), 343-350.

[38] Budinova, T., Savova, D., Tsyntsarski, B., Ania, C. O., Cabal, B., Parra, J. B., Petrov, N. (2009). Biomass waste-derived activated carbon for the removal of arsenic and manganese ions from aqueous solutions. *Applied surface science, 255*(8), 4650-4657.

[39] Yang, T., Lua, A. C. (2006). Textural and chemical properties of zinc chloride activated carbons prepared from pistachio-nut shells. *Materials chemistry and physics, 100*(2), 438-444.

[40] Imamoglu, M., Tekir, O. (2008). Removal of copper (II) and lead (II) ions from aqueous solutions by adsorption on activated carbon from a new precursor hazelnut husks. *Desalination, 228*(1), 108-113.

[41] Wu, F. C., Tseng, R. L., Juang, R. S. (2001). Kinetic modeling of liquid-phase adsorption of reactive dyes and metal ions on chitosan. *Water research, 35*(3), 613-618.

[42] Han, R., Zhang, J., Zou, W., Shi, J., Liu, H. (2005). Equilibrium biosorption isotherm for lead ion on chaff. *Journal of hazardous materials, 125*(1), 266-271.

[43] Wu, Q., Chen, J., Clark, M., Yu, Y. (2014). Adsorption of copper to different biogenic oyster shell structures. *Applied surface science, 311*, 264-272.

[44] Ghassabzadeh, H., Mohadespour, A., Torab-Mostaedi, M., Zaheri, P., Maragheh, M. G., Taheri, H. (2010). Adsorption of Ag, Cu and Hg from aqueous solutions using expanded perlite. *Journal of hazardous materials, 177*(1), 950-955.

[45] Hamdaoui, O., Naffrechoux, E. (2007). Modeling of adsorption isotherms of phenol and chlorophenols onto granular activated carbon: Part I. Two-parameter models and equations allowing determination of thermodynamic parameters. *Journal of hazardous materials, 147*(1), 381-394.

[46] Dang, V. B. H., Doan, H. D., Dang-Vu, T., Lohi, A. (2009). Equilibrium and kinetics of biosorption of cadmium (II) and copper (II) ions by wheat straw. *Bioresource technology, 100*(1), 211-219.

11

Removal of As(V), Cr(VI) and Pb(II) from aqueous solution using surfactant-modified Sabzevar nanozeolite

Kourosh Razmgar, Zahra Beagom Mokhtari Hosseini [*]

Department of Chemical Engineering, Hakim Sabzevari University, Sabzevar, Iran

Keywords:
Adsorption
Cationic surfactant
Heavy metal
Modification
Natural nanozeolite

ABSTRACT

The pollution of water environments is a challenging issue especially in developing countries. Contamination of drinking water with heavy metals has been reported in many parts of the world. Arsenic, chromium and lead are dangerous heavy metals and also common contaminants of drinking water. In this study, the capacity and performance of the surfactant-modified Sabzevar natural nanozeolite (SMSNZ) on the removal of heavy metals from an aqueous solution was investigated. Initially, the appropriate concentration of hexadecyltrimethylammonium bromide HDTMA-Br solution for modification was investigated; it was found that it must be higher than the critical concentration micelle (CMC). Then, the removal of As (V), Cr (VI), and Pb(II) from an aqueous solution was studied using SMSNZ. The results indicated that the removal efficiency was very high in different initial concentrations of heavy metals. The Linear, Langmuir and Freundlich isotherm models were used to investigate the adsorption equilibrium of the surfactant-modified natural zeolite for heavy metals adsorption. The results showed that the Linear isotherm is a better fit for the three studied heavy metals.

1. Introduction

Arsenic, chromium and lead are heavy metals which are potentially toxic environmental pollutants. One important feature that distinguishes heavy metals from other pollutants is that the former are non-biodegradable. Once metal ions enter the environment, their chemical form largely determines their potential toxicity [1]. Arsenic has a serious impact on human health and causes problems such as skin, lung, urinary bladder, liver and kidney cancers [2]. Dangerous concentrations of arsenic in natural water is now a worldwide problem and is often referred to as a 20th-21st century calamity [3]. This heavy metal exists in the environment in different oxidation states and in various forms, e.g., As, As(III), As(V), As(0), and As(-III). This element presents a major problem as it cannot be easily destroyed and can only be converted into different forms or transformed into insoluble compounds in combination with other elements such as iron [4]. The Maximum Contamination Level (MCL) of arsenic in water is 10 µg/L as

established by the World Health Organization (WHO) [5].Chromium is a redox active metal element in the environment which usually exists as Cr(III) or Cr(VI). Cr(VI) may be found in the form of dichromate ($Cr_2O_7^{2-}$), hydrochromate ($HCrO_4^-$), or chromate (CrO_4^{2-}). Cr(III) in aqueous solutions, however, may take the form of Cr^{3+}, $Cr(OH)^{2+}$, or $Cr(OH)_2^+$. Hexavalent chromium is more toxic to living organisms than Cr(III). Chromium(VI) is on the US EPA priority list of toxic pollutants and is present in electroplating wastewater and many other industrial discharges [1]. In epidemiological studies, an association has been found between lung cancer and exposure to Cr(VI) through inhalation. The International Agency for Research on Cancer has categorized pentavalent chromium in Group 1 which includes substances that are carcinogenic to humans (World Health Organization, 2004) [6]. The MCL of hexavalent chromium in drinking water is 100 µg/L [5]. Another contaminant investigated in this research is lead, which is not only an issue in developing countries. In Japan, lead piping has been widely used

*Corresponding author
E-mail address: z.mokhtari@hsu.ac.ir

because it is inexpensive and easy to fabricate. This has resulted in lead seeping into tap water and has raised concerns in the society [7]. Lead disturbs hemoglobin synthesis and renal function as well as causing neurological and behavioral disturbances in children. Even low blood lead concentrations have been associated with intellectual impairment in children [8]. The MCL of lead in drinking water established by the WHO is 10 μg/L [5]. Zeolites are a widely used adsorbent for water treatment. Zeolites are hydrated alumina silicate made from the interlinked tetrahedral of alumina (AlO_4) and silica (SiO_4) and form with many different crystalline structures which have large open pores (sometimes referred to as cavities) in a very regular arrangement. The framework structure and high surface area of zeolites traps molecules in a cage-like structure. The widespread use of zeolites involves both types, natural and synthetic. Zeolites were first discovered at the end of the 18th century and came to prominence in the 1960's when synthetically produced zeolites were employed in large industrial applications as molecular sieves for filtration purposes and as catalysts in the cracking of crude oil [9]. In this study, we utilized natural zeolite from the Chahtalkh region surrounding the Sabzevar area in Iran as an adsorbent. Natural zeolites are abundant and low cost resources, which are crystalline hydrated aluminosilicates with a framework structure containing pores occupied by water, alkali and alkaline earth cations. Due to their high cation-exchange ability as well as their molecular sieve properties, natural zeolites have been widely used as adsorbents in separation and purification processes in past decades [10]. More than 40 natural zeolites have been identified during the past 200 years. Zeolites are distinguished by the differences in their chemical compositions and the size and arrangement of their crystal structures [11]. Clinoptilolite, mordenite, phillipsite, chabazite, stilbite, analcime and laumontite are very common forms whereas offretite, paulingite, barrerite, and mazzite are much rarer [16]. Working with zeolites in nanoscale depicts the significant properties of nanozeolites in comparison to typical zeolites. Nanoscience and nanoporous materials are currently attracting attention from many researchers. Microporous materials with nanometer particle size (nanozeolites) are being studied because of their outstanding properties that could not be found in the micrometer zeolites. Reducing the particle size from a micrometer to a nanometer scale leads to a significant change in material characteristics and their applications in catalysis and adsorption. The number of atoms in the unit cell increases when particle sizes decrease and nanozeolites have a large external surface area. The diffusion path length in nanozeolites is shortened as compared to that in the conventional micrometer zeolites [12]. The use of nanoscale zeolites means a reduction in weight, smaller cross sections, enhanced adsorption/desorption kinetics, and a higher

overall efficiency of the entire system [9]. Although natural zeolites have been widely used to remove cationic heavy metals in aqueous solutions, they are not useful for removing inorganic oxyanions such as chromate (CrO_4^{2-}). Morever, As, Cr and Pb exist in an anion form over a broad range of pH. The external surface charge of natural zeolites could be altered with a surfactant to make it possible to adsorb heavy metals and their oxyanions as well [13, 19]. In this study, HDTMA-Br was chosen as a cationic surfactant due to its high stability and proper interactions with the SMSNZ surface. Since the performance of surfactant on a zeolite surface is directly related to Critical Micelle Concentration (CMC), it was important to use the proper concentration of surfactant with respect to its CMC [15]. The structural formula of the used surfactant is shown in Figure 1.

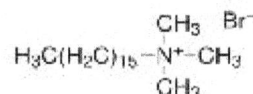

$$H_3C(H_2C)_{15}-\overset{\overset{\displaystyle CH_3}{|}}{\underset{\underset{\displaystyle CH_3}{|}}{N^+}}-CH_3 \quad Br^-$$

Fig. 1. Structural formula of HDTMA-Br

2. Materials and methods

2.1. Materials

The hexadecyltrimethylammonium bromide (HDTMA-Br) with a purity>99% was supplied by A-3, Okhala Industrial Area, New Delhi. The $Na_2HAsO_4.7H_2O$ (>98%) was purchased from Sigma-Aldrich while the $Pb(NO_3)_2$ was provided by Merck, KGaA, Darmstadt, Germany and the K_2CrO_4 was obtained from E. Merck, Darmstadt, Germany. Natural nanozeolite was prepared from the area of Sabzevar, Iran. All the chemicals used in this research were of analytical grade.

2.2. Surfactant concentration

Before choosing the best concentration of surfactant for removal, a pretest was conducted with three concentrations of surfactant. The CMC (Critical Micelle Concentration) of HDTMA-Br was 1.8 mmol/L [15]. The three concentrations included the following: lower than the CMC (0.5 mmol/L), nearly equal to CMC (2 mmol/L), and higher than CMC (30 mmol/L). The natural nanozeolite modified with these concentrations were used for treating the 1 mg/L arsenate solution. The results of the pretest are shown in Table 1.

Table 1. Removal efficiency of As from aqueous solution using nanozeolites that were modified with different concentrations of surfactant

Surfactant Concentration	Removal Efficiency (%)
0.5	47.5
2	57.0
30	79.0

2.3. Preparation of nanozeolite

The nanozeolite was manufactured using a ball mill. 10 g of nanozeolite powder was dissolved in 100 mL of 0.03 mol/L HDTMA-Br solution for surface modification. The resulting suspension was shaken for 24 h at 30°C and 150 rpm. The solid phase was separated by centrifuging and was rinsed in distilled water. The nanozeolites were dried for 48 h at 100°C and kept in the desiccator to prevent moisture adsorption. Figure 2 shows the nanozeolite after modification.

Fig. 2. Surfactant-modified Sabzevar nanozeolite

2.4. Removal procedure

Different concentrations of As, Pb and Cr solution were prepared using $Na_2HAsO_4.7H_2O$, $Pb(NO_3)_2$ and K_2CrO_4, respectively. For instance, the initial arsenate solution with a concentration of 10 mg/L was prepared by dissolving a sufficient amount of $Na_2HAsO_4.7H_2O$ in 1000 mL of deionized water. Then the solutions with concentrations of 0.2, 0.5, 1, 2 and 5 mg/L were prepared by dilution; 0.5 g of surface modified natural nanozeolite was added to 10 mL of each solution. The batch process of heavy metal removal by SMSNZ was conducted under the following conditions: SMSNZ dosage of 50 g/L, 30°C, 150 rpm, and 24h. After the separation of nanozeolite by centrifuging at 3000 rpm for 10 minutes, the heavy metal concentration in supernatant was measured by atomic adsorption.

3. Results and discussion

In this study, we increased the interaction time and amount of nanozeolite in comparison to similar works in this field. Our aim was to attain the highest removal for each pollutant concentration. The adsorption percent showed that this technique was able to adsorb high levels of the mentioned ions. As(V), Cr(VI) and Pb(II) exist mainly in anion form over a wide range of pH [13]. The surface of natural zeolite has a negative charge because of its structure. For the adsorption of anion, it must be modified by cationic agents [14]. In this study, HDTMA-Br was used for modification. Three concentrations of surfactant were used for the modification of zeolite and their removal efficiency of As was investigated. The results (Table 1) showed that the removal efficiency was higher at a

concentration of the higher CMC (>CMC). Modification with this concentration formed a bilayer on the surface of zeolite and the charge on the zeolite surface was reversed from negative to positive [15]. As shown in the Figure 3, in modifications with concentrations equal to or lower than the CMC, monolayer forms on the surface of zeolite. Also, its surface charge was negative and caused a decrease in the the amount of removal efficiency. Therefore, modification with higher CMC of surfactant was selected for the following study. The adsorption of heavy metal from an aqueous solution by SMSNZ in different initial concentrations of heavy metal (0.2-5 mg/L) was investigated. Figure 4 shows the results of this survey in terms of removal efficiency and adsorption capacity. According to this figure, SMSNZ can significantly remove heavy metal in different initial concentrations and the adsorption capacity of SMSNZ was about 0.1 mg/g for each of the heavy metals. Also, it indicated that the removal efficiency of Pb was higher than the two other studied heavy metals under the same condition. The equilibrium experimental data were fitted with three isotherm models, Linear, Langmuir and Freundlich. The parameters of these models can be determined by linear regression of the experimental data. The correlation coefficients (R^2) of models for three heavy metals are summarized in Table 2. The higher values of R^2 for the Linear isotherm indicated a better fitness of the Linear isotherm to the experimental data. First, q_e (mg/g) is defined as the amount of adsorption at equilibrium. In fact, q_e depicts the mg of adsorbate removed by 1gr of adsorbent. It was calculated by the mass balance equation [18]:

$$q_e = C_0 - C_e \frac{V}{m} \qquad (1)$$

where C_0 and C_e are the initial and equilibrium concentrations (mg/L), V is the volume of solution (mL), and m is the adsorbent weight (g). The adsorption percent was calculated using the equation below [18]:

$$Adsorption\% = \frac{C_0 - C_e}{C_0} \times 100 \qquad (2)$$

The logarithmic form of the Freundlich equation could be written as:

$$q_e = K_F C_e^{\frac{1}{n}} \qquad (3)$$

where q_e is the adsorption amount at equilibrium (mg/g), K_F is the Freundlich constant related to adsorption capacity $(mg/g)(L/g)^{nF}$, C_e is the equilibrium concentration in solution (mg/L), and n is reaction energy. The term 1/n shows the curvature in the isotherm and may represent the energy distribution of the adsorption site [6, 18]. The linear form of Freundlich could be written as:

$$\ln q_e = \ln K_F + \frac{1}{n} \ln C_e \qquad (4)$$

The Linear isotherm would be written as:

$$q_e = K_L C_e \qquad (5)$$

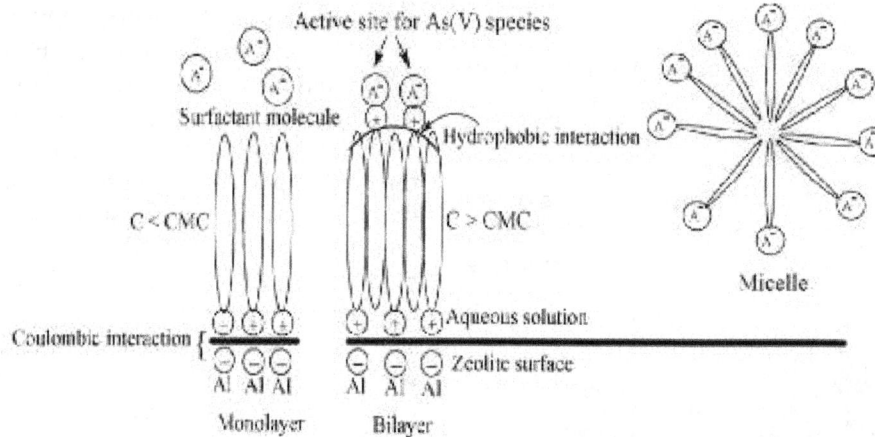

Fig. 3. The effect of surfactant on zeolite surface for higher and lower CMC [17]

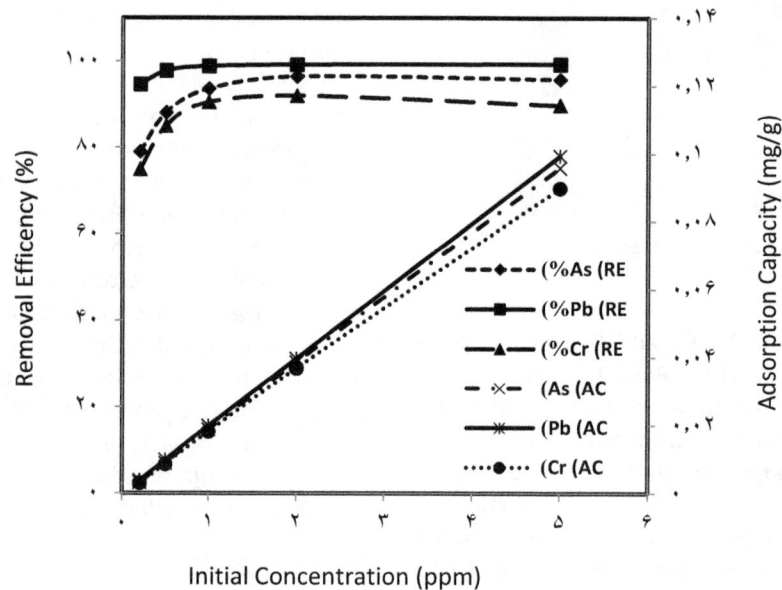

Initial Concentration (ppm)

Fig. 4. Removal efficiency and adsorption capacity results for different initial concentration of arsenic, chromium, and lead solutions

Table 2. Correlation coefficients (R^2) of Linear, Langmuir and Freundlich isotherms for heavy metal adsorption from aqueous solution by surfactant-modified Sabzevar nanozeolite

Heavy Metals	Isotherm Models	Correlation Coefficients (R^2)		
		Linear	Langmuir	Freundlich
As (V)		0.9411	0.713	0.792
Cr (VI)		0.9746	0.8463	0.9017
Pb (II)		0.977	0.4321	0.7664

4. Conclusions

Natural nanozeolite from the Sabzevar area was used for the removal of As, Cr and Pb from aqueous solutions. It was modified by HDTMA-Br. The results showed that

modification with a surfactant dosage higher than CMC had a higher removal efficiency because of the formation of bilayer on the surface of the zeolite and a change in the surface charge of the zeolite. The study of heavy metal adsorption by surfactant modified Sabzevar nanozeolite in different initial concentrations of heavy metal resulted in

the significant adsorption of heavy metal from aqueous solutions. Considering the accessibility of natural zeolites and their capability to remove heavy metals, the use of these adsorbents in future research is recommended.

References

[1] Namasivayam, C., Sureshkumar, M. V. (2008). Removal of chromium (VI) from water and wastewater using surfactant modified coconut coir pith as a biosorbent. *Bioresource technology*, *99*(7), 2218-2225.

[2] Vu, D., Li, X., Wang, C. (2013). Efficient adsorption of As (V) on poly (acrylo-amidino ethylene amine) nanofiber

membranes. *Chinese science bulletin, 58*(14), 1702-1707.

[3] Mohan, D., Pittman, C. U. (2007). Arsenic removal from water/wastewater using adsorbents—a critical review. *Journal of hazardous materials, 142*(1), 1-53.

[4] Choong, T. S., Chuah, T. G., Robiah, Y., Koay, F. G., Azni, I. (2007). Arsenic toxicity, health hazards and removal techniques from water: an overview. *Desalination, 217*(1), 139-166.

[5] Momodu, M. A., Anyakora, C. A. (2010). Heavy metal contamination of groundwater: the Surulere case study. *Research journal* of e*nvironmental* and e*arth* sciences, *2*(1), 39-43.

[6] Tashauoei, H. R., Attar, H. M., Amin, M. M., Kamali, M., Nikaeen, M., Dastjerdi, M. V. (2010). Removal of cadmium and humic acid from aqueous solutions using surface modified nanozeolite A. *International journal of environmental science and technology,7*(3), 497-508.

[7] Tokimoto, T., Kawasaki, N., Nakamura, T., Akutagawa, J., Tanada, S. (2005). Removal of lead ions in drinking water by coffee grounds as vegetable biomass. *Journal of colloid and interface science, 281*(1), 56-61.

[8] Halttunen, T., Salminen, S., Tahvonen, R. (2007). Rapid removal of lead and cadmium from water by specific lactic acid bacteria. *International journal of food microbiology, 114*(1), 30-35.

[9] Daniell, W., Sauer, J., Kohl, A. (2008). NanoZeolites, porous nanomaterials for cleanTech, Encapsulation, and triggered-release applications. NanoEurope congress and exhibition (Switzerland), 8-19

[10] Wang, S., Peng, Y. (2010). Natural zeolites as effective adsorbents in water and wastewater treatment. *Chemical engineering journal, 156*(1), 11-24.

[11] Keane, C., Sever, M., 2014. Earth's consumer's guide to minerals, The American geosciences institute, 4220 King Street, Alexandria, USA

[12] Ngoc, D. T., Pham, T. H., Nguyen, K. D. H. (2013). Synthesis, characterization and application of nanozeolite NaX from Vietnamese kaolin. *Advances in natural sciences: Nanoscience and nanotechnology, 4*(4), 1-13

[13] Yusof, A. M., Malek, N. A. N. N. (2009). Removal of Cr (VI) and As (V) from aqueous solutions by HDTMA-modified zeolite Y. *Journal of hazardous materials, 162*(2), 1019-1024.

[14] Mohammadi, A., Bina, B., Ebrahimi, A., Hajizadeh, Y., Amin, M. M., Pourzamani, H. (2012). Effectiveness of nanozeolite modified by cationic surfactant in the removal of disinfection by-product precursors from water solution. *International journal of environmental health engineering, 1*(1), 3.

[15] Samadi, M. T., Saghi, M. H., Ghadiri, K., Hadi, M., Beikmohammadi, M. (2010). Performance of Simple Nano Zeolite Y and Modified Nano zeolite Y in

Phosphor removal from aqueous solutions. *Iranian journal of health and environment, 3*(1), 27-36.

[16] Wang, S., Peng, Y. (2010). Natural zeolites as effective adsorbents in water and wastewater treatment. *Chemical engineering journal, 156*(1), 11-24.

[17] Chutia, P., Kato, S., Kojima, T., Satokawa, S. (2009). Adsorption of As (V) on surfactant-modified natural zeolites. *Journal of hazardous materials, 162*(1), 204-211.

[18] Kumar, K. V., Sivanesan, S. (2005). Comparison of linear and non-linear method in estimating the sorption isotherm parameters for safranin onto activated carbon. *Journal of hazardous materials, 123*(1), 288-292.

[19] Neupane, G., Donahoe, R. J. (2009). Potential use of surfactant-modified zeolite for attenuation of trace elements in fly ash leachate. In *world of coal ash (WOCA) conference* (pp. 2-7).

12

Preparation of Kissiris/TiO₂/Fe₃O₄/GOx biocatalyst: Feasibility study of MG decolorization

Vahide Elhami[1], Afzal Karimi*,[2]

[1]*Department of Chemical Engineering, Faculty of Chemical and Petroleum Engineering, University of Tabriz, Tabriz, Iran*
[2]*Faculty of Advanced Technologies in Medicine, Iran University of Medical Sciences, Tehran, Iran*

Keywords:
Decolorization
Glucose oxidase
Kissiris
Heterogeneous Bio-Fenton

ABSTRACT

Titanium dioxide (TiO₂) and Fe₃O₄ magnetite particles were coated on spherical Kissirises; glucose oxidase (GOx) enzyme was immobilized on Kissiris/Fe₃O₄/TiO₂ by physical adsorption. This catalyst was analyzed by a scanning electron microscopy (SEM), Fourier transform infrared spectroscopy (FTIR), and energy dispersive X-ray (EDX) measurements. The performance of the prepared biocatalyst in the decolorization of Malachite Green dye was investigated. The optimal operation parameters were 20 mg/L, 20 mM, 5.5 and 40 °C for initial dye concentration, initial glucose concentration, pH and temperature, respectively. Under these conditions, a 95% Malachite Green decolorization efficiency was obtained after 150 min of reaction by using 1 g of prepared heterogeneous bio-Fenton catalyst. In this process, in contrast to a conventional Fenton's reaction, external hydrogen peroxide and ferrous ion sources were not used. The effect of various reaction parameters such as initial concentration of dye, amount of catalyst, concentration of glucose, pH value and temperature on MG decolorization efficiency was studied.

1. Introduction

Synthetic dyes are used in many industries such as textile, leather tanning, paper, plastics, pharmaceuticals, and foods[1]. There are more than 100,000 types of mercantile dyes and over 7×10^5 tons of dyestuff are produced per year [2, 3]. Malachite green (MG) is a triphenylmethane dye that is extensively used in the above mentioned industries. MG has some damaging effects on the ecosystem and its contact with skin leads to irritation, redness, and pain [4, 5]. Therefore, the elimination of MG from industrial wastewaters has become environmentally important. Many methods such as advanced oxidation, ozonation, adsorption, reverse osmosis, ion exchange, and membrane filtration have been utilized to remove dyes from waste effluents [6-10]. Fenton's reaction is a well-known method for the elimination of organic pollutants [11]. Fenton and Fenton like reactions can be demonstrated according to Equations (1) and (2) [10]:

$$Fe^{2+} + H_2O_2 \rightarrow Fe^{3+} + OH^. + OH^- \qquad (1)$$

$$Fe^{3+} + H_2O_2 \rightarrow Fe^{2+} + OOH^. + H^+ \qquad (2)$$

Hydroxyl radicals have high oxidizing potential (E⁰ =2.8 v) in terms of oxidizing various organic materials such as dyestuffs [12]. H₂O₂ is widely used as a source of hydroxyl radicals. Recently, in-situ production of hydrogen peroxide within the reaction medium has been developed in order to increase the efficiency of wastewater treatment and decrease the risk of hydrogen peroxide transportation and storage [13, 14]. Glucose oxidation, a simple enzymatic reaction, is catalyzed by glucose oxidase and applied for the in-situ generation of hydrogen peroxide. Therefore, simultaneous bio-Fenton's reactions can be done to decolorize dyes via Eqs. (1-4) [15].

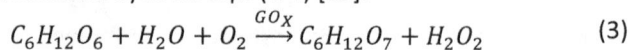

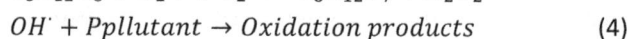

$$C_6H_{12}O_6 + H_2O + O_2 \xrightarrow{GOx} C_6H_{12}O_7 + H_2O_2 \qquad (3)$$

$$OH^. + Ppllutant \rightarrow Oxidation\ products \qquad (4)$$

To increase the efficiency of the bio-Fenton process, a proper carrier should be prepared to enhance GOx loading, activity, and stability as well as to decrease the enzyme usage cost in continuous reactors [17]. The high surface

*Corresponding author
E-mail address: akarimi@tabrizu.ac.ir

area to volume ratio provided by the nanoparticles, such as TiO_2 and Fe_3O_4, favors high binding capacity and high catalytic specificity of the conjugated enzyme [16]. However, the main problem of the suspended biocatalyst is in the separation of nanoparticles after the treatment [18]. To solve the separation problem of carriers, this study used Kissirisfor the $TiO_2/Fe_3O_4/GOx$ immobilization. Kissiris is formed by foam thickening of volcanic lava and has micro-pores which are irregularly distributed throughout the surface. This natural mineral with a highly porous structure, good mechanical strength and stability toward chemical agents could be a very attractive material for enzyme immobilization [19-22]. In order to provide a better medium for the GOx enzyme, this research used TiO_2 nanoparticles due to their chemical internees, rigidity, thermal stability, good adhesion to carriers, and high surface area [23]. Fe_3O_4 was coupled with TiO_2 to make a TiO_2/Fe_3O_4 adequate composite for GOx immobilization, the capacity of which is usually higher than pure TiO_2 and pure Fe_3O_4 [24]. Moreover, Fe_3O_4 was a ferrous source for the heterogenic Fenton's reaction and the adherence property of TiO_2 nanoparticles caused Fe_3O_4 nanoparticles to be strongly linked to Kissiris carriers. Therefore, the prepared carrier had the advantage of being enzyme compatible and included a ferrous source for the commencement of bio-Fenton reactions.

2. Materials and methods

2.1. Materials and equipment

TiO_2 nanoparticles (commercial Degussa P25) were a mixed phase containing 80% anatase and 20% rutile with an average crystal size of 21 nm. Methanol (99.9%), glucose oxidase (EC 1.1.3.4, from Aspergillus niger), β-D-glucose, and Malachite Green oxalate (MG) were obtained from Sigma Aldrich.Iron (III) chloride tetrahydrate, iron (II) sulfate heptahydrate, ammonia trihydrate (Merck) and nitrogen gas were used to prepare magnetite nanoparticles. All the solutions were prepared using distilled water. A UV–vis spectrophotometer (1700 UV–vis Shimadzu, Japan) was used to determine dye concentration. Scanning electron microscopy (SEM) images and EDX analysis were taken by MIRA3FEG-SEM (TescanBrno, Czech Republic). FTIR spectrums were obtained by Tensor 27 (Bruker, Germany). A Sonoplus Ultrasonic Homogenizer HD 2200 (Germany) was used for sonication.

2.2. Coating TiO₂ and Fe₃O₄ on Kissiris

A suspension of 0.5 g TiO_2 nanoparticles in 25 mL of methanol was sonicated for 15 min. The desired amount of HCl (1 N) was added to the solution to reach apH value of 5. The suspended solution was poured onto 5 g of Kissirises at 90°C and was allowed to dry at the mentioned temperature for 5 h [25]. The coated Kissirises were washed with distilled water to remove weakly attached particles. The deposited

amount of TiO_2 was measured by the difference in the mass of Kissirises before and after TiO_2 deposition. Fe_3O_4 nanoparticles were synthesized using a chemical co-precipitation method [26]. Kissirises, coated with TiO_2 were dispersed in 100 mL of deionized water and then 2.7 g of $FeSO_4.7H_2O$ and 5.4 g of $FeCl_3.4H_2O$ were added. The mixed solution was continuously stirred under nitrogen gas at 80°C for 1 h. Subsequently, 30 mL of $NH_3.3H_2O$ was rapidly added to the mixture and stirred for another 1 h. Finally, the carriers were washed with 100 mL of deionized water to remove impurities and allowed to dry in a vacuum condition for 1 h.

2.3. Immobilization of GOx on Kissiris/TiO₂/Fe₃O₄

1 g of the carrier was dispersed in 10 ml of GOx solution in a phosphate buffer at a pH value of 5.5 by stirring in a shaker incubator for 2 h at 30°C. The prepared heterogeneous bio-Fenton catalyst was separated from the solution and washed with distillated water to remove weakly attached GOx [17].

2.4. Decolorization

Experiments were carried out at a certain temperature and a constant stirring rate of 160 rpm in a 100 mL Erlenmeyer flask, which contained 20 mL of reaction mixture. The reaction mixture contained a certain concentration of MG and glucose with the desired amount of bio-Fenton catalyst to measure MG concentration in the sample solutions. The dyeabsorbance was measured at a maximum wavelength (λ_{max}=617 nm). The decolorization percentage was calculated by the following equation:

$$\text{Decolorization (\%)} = \left(1 - \frac{C}{C_0}\right) \times 100 \qquad (5)$$

Where C_0 and C are the concentrations of the sample solution at times 0 and t, respectively.

3. Results and discussion

3.1. Characterizing of the catalysts

Mineral composition of the prepared support was analyzed using EDX (energy dispersion X-ray) patterns (Figure 1). According to the figure, the applied Kissiris contained Al_2O_3, SiO_2, CaO, etc. The sharp peak was assigned to Au since the samples were coated by gold in the analysis procedure. The appearance of Ti and Fe can be clearly observed in the Figure 1, implying the successful deposition of TiO_2 and the synthesis of Fe_3O_4.

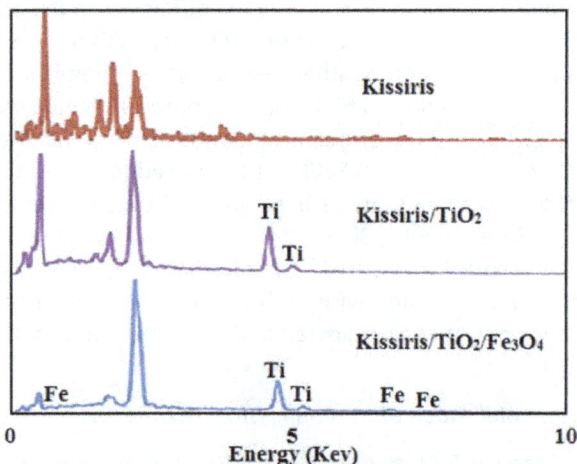

Fig. 1. EDX spectrum of Kissiris, Kissiris/TiO$_2$ and Kissiris/TiO$_2$/Fe$_3$O$_4$

Figure 2 presents the FTIR spectrum of Kissiris, Kissiris/TiO$_2$, Kissiris/TiO$_2$/Fe$_3$O$_4$, and Kissiris/TiO$_2$/Fe$_3$O$_4$/GOx. In the measured spectra, the bonds at about 3400-3500 cm^{-1} were attributed to the vibration of the OH group. The bending vibration of H$_2$O molecules was observed within 1400-1700 cm^{-1}[27]. In the spectrum of volcanic Kissiris, the peaks at 400-600 cm^{-1} were assigned to the vibration of Al-O bonds of Al$_2$O$_3$ and those at 600-900 cm^{-1} and 1100-1250 cm^{-1} corresponded to the vibration of Si-O-Si and Si = O, respectively [28]. The peaks at 400-700 cm^{-1} in the spectra of Kissiris/TiO$_2$ corresponded to the stretching vibration of Ti-O and Ti-O-Ti [28]. The peaks at 400-700 cm^{-1} were assigned to the vibration of Fe-O-Fe and Fe-O in the spectra of synthesized Fe$_3$O$_4$ [27]. In Figure 2, the Kissiris/TiO$_2$/Fe$_3$O$_4$/GOx represents the FTIR spectra of immobilized GOx on the carrier, in which the peaks at 1645.82 and 1513.17cm^{-1} correspond to amide I and amide II bonds of GOx, respectively [29]. Thus, the results proved the successful preparation of the bio-Fenton catalyst of Kissiris/TiO$_2$/Fe$_3$O$_4$/GOx. Scanning electron microscopy (SEM) images of the Kissiris, thin deposited layers of TiO$_2$/Fe$_3$O$_4$, and immobilized GOx were taken. Figure 3(b) and (c) clearly shows the attachment of TiO$_2$ and Fe$_3$O$_4$ nanoparticles on the Kissirises. Figure 3d shows that GOx was successfully immobilized on the carrier.

3.2. Catalytic activity of the Kissiris/TiO$_2$/Fe$_3$O$_4$/GOx

To ensure that the enzyme was acting, the new catalyst was used to decolorize 5 mg/L of malachite green aqueous solution with and without adding glucose. In the absence of glucose, the decolorization was only 33% at 60 min. The GOx enzyme did not produce H$_2$O$_2$ without substrate (e.g., glucose) and it was physical adsorption that caused the disappearance of dye in this case. When glucose was added to the reaction medium, the decolorization percentage increased to 99.9% at 60 min.

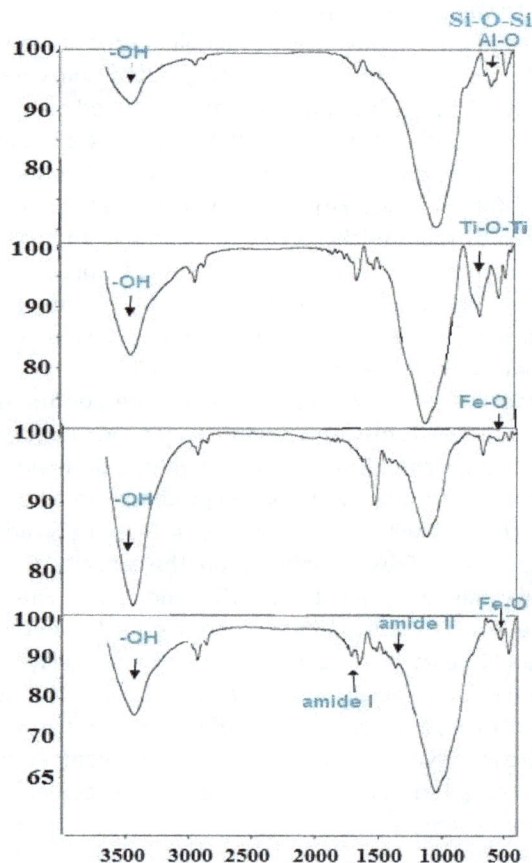

Fig. 2. FT-IR spectrum of Kissiris, Kissiris/TiO$_2$, Kissiris/TiO$_2$/Fe$_3$O$_4$, and Kissiris/TiO$_2$/Fe$_3$O$_4$/GOx.

A possible mechanism of decolorization can be outlined as the following reactions [30]. First, MG was adsorbed onto the catalyst surface, Eq. (6), where Me refers to the Fe or Ti on the surface of catalyst. The enzymatic reaction produced in-situ hydrogen peroxide according to Eq. (3). Then, Fenton and Fenton-like processes took place and active radicals such as hydroxyl and perhydroxyl radicals (˙OH and ˙OOH) were generated, as shown in Eqs. (1) and (2), respectively. Finally, the produced radicals oxidized the organic molecules of dye, leading to MG decolorization (Eqs. (7) and (8)).

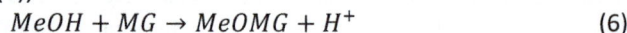

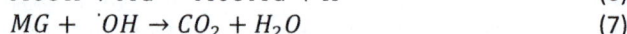

$$MeOH + MG \rightarrow MeOMG + H^+ \qquad (6)$$
$$MG + \dot{O}H \rightarrow CO_2 + H_2O \qquad (7)$$
$$MG + \dot{O}OH \rightarrow CO_2 + H_2O \qquad (8)$$

3.2.1. Effect of dye concentration

The experiments were conducted at an initial concentration of MG in the range of 20 – 100 mg/L, while maintaining agitation rate, catalyst loading, initial glucose concentration and temperature at 160 rpm, 1 g, 20 mM and 35°C, respectively. Figure 4 shows that MG disappearance efficiency decreased with a decrease in initial dye concentration. It can be seen that the decolorization rate was faster at the start of the reaction for each initial concentration, which was mainly due to the higher number of available dye molecules for reaction and the rapid engagement of active radicals with organic molecules decreased the chance of undesired reactions between

radicals in the first minutes of the reaction [31]. Table 1 shows the overall decolorization rate.

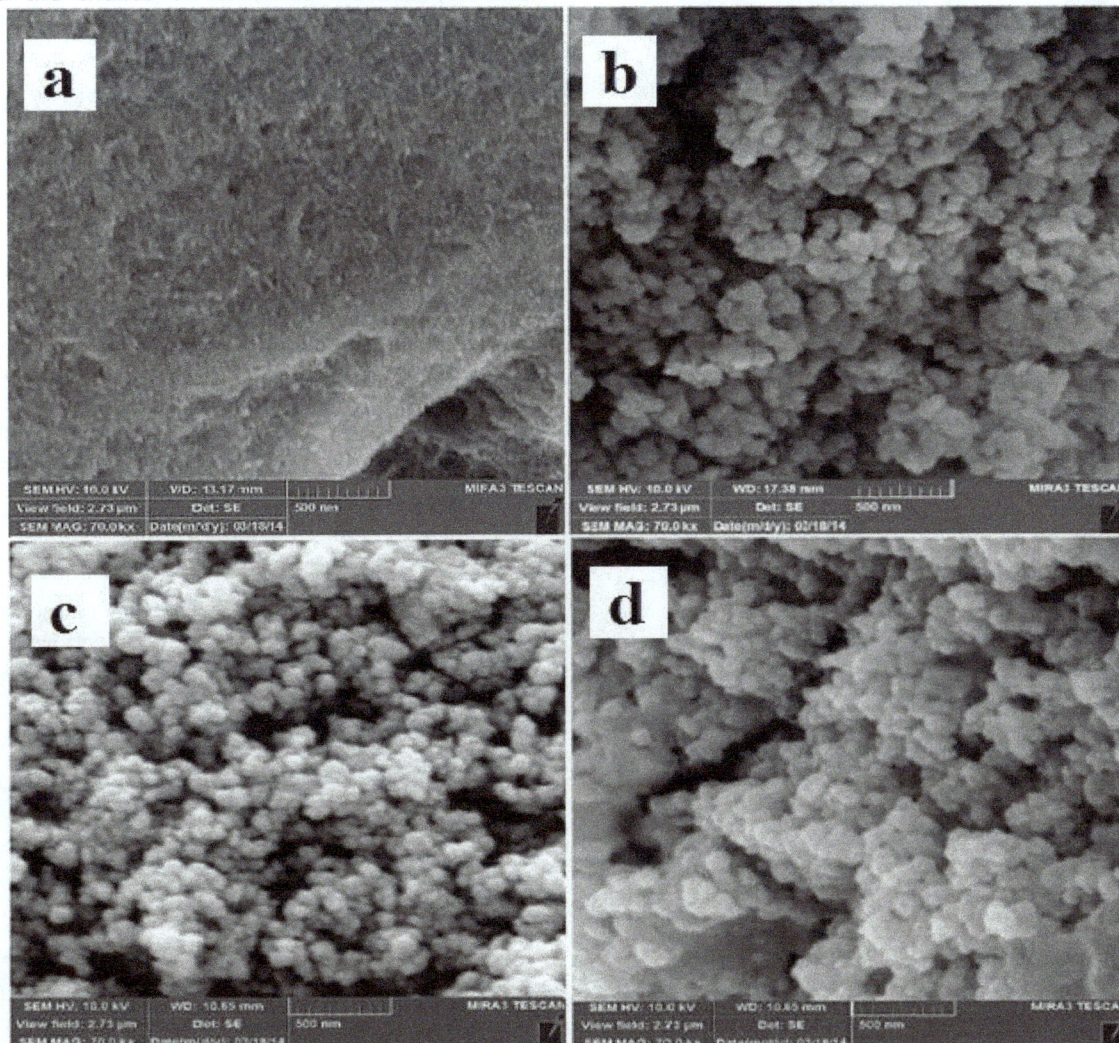

Fig. 3. SEM images of (a) Kissiris, (b) Kissiris/TiO$_2$, and (c) Kissiris/TiO$_2$/Fe$_3$O$_4$, (d) Kissiris/TiO$_2$/Fe$_3$O$_4$/GOx.

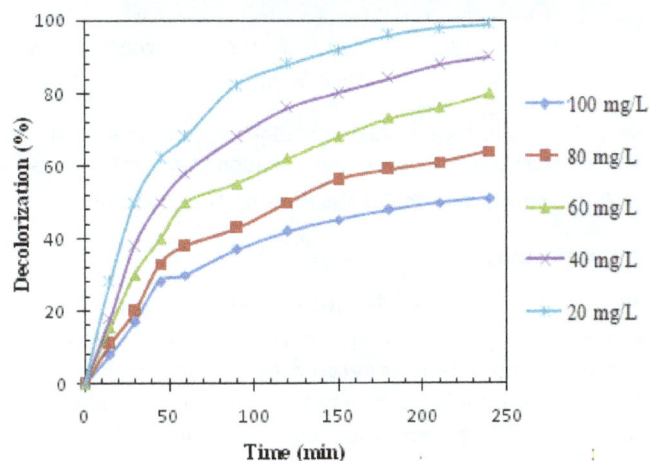

Fig. 4. Effect of initial concentration of dye on pseudo-first order rate constant during heterogeneous bio-Fenton oxidation treatment; [glucose]$_0$=20 mM, 1 g Kissiris/Fe$_3$O$_4$/TiO$_2$/GOx, T= 35°C.

Table 1. Overall pseudo-first order kinetic rate constant for different dye concentration

MG concentration	k_{app}(min^{-1})	R^2
20	0.0029	0.9074
40	0.0041	0.9439
60	0.0063	0.9714
80	0.0093	0.9802
100	0.0182	0.9935

3.2.2. Effect of glucose concentration

To study the effect of glucose concentration on MG degradation, different concentrations of glucose (5-25) mM were tested (Fig. 5). The experiments were performed at a fixed initial Malachite Green concentration of 20 mg/L, 1 g of heterogeneous bio-Fenton catalysts, and a temperature of 35°C. The increase in decolorization rate, shown in Figure 5, was related to the H$_2$O$_2$ generation rate, which was produced by enzymatic reaction using immobilized GOx. This in-situ production of H$_2$O$_2$ was the main advantage of the immobilized enzyme. By the increase of glucose amount, the in-situ production and consumption rate of hydrogen peroxide increased. The decolorization efficiency was raised, as a result of the increased glucose concentration from 5 to 20 mM. However, at 25 mM, the degradation efficiency decreased; this can be due to the adverse reactions caused by hydrogen peroxide scavenging (Eqs. (10)- (11)) at such a high concentration of H_2O_2 [32].

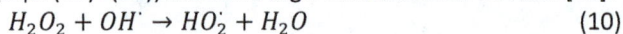

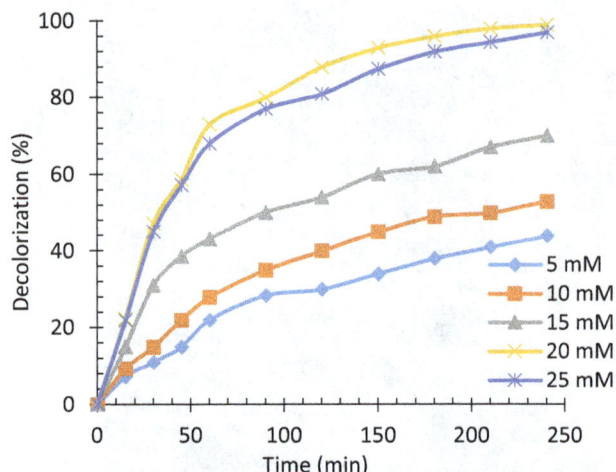

$$H_2O_2 + OH^{\cdot} \rightarrow HO_2^{\cdot} + H_2O \qquad (10)$$
$$HO_2^{\cdot} + OH^{\cdot} \rightarrow H_2O + O_2 \qquad (11)$$

Fig. 5. Effect of initial glucose concentration on pseudo-first order rate constant during heterogeneous bio-Fenton oxidation treatment; $[MG]_0$= 20 mg/L, 1 g Kissiris/Fe₃O₄/TiO₂/GOx, T= 35°C.

3.2.3. Effect of pH

As can be seen in Figure 6, the MG disappearance efficiency was influenced by the pH value of the solution. The efficiency increased by raising the pH value from 3.5 to 5.5 and decreased significantly above a pH of 5.5. These changes of decolorization efficiency can be attributed to the stability of H_2O_2 in the acidic medium. The decrease in efficiency at pH values higher than 5.5 was due to Fe (OH)₃ formation; in this form, iron disintegrated hydrogen peroxide to water and oxygen [31]. On the other hand, the higher and lower pH values disrupted the enzymatic reaction, so that there was an optimal pH value of 5.5 for enzyme activity and decolorization process.

3.2.4. Effect of temperature

To determine the effect of temperature on MG decolorization, experiments were conducted at a temperature range of 298 to 318 K. According to Figure 7, the decolorization efficiency increased by increasing temperature to 20°C; this was due to an increase in GOx activity by temperature which consequently enhanced the hydrogen peroxide production rate. Also, the reaction of hydrogen peroxide with Fe ions was accelerated by an increase in temperature. Decolorization efficiency decreased above 20°C, which was due to the denaturation

of the protein molecules in the structure of the enzyme. Thus, excessive temperature deactivated glucose oxidase and disrupted its bio-catalytic activity; as a result, the in-situ generation of hydrogen peroxide decreased and consequently, decoloirzation efficiency decreased.

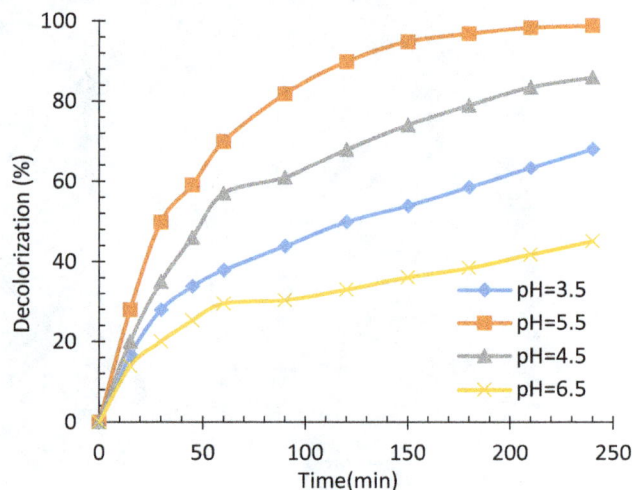

Fig. 6. Effect of pH value on pseudo-first order rate constant during heterogeneous bio-Fenton oxidation treatment; $[MG]_0$=20 mg/L, $[glucose]_0$=20 mM, 1 g Kissiris/Fe₃O₄/TiO₂/GOx, T= 35°C.

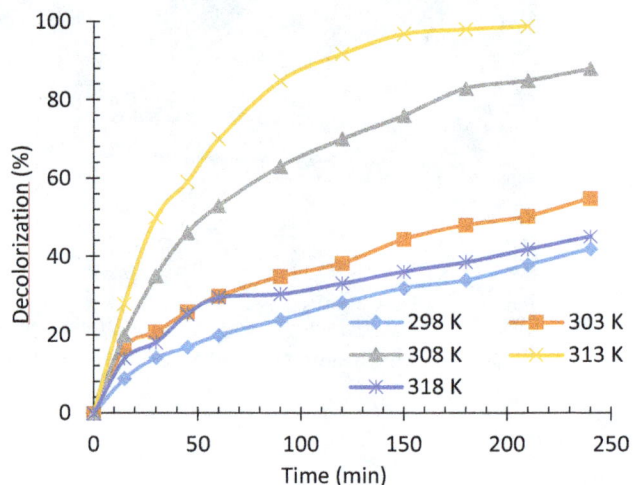

Fig. 7. Effect of temperature on pseudo-first order rate constant during heterogeneous bio-Fenton oxidation treatment; $[MG]_0$=20 mg/L, $[glucose]_0$=20 mM, 1 g Kissiris/Fe₃O₄/TiO₂/GOx, pH=5.5.

4. Conclusions

Bio-Fenton is a newly introduced method for the elimination of organic pollutants. In this technique, hydrogen peroxide is generated by an enzymatic reaction and reacts with a Fe^{2+} ion to produce a hydroxyl radical. The glucose oxidase enzyme was immobilized on the Kissiris/TiO₂/Fe₃O₄ carrier to enhance enzyme reusability. The results of EDX, FTIR, and SEM analyses showed that the synthesis of the hybrid heterogeneous bio-Fenton catalyst (e.g.,Kissiris/TiO₂/Fe₃O₄/GOx) was successful. The best MG decolorization efficiency, by using 1 g of prepared heterogeneous bio-Fenton catalyst, was 95% after 150 min at an optimal operating condition of 20 mg/L MG initial concentration, 20 mM glucose, a pH value of 5.5 and a temperature of 40 °C. It was observed that Kissiris/TiO$_2$/Fe$_3$O$_4$/GOx is a promising catalyst for the bio- P-values of 0.0020 and 0.0066, respectively. In other words, increasing pH had a greater influence on the removal efficiency.

References

[1] Doğan, M., Abak, H., Alkan, M. (2009). Adsorption of methylene blue onto hazelnut shell: kinetics, mechanism and activation parameters. *Journal of hazardous materials*, *164*(1), 172-181.

[2] Gupta, V. K., Mittal, A., Krishnan, L., Gajbe, V. (2004). Adsorption kinetics and column operations for the removal and recovery of malachite green from wastewater using bottom ash. *Separation and purification technology*, *40*(1), 87-96.

[3] Jasińska, A., Różalska, S., Bernat, P., Paraszkiewicz, K., Długoński, J. (2012). Malachite green decolorization by non-basidiomycete filamentous fungi of Penicillium pinophilum and Myrothecium roridum. *International biodeterioration and biodegradation*, *73*, 33-40.

[4] Srivastava, S., Sinha, R., Roy, D. (2004). Toxicological effects of malachite green. *Aquatic toxicology*, *66*(3), 319-329.

[5] Nethaji, S., Sivasamy, A., Thennarasu, G., Saravanan, S. (2010). Adsorption of Malachite Green dye onto activated carbon derived from Borassus aethiopum flower biomass. *Journal of hazardous materials*, *181*(1), 271-280.

[6] Khataee, A. R., Vatanpour, V., Ghadim, A. A. (2009). Decolorization of CI Acid Blue 9 solution by UV/Nano-TiO 2, Fenton, Fenton-like, electro-Fenton and electrocoagulation processes: a comparative study. *Journal of hazardous materials*, *161*(2), 1225-1233.

[7] Aleboyeh, A., Kasiri, M. B., Olya, M. E., Aleboyeh, H. (2008). Prediction of azo dye decolorization by UV/H 2 O 2 using artificial neural networks. *Dyes and pigments*, *77*(2), 288-294.

[8] Rauf, M. A., Meetani, M. A., Hisaindee, S. (2011). An overview on the photocatalytic degradation of azo dyes in the presence of TiO$_2$ doped with selective transition metals. *Desalination*, *276*(1), 13-27.

[9] Xu, N., Zhang, Y., Tao, H., Zhou, S., Zeng, Y. (2013). Bio-electro-Fenton system for enhanced estrogens degradation. *Bioresource technology*, *138*, 136-140.

[10] Zhang, G., Qin, L., Meng, Q., Fan, Z., Wu, D. (2013). Aerobic SMBR/reverse osmosis system enhanced by Fenton oxidation for advanced treatment of old municipal landfill leachate. *Bioresource technology*, *142*, 261-268.

[11] Yalfani, M. S., Contreras, S., Medina, F., Sueiras, J. (2009). Phenol degradation by Fenton's process using catalytic in situ generated hydrogen peroxide. *Applied catalysis B: Environmental*, *89*(3), 519-526.

[12] Barreca, S., Colmenares, J. J. V., Pace, A., Orecchio, S., Pulgarin, C. (2014). Neutral solar photo-Fenton degradation of 4-nitrophenol on iron-enriched hybrid montmorillonite-alginate beads (Fe-MABs). *Journal of Photochemistry and Photobiology A: Chemistry*, *282*, 33-40.

[13] Osegueda, O., Dafinov, A., Llorca, J., Medina, F., Suerias, J. (2012). In situ generation of hydrogen peroxide in catalytic membrane reactors. *Catalysis today*, *193*(1), 128-136.

[14] Torabi, S. F., Khajeh, K., Ghasempur, S., Ghaemi, N., Siadat, S. O. R. (2007). Covalent attachment of cholesterol oxidase and horseradish peroxidase on perlite through silanization: activity, stability and co-immobilization. *Journal of biotechnology*, *131*(2), 111-120.

[15] Karimi, A., Aghbolaghy, M., Khataee, A., Shoa Bargh, S. (2012). Use of enzymatic bio-Fenton as a new approach in decolorization of malachite green. *The scientific world journal*, *2012*.

[16] Ansari, S. A., Husain, Q. (2012). Potential applications of enzymes immobilized on/in nano materials: a review. *Biotechnology advances*, *30*(3), 512-523.

[17] Karimi, A., Mahdizadeh, F., Salari, D., Niaei, A. (2011). Bio-deoxygenation of water using glucose oxidase immobilized in mesoporous MnO$_2$. *Desalination*, *275*(1), 148-153.

[18] Khataee, A. R., Fathinia, M., Aber, S., Zarei, M. (2010). Optimization of photocatalytic treatment of dye solution on supported TiO 2 nanoparticles by central composite design: intermediates identification. *Journal of hazardous materials*, *181*(1), 886-897.

[19] Ghasemzadeh, R., Kargar, A., Lotfi, M. (2011). Decolorization of synthetic textile dyes by immobilized white-rot fungus. In *international conference on chemical, ecology and environmental sciences, Pattaya* (pp. 434-438).

[20] Jamshidian, H., Khatami, S., Mogharei, A., Vahabzadeh, F., Nickzad, A. (2013). Cometabolic degradation of para-nitrophenol and phenol by Ralstonia eutropha in a Kissiris-immobilized cell bioreactor. *Korean Journal of chemical engineering*, *30*(11), 2052-2058.

[21] Karimi, A., Vahabzadeh, F., Bonakdarpour, B. (2006). Use of Phanerochaete chrysosporium immobilized on Kissiris for synthetic dye decolourization: involvement of manganese peroxidase. *World journal of microbiology and biotechnology*, *22*(12), 1251-1257.

[22] Tsoutsas, T., Kanellaki, M., Psarianos, C., Kalliafas, A., Koutinas, A. A. (1990). Kissiris: A mineral support for the promotion of ethanol fermentation by Saccharomyces cerevisiae. *Journal of fermentation and bioengineering*, *69*(2), 93-97.

[23] Xiao, P., Zhang, Y., Cao, G. (2011). Effect of surface defects on biosensing properties of TiO$_2$ nanotube

arrays.*sensors and actuators B: Chemical, 155*(1), 159-164.

[24] Meng, H., Wang, B., Liu, S., Jiang, R., Long, H. (2013). Hydrothermal preparation, characterization and photocatalytic activity of TiO_2/Fe–TiO_2 composite catalysts. *Ceramics international, 39*(5), 5785-5793.

[25] Shoaebargh, S., Karimi, A., Dehghan, G. (2014). Performance study of open channel reactor on AO_7 decolorization using glucose oxidase/TiO_2/polyurethane under UV–vis LED. *Journal of the Taiwan institute of chemical engineers, 45*(4), 1677-1684.

[26] Ozmen, M., Can, K., Arslan, G., Tor, A., Cengeloglu, Y., Ersoz, M. (2010). Adsorption of Cu (II) from aqueous solution by using modified Fe_3O_4 magnetic nanoparticles. *Desalination, 254*(1), 162-169.

[27] Mesgari, Z., Gharagozlou, M., Khosravi, A., Gharanjig, K. (2012). Spectrophotometric studies of visible light induced photocatalytic degradation of methyl orange using phthalocyanine-modified Fe-doped TiO_2 nanocrystals. *Spectrochimica acta part A: Molecular and biomolecular spectroscopy, 92*, 148-153.

[28] Fan, Y., Ma, C., Li, W., Yin, Y. (2012). Synthesis and properties of Fe_3O_4/SiO_2/TiO_2 nanocomposites by hydrothermal synthetic method. *Materials science in semiconductor processing, 15*(5), 582-585.

[29] Zuo, S., Teng, Y., Yuan, H., Lan, M. (2008). Direct electrochemistry of glucose oxidase on screen-printed electrodes through one-step enzyme immobilization process with silica sol–gel/polyvinyl alcohol hybrid film. *Sensors and actuators B: Chemical, 133*(2), 555-560.

[30] Abbas, M., Rao, B. P., Reddy, V., Kim, C. (2014). Fe_3O_4/TiO_2 core/shell nanocubes: Single-batch surfactantless synthesis, characterization and efficient catalysts for methylene blue degradation. *Ceramicsi international, 40*(7), 11177-11186.

[31] Hameed, B. H., Lee, T. W. (2009). Degradation of malachite green in aqueous solution by Fenton process. *Journal of hazardous materials, 164*(2), 468-472.

[32] Romanias, M. N., El Zein, A., Bedjanian, Y. (2012). Heterogeneous interaction of H_2O_2 with TiO_2 surface under dark and UV light irradiation conditions. *The journal of physical chemistry A, 116*(31), 8191-8200.

13

Employing response surface analysis for photocatalytic degradation of MTBE by nanoparticles

Hossein Lotfi[1], Mohsen Nademi[1], Mohsen Mansouri[2],* , Mohammad Ebrahim Olya[3]
[1]*Department of Chemical Engineering, North Tehran Branch, Islamic Azad University, Tehran, Iran*
[2]*Department of Chemical Engineerin, Ilam University, Ilam, Iran*
[3]*Department of Environmental Research, Institute for Color Science and Technology, Tehran, Iran*

Keywords:
MTBE
Photocatalytic degradation
Response surface method
TiO₂-ZnO-CoO nanoparticles

ABSTRACT

Since groundwaters are a major source of drinking water, their pollution with organic contaminants such as methyl tertiary-butyl ether (MTBE) is a very significant issue. Hence, this research investigated the photocatalytic degradation of MTBE in an aqueous solution of TiO_2-ZnO-CoO nanoparticle under UV irradiation. In order to optimize photocatalytic degradation, response surface methodology was applied to assess the effects of experimental variables such as catalyst loading, initial concentration of MTBE and pH on the dye removal efficiency. The optimal condition to achieve the best degradation for the initial concentration of 30.58 mg/L of MTBE was found at a pH of 7.68 and a catalyst concentration of 1.68 g/L after 60 min.

1. Introduction

Methyl tertiary-butyl ether (MTBE) has received considerable attention over the last decade due to its widespread detection in indoor environments [1, 2]. MTBE is used as an additive in gasoline in order to elevate the octane number, improve combustion, and reduce CO_2 production. MTBE is a colorless clear liquid with an inherent smell and has a molecular formula and a molecular weight of C4H9OCH3 ($C_5H_{12}O$) and of 88.15 g/mole, respectively. It is widely used as a solvent in many industries along with its application as a fuel oxygenate. MTBE is equally flammable and has high vapor pressure (204 mm Hg at 20°C). The solvent has a boiling point of 55 °C at atmospheric pressure and a freezing point of -109°C. Human exposure to MTBE is generally through inhalation upon contact at the workplace, consumer use of products containing MTBE, or through environmental release. MTBE is readily absorbed and metabolized by the body[3].MTBE also has a low tendency for adsorption to soil particles. As such, MTBE readily dissolves into groundwater and experiences little retardation resulting in migration nearly equal to the groundwater flow rate and the potential for widespread migration. MTBE also creates taste and odor problems in drinking water at relatively low concentrations. So far, many methods have been studied for removing MTBE. These include adsorption, air stripping, photocatalysis, ozone treatment, Fenton process, high energy electron beam irradiation, cavitation, biodegradation and electrochemical oxidation. Advanced oxidation processes (AOPs) employing heterogeneous catalysts have been used extensively for various types of degradation of organic pollutants in water. This is due to the ability of the catalyst to generate a strongly oxidizing hydroxyl radical with a high oxidative power of Eo = 2.8 eV and thus acts to degrade various organic pollutants [4]. Heterogeneous photocatalysis have shown great potential in the oxidation of organic compounds using a semiconductor material such as a catalyst [5]. This process generates holes that can react with water to produce OH radicals. Titania (TiO_2) and zinc oxide (ZnO) are two of the most widely explored semiconductor materials that have been studied and are currently the most commonly used in AOP because of their photochemical stability, minimal toxicity and high efficiency in the degradation of pollutants [6, 7].The main advantage of the UV/TiO_2-ZnO system is

*Corresponding author
E-mail address: m.mansouri@ilam.ac.ir

that the process can be conducted at a wavelength (300–380 nm) higher than other UV based AOPs [8]. It has also been demonstrated by numerous studies that modified titanium dioxide and zinc oxide can perform photocatalytic activity under visible light irradiation [6-9].The fast recombination of the electron-hole pairs can be countered by the presence of co-dopants. The photodegradation efficiency of co-dopants-ZnO is higher than a bare ZnO and a single dopant ZnO system [10]. This is due to the co-dopants ability to simultaneously trap the photogenerated electron from the conduction band of the ZnO and subsequently reduce the recombination rate of the electron-hole [11]. Response surface methodology (RSM) is a very useful, quick and cost effective statistical method that optimizes different parameters and exhibits the interaction of these parameters; the Box–Behnken design was applied to determine the optimum photodegradation of paraquat and also to explain the interaction among the parameters studied [12, 13]. In our previous study, MTBE photocatalytic degradation with UV/TiO_2-ZnO-CuO nanoparticles was investigated; the optimized values were obtained at a PH of (7), a catalyst concentration of (1.49 g/L), and the initial MTBE concentration of (31.46 mg/L) [14]. So, the aim of this paper was to optimize the degradation of MTBE by a TiO_2-ZnO-CoO nanoparticles system using RSM based on the Box–Behnken design. Important parameters such as the initial pH values of the solution, TiO_2-ZnO-CoO loading, and MTBE concentration were investigated in this study as well as the interactions between these different parameters. Furthermore, the sol-gel method was used to synthesize TiO_2-ZnO-CoO nanoparticles. Then, the nanoparticles were characterized by X-ray diffraction (XRD) and a scanning electron microscope (SEM).

2. Materials and methods

2.1. Materials and Equipment

TiO_2 nanoparticles (commercial Degussa P25) were a mixed phase containing 80% anatase and 20% rutile with an average crystal size of 21 nm. Methanol (99.9%), glucose oxidase (EC 1.1.3.4, from Aspergillus niger), β-D-glucose, and Malachite Green oxalate (MG) were obtained from Sigma Aldrich.Iron (III) chloride tetrahydrate, iron (II) sulfate heptahydrate, ammonia trihydrate (Merck) and nitrogen gas were used to prepare magnetite nanoparticles. All the solutions were prepared using distilled water. A UV–vis spectrophotometer (1700 UV–vis Shimadzu, Japan) was used to determine dye concentration. Scanning electron microscopy (SEM) images and EDX analysis were taken by MIRA3FEG-SEM (TescanBrno, Czech Republic). FTIR spectrums were obtained by Tensor 27 (Bruker, Germany). A Sonoplus Ultrasonic Homogenizer HD 2200 (Germany) was used for sonication.

2. 2. Instruments

The morphology and structure of the prepared samples were characterized using a field emission scanning electron microscope (Leo 1455 VP, England) and an X-ray diffractometer (Philips PW 1800, Netherlands). The MTBE concentrations were measured with a UV-Vis spectrophotometer (Model T80+, PG Instruments, UK) device. The gas chromatography was equipped with a helium ionization detector (HID) (Model GC-Acme 6100, Korea). A TRB-5 quartz capillary column of (30 m × 0.53 mm) with a 3-μm film thickness was used in the UV-Vis spectrophotometer.

2.3. Preparation of nanocatalyst

The TiO_2-ZnO-CoO nanocomposite tested in this study was prepared using the sol-gel procedure which was described in our previous work [14]. To prepare the Titania, hydroxyl propyl cellulose (HPC) was dissolved in ethanol under quick stirring for five minutes. Then, titanium tetraisopropoxide (TTIP) was added to the previous mixture and was stirred for fifteen minutes. Subsequently, the mixture of glacial acetic acid, pure alcohol and deionized water was added to the previous mixture. It was stirred for fifteen minutes to make sure it achieved a yellow transparent acidic TiO_2 sol. The sol was kept at room temperature for thirty minutes. The second component of the nanocomposite was ZnO. Initially, the Zn $(NO_3)_2.6H_2O$ was dissolved in pure alcohol and stirred for five minutes. Then the mixture of di-ethanol amine, pure alcohol, and deionized water was added to the solution under a vigorous and constant stirring condition. The solution was constantly stirred for fifteen minutes to reach a transparent sol ZnO. The third component of the nanoparticle was CoO. To begin with, the Co $(NO_3)_2.6H_2O$ was dissolvedin the pure alcohol and stirred for five minutes. Then, the mixture of ethanolamine, pure alcohol and distilled water was added to the solution under a vigorous, constant mixing condition. The solution was steadily mixed for fifteen minutes to reach an alkalinity transparent sol CoO. Finally, the sol of ZnO and CoO was mixed directly with the sol of TiO_2 to prepare the TiO_2-ZnO-CoO. This nano composite was dried at room temperature. Then, it was sintered at a temperature of 350 °C for 10 minutes and afterward, it was sintered at a temperature of 500 °C for five hours in order to calcinate (the temperature was increased at a rate of five degree Celsius per second); finally, the catalyst was prepared [15].

2.4. Experimental

All photochemical reactions for the destruction of MTBE with TiO_2-ZnO-CoO were fulfilled in a batch reactor made from cylindrical glass with a volume of three liters. A scheme of the reactor used in this study is presented in Figure 1. The reaction mixture in the reactor circulated in a closed cycle between the pump and the reactor. In addition, the temperature of the reaction was monitored. Three 15 W lamps from Phillips emitted UV light with a wavelength 254 nm, which were immersed in the solution and used to provide the UV radiation in the reactor. The volume ofthe reaction mixture for each of the tests, which was proposed by RSM experimental design, was 3L. In the end, the MTBE concentrations were measured with a UV-Vis spectrophotometer.

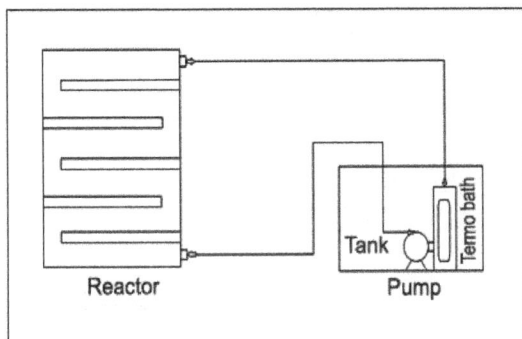

Fig. 1. Scheme of the reactor.

2.5. Analysis

The percentage of photocatalytic degradation (% MTBE removal) was calculated using Eq. 1.

$$MTBE\ removal\ (\%) = (C_0 - C_t)/\ C_0 \times 100 \qquad (1)$$

In the above mentioned equation, MTBE removal is the percentage of photocatalytic degradation, C_0 is the initial concentration of the sample in mg/L before irradiation under UV light, and C_t is the sample's concentration based on mg/L after irradiation under UV light at any time.

2.6. Experimental design and statistical analysis

Initially, preliminary experiments were conducted following the single factor study method to decide the most influential experimental parameters affecting the photocatalytic degradation of MTBE and to find their ranges. The selected factors were catalytic dose, initial concentration of MTBE, and pH of reaction mixture. The three selected experimental parameters were optimized by RSM as the independent variables and the percentage of degradation of MTBE as the response variables. The Box–Behnken design of experiments was employed to examine the combined effects of the three independent variables on the response through 15 sets of experiments. The ranges and levels of the independent variables are shown in Table 1. The Box–Behnken design was applied because it is highly efficient and does not involve any point at the peaks of the cubic region formed by the upper and lower limits of the variables. This design along with RSM has been widely used to optimize various physical, chemical, and biological processes [12-16]. By using RSM, the results were matched to an empirical quadratic polynomial model for the three parameters expressed in Equation 2:

$$Y=\beta_0+\beta_1 A+\beta_2 B+\beta_3 C+\beta_4 D+\beta_{11}A^2+\beta_{22}B^2+\beta_{33}C^2+\beta_{44}D^2+\beta_{12}AB+ \qquad (2)$$
$$\beta_{23}BC+\beta_{31}CA+\beta_{14}AD+\beta_{24}BD+\beta_{34}CD$$

where Y denotes the response variable; β_0 the intercept; β_1, β_2, β_3 the coefficients of the independent variables; β_{11}, β_{22}, β_{33} quadratic coefficients; β_{12}, β_{23}, β_{31}, β_{14}, β_{24}, β_{34} the interaction coefficients; and A, B, C are the independent variables. The multivariate regression analysis and optimization process were performed by means of RSM via Design Expert software (version 7, Stat Ease Inc., USA). The values obtained from the analysis of variance (ANOVA) were found to be significant at $p < 0.05$. The optimum values for the independent variables were found using three-dimensional response surface analysis of the independent and dependent variables. The designed experiments and the actual and predicted values of the response are detailed in Table 2. Also, the variations are shown in Figure 2 d. The optimum conditions for the maximum degradation of MTBE are shown in Table 3, and the effect of the independents variable on the degradation of MTBE are shown in Figure 2 (a,b,c).

Table 1. The levels and ranges of variables in Box–Behnken experiment design

Coded variable level	Symbol	Independent variables
low		
-1		
4	A	pH
30	B	MTBE concentration (mg/L)
1	C	Catalytic loading (g/L)

Table2. Box–Behenken experiments along with actual and predicted values of responses

Run	A, pH	B, MTBE concentration	C, Catalyst loading	MTBE removal% Actual	MTBE removal% Predicted
1	10	40	3	63.68	64.03
2	7	50	3	64.78	64.68
3	4	30	2	83.35	83.60
4	7	40	2	81.78	81.62
5	4	40	1	51.27	50.92
6	7	40	2	81.67	81.62
7	7	30	1	83.63	83.73
8	7	30	3	88.79	88.66
9	7	50	1	57.36	57.50
10	10	40	1	58.68	58.79
11	7	40	2	81.39	81.62
12	4	50	2	57.38	57.59
13	10	30	2	89.98	89.76
14	10	50	2	65.81	65.56
15	4	40	3	57. 90	57.78

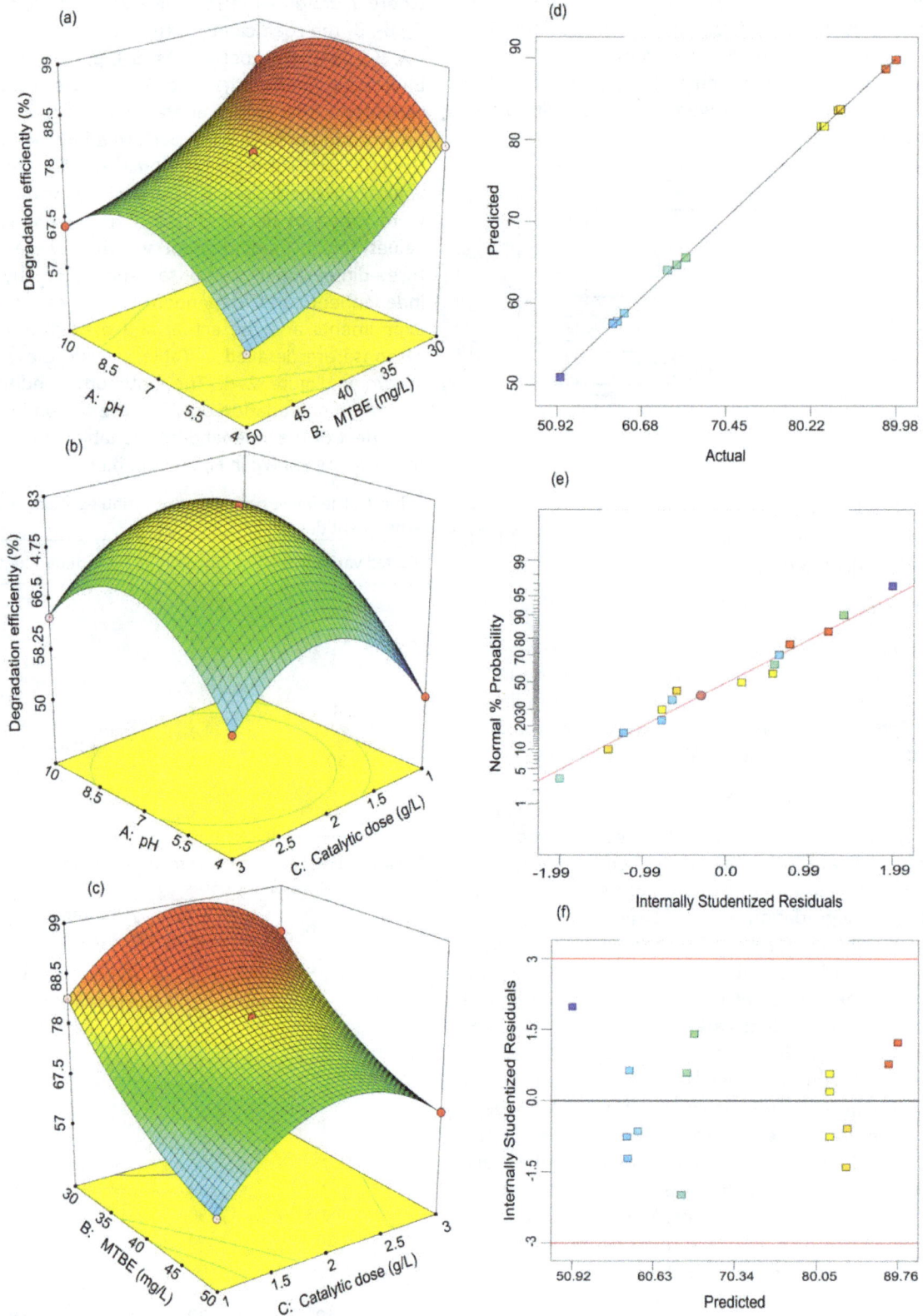

Fig. 2. Effects of catalyst loading, initial MTBE and pH on degradation efficiency (%):(a) catalytic dose: 2(g/L); (b) initial MTBE concentration: 40(mg/L); (c) pH: 7; (d) plot of the actual and predicted values for degradation efficiency (%); (e) Normal probability plot of residual for degradation efficiency %; (f) normal probability plot of residual for degradation efficiency (%).

Table 3. The Optimum conditions selected for the maximum possible MTBE removal (%)

	Num.	A, pH	B, initial MTBE Concentration (mg/L)	C, catalytic dose (g/L)	MTBE removal (%) (predict)	MTBE removal (%) (actual)	Desirability	
Solutions	1	7.68	30.58	1.63	94.9298	92.3681	1	Selected

3. Results and Discussion

3.1. Scanning electron microscopy/ X-ray Spectroscopy (SEM/XRD)

Figure 3a shows the SEM image of TiO$_2$-ZnO-CoO nanocomposites which were prepared in this research. As can be seen, the TiO$_2$-ZnO-CoO nanoparticles were well distributed on the surface and sphere-shaped particles were formed in good resemblance to each other. Relying on the SEM images, the average particle size of the TiO$_2$-ZnO-CoO nanoparticles was approximately 16 nm. Moreover, it can be seen that there was a difference between the crystal size evaluated by XRD and by SEM. This difference could originate from the fact that the outcome of an XRD pattern revealed the crystal size of a particle, whereas the result of a SEM image represented

the particle size itself which was the accumulation of several crystals [17, 18]. The XRD patterns of TiO$_2$-ZnO-CoO and TiO$_2$ are shown in Fig. 3b. According to Fig. 3a, it can be seen that all peaks are found as follows: 25.26°, 37.98°, 47.97°, 53.56° and 62.53° for TiO$_2$-ZnO-CoOand 25.38°, 37.94°, 48.04°, 54.69° and 62.93° for TiO$_2$. It is recognized that the 2 theta values of the X-ray patterns of TiO$_2$ and TiO$_2$-ZnO-CoOwerecompatible with anatase for both of them. The XRD patterns illustrate that the composition of TiO$_2$-ZnO-CoOdoes not change the catalyst structure of TiO$_2$. This may result from the low concentration of CoO and ZnO in the composition. The particle size of the samples can be calculated by the Debye-Scherrer formula. The particle size calculated value for TiO$_2$ and TiO$_2$-ZnO-CoOnanoparticles was 13.22 nm and 14.15 nm, respectively.

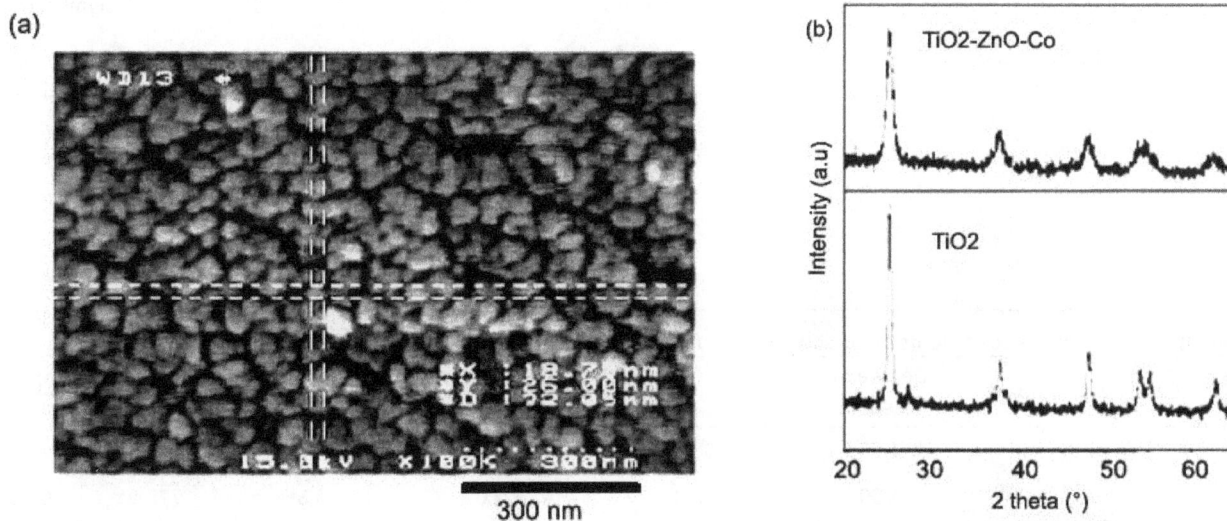

Fig. 3. The catalyst characterization (a) SEM image of TiO$_2$–ZnO-CoO and (b) XRD pattern of TiO$_2$ and TiO$_2$–ZnO-CoO.

3.2. Statistical analysis

In order to achieve an appropriate model, the tests of importance for the regression model and for each coefficient of the model similar to the test for lack-of-fit had to be performed. The test results were summarized in a normal ANOVA table. The table of the ANOVA test for MTBE's destruction reply is provided in Table 4. As presented in Table 4, the predicted decolorization efficiencies by the mentioned equation were in good agreement with the experimental values. The correlation coefficient (R^2) is a quantitative criterion for evaluating the solidarity between the experiential data and the predicted

values. By comparing the empirical outcomes and the predicted data, it was justified that there was a sensible relationship between the predicted values and the empirical data with R^2 = 0.9997 %. Moreover, the modified R^2 (Adj-R^2) was 0.9993,which is near to the corresponding value of R^2 (Table 4) and indicated a measure of fit goodness as well as being more suitable for comparing models with different numbers of independent variables. Sufficient accuracy compared the range of the computed value at the design points with the average prediction error. The ratios that were greater than 4 indicated the sufficient differentiation power of the model. The outcomes of the above comparison was greater than 4 (132.625), implying the adequate discrimination power of the model. The absence of a fit P-value of 0.1807 suggested that the lack-of-fit was not considerable relative to pure error; this was suitable, since we were looking for a sample that matches. According to the experimental plan (Table 2), empirical second order multinomial equations were expanded for the destruction percentage of MTBE in terms of the three independent variables as described in Equation 3.

$$\text{MTBE removal (\%)} =+ 81.62 + 3.53\,A - 12.55\,B + 3.03\,C + 0.45\,A\,B - 0.41\,A\,C + 0.56\,B\,C - 11.62\,A^2 + 4.14\,B^2 - 12.11\,C^2 \quad (3)$$

The empiric information for the destruction of MTBE was statistically analyzed using the ANOVA test and the results of are presented in Tables 2 and 4. The ANOVA of the second order multinomial model (F-value = 2183.83, p-value < 0.0001) indicated that the model was notable, i.e., there was only a chance of 0.01% for occurrence of the model's F-value because of the noise. The F-value of the sample was much more than the tabular F-value with similar numeral of degrees of freedom which showed that the model was suitable for prediction of MTBE removal. The regression model's coefficient of MTBE removal is presented in Table 5 as an additional tool to evaluate the sufficiency of the final model by the ANOVA test. The ordinary possibility plan (Scatter Diagram) for the studentized residuals is shown in Figure 2e.The points on this plot lie reasonably close to the straight line, confirming that the errors had normal distribution with a zero mean and a constant. The curvature P value < 0.0001 indicated that there was significant curvature (as measured by the difference between the mean center points and the mean factorial points) in the design space. As a result, a linear model along with the interaction terms giving a twisted plane was not sufficient to describe the response. Likewise, the patterns of the residuals in Figure 2f showed that they had no clear plan and their structure was relatively eccentric. Moreover, they indicated equal scatter above and below the x-axis, implying the adequacy of the proposed model; thus, there was no reason to suspect any violation.

Table 4. ANOVA for response surface reduced quadratic model-aAnalysis of variance table

Source	Sum of Squares	Degrees of freedom	Mean Square	F Value	P-value Prob> F	
Model	2529.58	9	281.06	2183.83	< 0.0001	significant
Residual	0.64	5	0.13			
Lack of Fit	0.56	3	0.19	4.69	0.1807	not significant
Pure Error	0.080	2	0.040			
Cor Total	2530.23	14		R-Squared	0.9997	
				Adj R-Squared	0.9993	
				Adeq Precision	132.625	

Table 5. ANOVA results for the coefficients of quadratic model for MTBE removal

Factor	Coefficient estimate	Degree of freedom	Standard error	95% confidence interval low	95% confidence interval high	F-value	p-Value
Intercept	81.62	1.00	0.21	81.08	82.15	-	-
A-pH	3.53	1.00	0.13	3.21	3.86	775.81	<0.0001
B-MTBE	-12.55	1.00	0.13	-12.88	-12.23	9793.84	<0.0001
C-Catalyst	3.03	1.00	0.13	2.70	3.35	569.45	<0.0001
AB	0.45	1.00	0.18	-0.011	0.91	6.29	0.0539
AC	-0.41	1.00	0.18	-0.087	0.055	5.12	0.0732
BC	0.56	1.00	0.18	0.10	1.02	9.84	0.0257
A^2	-11.62	1.00	0.19	-12.10	-11.14	3876.11	<0.0001
B^2	4.14	1.00	0.19	3.66	4.62	490.72	<0.0001
C^2	-12.11	1.00	0.19	-12.59	-11.63	4208.57	<0.0001

3.3. Optimization of MTBE removal by RSM

3.3.1. Effect of Initial pH

The effect of pH on the photocatalytic degradation rate of organic compounds is a complex issue because this variable can modify the electrostatic interactions between the catalyst surface and substrate molecules as well as the formation of hydroxyl radicals by the reaction between hydroxide ions/$H_2O$and the generated positive holes in the catalyst surface [19].The surface charge of TiO_2 changed from positive to negative as the pH increased at values higher than the point of zero charges. Initially, the pH of the solution was adjusted and it was not controlled during the course of the reaction. The impacts of pH on the photocatalytic degradation of MTBE were assessed with the initial pH at three diverse values of 4, 7 and 10, as illustrated in Figure 4a. The destruction of MTBE occurred as the pH of solution increased from 4 to 7. Then, the percentage of MTBE degradation went up whereas the solution's pH value rises from 7 to 10. The electrostatic interaction between the semiconductor surface, solvent molecules, substrate, and charged radicals formed during photocatalytic oxidation was strongly dependent on the pH of the solution. In addition, protonation and deprotonation of the organic pollutants can take place depending on the solution pH [16]. The phenomenon can be represented in terms of the location of the point of zero charge (isoelectric point) of the TiO_2-ZnO-CoO. According to the results in Figure 2, it can be concluded that the best pH value for the degradation of MTBE under the mentioned condition was 7.68.

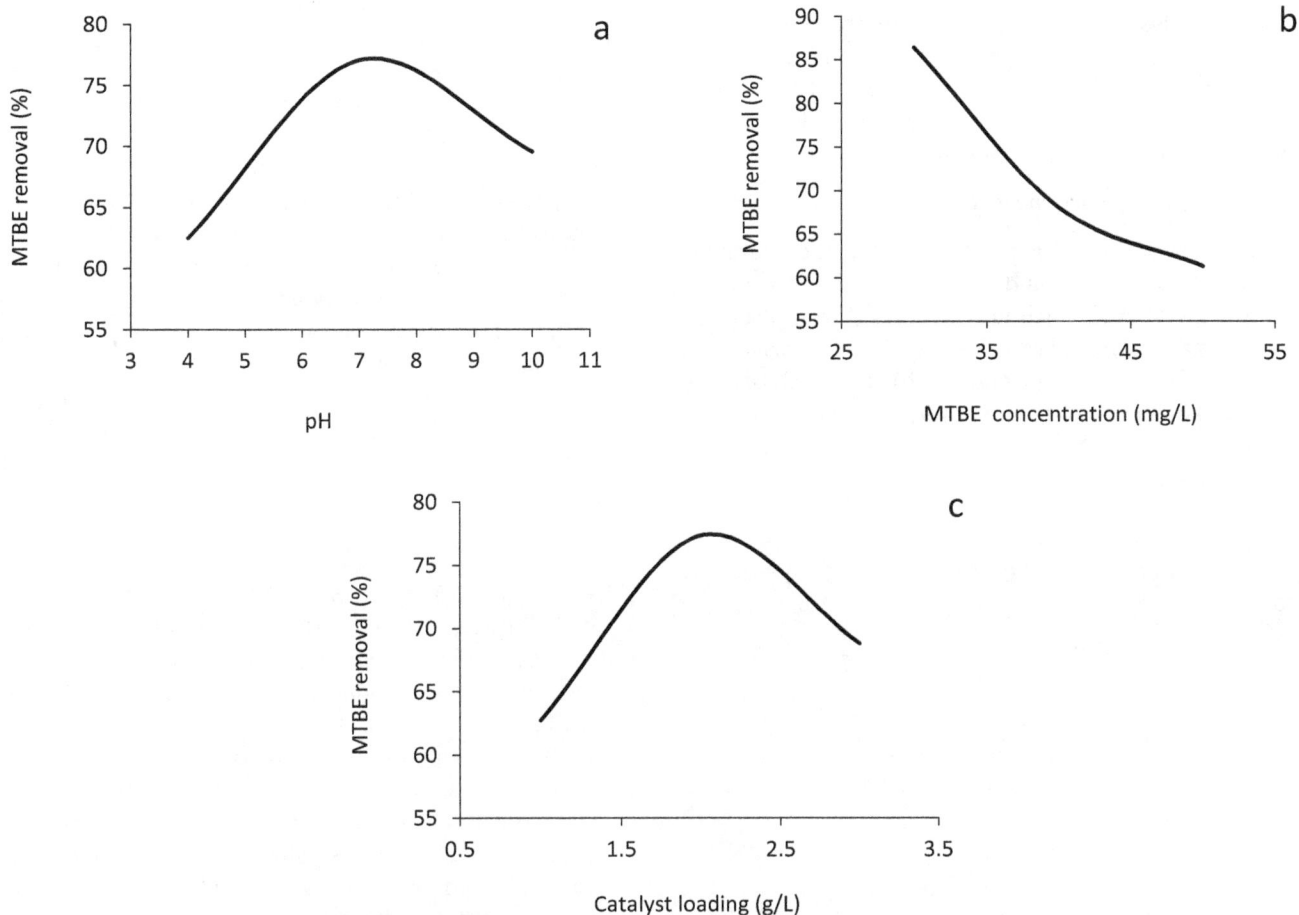

Fig.4. Effect of pH, initial MTBE concentration and catalytic dose on degradation efficiency (%): (a) initial MTBE concentration: 40 (mg/L), catalytic dose: 2 (g/L); (b) pH :7, catalytic dose: 2(g/L); and (c) pH :7, initial MTBE concentration: 40 (mg/L).

3.3.2. Effect of nanocatalyst loading

The effect of TiO_2–ZnO-CoO nanocatalyst loading on the photocatalytic degradation of MTBE under conditions 1, 2 and 3g/L of catalyst loading was studied. The results are presented in Figure 4b. It is obvious that the photodegradation rate increased with the increase of the catalyst's concentration up to a level which corresponded to the optimum activation of the catalyst particles by the incident light. In our case, this limit corresponded to 2g/L of TiO_2 – ZnO-CoO; the increase of the reaction rate that was observed up to this concentration level was attributed to an increase in the photo generated active sites on the catalyst surface and consequently, the formation of greater amounts of reactive oxygen species (ROS).This behavior can be attributed to the fact that some photocatalyst particles may not get sufficient energy to produce hydroxyl radical and started MTBE oxidation [12].

3.3.3. Effect of initial concentration of MTBE

The effects of the initial MTBE concentration on the photocatalytic rate were also investigated by carrying out a series of experiments at different initial MTBE concentrations. The experimental data showed that the photodegradation rate decreased with the rise of the initial MTBE concentration (Figure 4c). A reduction of the photodegradation rate to about 75% was observed when the initial MTBE concentration was scaled up from 30 mg/L to 50 mg/L. This change of the photodegradation rate was explained with the increased light absorption due to the MTBE molecules which reduced the excitation density at the photocatalyst surface and there with, the formation of reactive hydroxyl and superoxide radical which are required for the photo degradation of MTBE. Similar results have been reported on the photocatalytic oxidation of other organic compounds interface [13, 20, 21].

3.4. Photocatalytic reaction kinetic

Figure 5 shows the kinetic study of the photocatalytic degradation of MTBE and was appraised based on optimum conditions, which were obtained from previous sections, at a catalyst concentration of 1.63 g/L, pH = 7.68, and an initial MTBE concentration of 30.58 (mg/L). Usually, first-order kinetics is suitable for photocatalytic reactions [22, 23].The kinetics model is as follows:

$$-r_A = -\frac{dc}{dt} = KC \tag{4}$$

After integration of Eq. (4), the following equation is obtained:

$$Ln\left(\frac{C_0}{C}\right) = K\,t \tag{5}$$

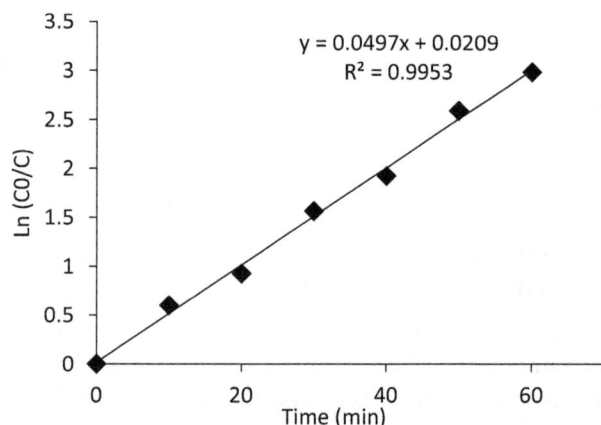

Fig.5. Effect of different initial concentrations of the MTBE on photocatalytic degradation.

where r_A is the oxidation rate of the MTBE (ppm min^{-1}), K is the rate constant (min^{-1}), C_0 is the initial concentration of MTBE, and C_t is the concentration of MTBE at the irradiation time. The linear relation of $Ln(C_0/C_t)$ versus irradiation time (t) for MTBE is presented in Fig. 5. The values of the first-order degradation constants (K) as well as the linear regression (R^2) values are reported in Table 6.

Table 6. Kinetic constant of MTBE degradation

A, pH	B, initial MTBE Concentration (mg/L)	C, catalytic dose (g/L)	R^2	K_{app}(min^{-1})
7.68	30.58	1.63	0.9953	0.0497

4. Conclusions

In this research, the photocatalytic degradation of MTBE from aqueous solutions using TiO$_2$-ZnO-CoO nanoparticles was studied.Response surface methodology was used for the assessment of the effects of experimental variables on the fading efficiency of MTBE. An empirical relationship between the decolorization efficiency (response) and independent variables (nanoparticles loading, initial concentration of MTBE, and pH) based on the experimental results was obtained and expressed by the second order polynomial equation. The maximum vital operation parameters were determined by the RSM method at a pH of 7.68, a TiO$_2$-ZnO-CoO concentration of 1.63 g/L, and an initial concentration of MTBE of 30.58 mg/L. In the mentioned conditions, photocatalytic degradation reached 99.53% in 60 minutes and the rate constant of degradation of MTBE was 0.0497 (min^{-1}).

References

[1] Belpoggi, F., Soffritti, M., Maltoni, C. (1995). Methyl-tertiary-butyl ether (MTBE)—a gasoline additive—causes testicular and lympho haematopoietic cancers in rats. *Toxicology and industrial health*, *11*(2), 119-149.

[2] Smith, A. E., Hristova, K., Wood, I., Mackay, D. M., Lory, E., Lorenzana, D., Scow, K. M. (2005). Comparison of biostimulation versus bioaugmentation with bacterial strain PM1 for treatment of groundwater contaminated with methyl tertiary butyl ether (MTBE). *Environmental health perspectives*, 113(3), 317-322.

[3] Danmaliki, G. I., Shamsuddeen, A. A., Usman, B. J. (2016). The effect of temperature, turbulence, and Ph on the solubility of MTBE. *European journal of earth and environment*, *3*(2), 31-39.

[4] El Madani, M., Harir, M., Zrineh, A., El Azzouzi, M. (2015). Photodegradation of imazethapyr herbicide by using slurry and supported TiO$_2$: Efficiency comparison. *Arabian journal of chemistry*, *8*(2), 181-185.

[5] Pirkanniemi, K., Sillanpää, M. (2002). Heterogeneous water phase catalysis as an environmental application: a review. *Chemosphere, 48*(10), 1047-1060.

[6] Siuleiman, S., Kaneva, N., Bojinova, A., Papazova, K., Apostolov, A., Dimitrov, D. (2014). Photodegradation of Orange II by ZnO and TiO_2 powders and nanowire ZnO and ZnO/TiO_2 thin films. *Colloids and surfaces A: Physicochemical and engineering aspects, 460*, 408-413.

[7] Evgenidou, E., Fytianos, K., Poulios, I. (2005). Semiconductor-sensitized photodegradation of dichlorvos in water using TiO_2 and ZnO as catalysts. *Applied catalysis B: Environmental, 59*(1), 81-89.

[8] Cui, L., Wang, Y., Niu, M., Chen, G., Cheng, Y. (2009). Synthesis and visible light photocatalysis of Fe-doped TiO_2 mesoporous layers deposited on hollow glass microbeads. *Journal of solid state chemistry, 182*(10), 2785-2790.

[9] Moustakas, N. G., Kontos, A. G., Likodimos, V., Katsaros, F., Boukos, N., Tsoutsou, D, Falaras, P. (2013). Inorganic–organic core–shell titania nanoparticles for efficient visible light activated photocatalysis. *Applied catalysis B: Environmental, 130*, 14-24.

[10] Zhang, G., Qin, L., Meng, Q., Fan, Z., Wu, D. (2013). Aerobic SMBR/reverse osmosis system enhanced by Fenton oxidation for advanced treatment of old municipal landfill leachate. *Bioresource technology, 142*, 261-268.

[11] Lee, K. M., Lai, C. W., Ngai, K. S., Juan, J. C. (2016). Recent developments of zinc oxide based photocatalyst in water treatment technology: a review. *Water research, 88*, 428-448.

[12] Zhang, J., Fu, D., Xu, Y., Liu, C. (2010). Optimization of parameters on photocatalytic degradation of chloramphenicol using TiO_2 as photocatalyst by response surface methodology. *Journal of environmental sciences, 22*(8), 1281-1289.

[13] Liu, P., Xu, Z., Ma, X., Peng, Z., Xiao, M., Sui, Y. (2016). Removal of Methyl Tertiary-Butyl Ether via ZnO-AgCl Nanocomposite Photocatalyst. *Materials research, 19*(3), 680-685.

[14] Mansouri, M., Nademi, M., Olya, M. E., Lotfi, H. (2017). Study of Methyl tert-butyl Ether (MTBE) Photocatalytic Degradation with UV/TiO_2-ZnO-CuO Nanoparticles. *Journal of chemical health risks, 7*(1). 19-32

[15] Pirkarami, A., Olya, M. E., Farshid, S. R. (2014). UV/Ni–TiO_2 nanocatalyst for electrochemical removal of dyes considering operating costs. *Water Resource and industry, 5*, 9-20.

[16] Saien, J., Khezrianjoo, S. (2008). Degradation of the fungicide carbendazim in aqueous solutions with UV/TiO_2 process: optimization, kinetics and toxicity studies. *Journal of hazardous materials, 157*(2), 269-276.

[17] Eslami, A., Nasseri, S., Yadollahi, B., Mesdaghinia, A., Vaezi, F., Nabizadeh, R., Nazmara, S. (2008). Photocatalytic degradation of methyl tert-butyl ether (MTBE) in contaminated water by ZnO nanoparticles. *Journal of chemical technology and biotechnology, 83*(11), 1447-1453.

[18] Zhou, M., Yu, J., Cheng, B. (2006). Effects of Fe-doping on the photocatalytic activity of mesoporous TiO_2 powders prepared by an ultrasonic method. *Journal of hazardous materials, 137*(3), 1838-1847.

[19] An, T., An, J., Yang, H., Li, G., Feng, H., Nie, X. (2011). Photocatalytic degradation kinetics and mechanism of antivirus drug-lamivudine in TiO_2 dispersion. *Journal of hazardous materials, 197*, 229-236.

[20] Samaei, M. R., Maleknia, H., Azhdarpoor, A. (2016). A comparative study of removal of methyl tertiary-butyl ether (MTBE) from aquatic environments through advanced oxidation methods of $H_2O_2/nZVI$, $H_2O_2/nZVI/ultrasound$, and $H_2O_2/nZVI/UV$. *Desalination and water treatment, 57*(45), 21417-21427.

[21] Moradi, H., Sharifnia, S., Rahimpour, F. (2015). Photocatalytic decolorization of reactive yellow 84 from aqueous solutions using ZnO nanoparticles supported on mineral LECA. *Materials chemistry and physics, 158*, 38-44.

[22] Hu, Q., Zhang, C., Wang, Z., Chen, Y., Mao, K., Zhang, X., Zhu, M. (2008). Photodegradation of methyl tert-butyl ether (MTBE) by UV/H_2O_2 and UV/TiO_2. *Journal of hazardous materials, 154*(1), 795-803.

[23] Safari, M., Nikazar, M., Dadvar, M. (2013). Photocatalytic degradation of methyl tert-butyl ether (MTBE) by Fe-TiO_2 nanoparticles. *Journal of industrial and engineering chemistry, 19*(5), 1697-1702.

14

Hydrodynamics and mass transfer investigation in three-phase airlift reactors for sewage and water treatment process by using activated carbon and sludge

Mohammad Ali Salehi*, Nasrin Hakimghiasi

Department of Chemical Engineering, University of Guilan, Rasht, Iran

Keywords:
Activated carbon
Activated sludge
Airlift reactor
Gas superficial velocity
Mass transfer

ABSTRACT

A bioreactor refers to any manufactured or engineered device that supports a biologically active environment. These kinds of reactors are designed to treat wastewater treatment. Volumetric mass transfer coefficient and the effect of superficial gas velocity, as the most important operational factor on hydrodynamics, in three-phase airlift reactors are investigated in this study. The experiments for the external airlift reactor were carried out at a 0.14 downcomer to riser cross-sectional area ratio, and for the internal reactor at 0.36 and 1. Air and water were used as the gas and liquid phases, respectively, as well as activated carbon/sludge particles as the solid phase. Increasing the superficial gas velocity resulted in greater liquid circulation velocity, gas hold-up, and volumetric mass transfer coefficient; increasing the suspended activated carbon particles resulted in a decreased concentration of activated sludge, downcomer to riser cross sectional area ratio, liquid velocity, gas hold-up and volumetric mass transfer coefficient. The maximum gas hold-up was 0.178 which was attained in the external airlift reactor with a 1 Wt. % of activated sludge at a gas superficial velocity of 0.25 (m/s). The maximum volumetric mass transfer coefficient was 0.0485 (l/s) that was observed in the external airlift reactor containing activated carbon with a 0.00032 solid hold-up. A switch was observed in the activated sludge airlift reactor flow regime at gas velocities higher than 0.15 (m/s) and 0.18 (m/s) in the activated carbon airlift reactors.

1. Introduction

Due to advantages such as low shear stress, low cost, high mass transfer and simple structure, the application of airlift reactors in biotechnology procedures has expanded significantly over the past years. These include aerobic fermentation for food production, sewage treatment, and similar operations [1,2]. Airlifts, like bubble columns, are reactors in which liquids are mixed as air bubbles move through them. This type of reactor is suitable for procedures in which a uniform and fast dispersion of reactants is necessary as well as for multi-phase (liquid-gas-solid) systems that require high mass and heat transfer [1]. Generally speaking, the design and operation of two-phase gas-liquid and three-phase gas-liquid-solid reactors depend on gas hold-up, overall volumetric mass transfer coefficient, and the liquid circulation intensity [3]. These parameters are also functions of gas sparger type, fluid properties, and gas superficial velocity. There have been numerous studies on the effects of the mentioned factors on gas hold-up, liquid circulation velocity and the overall volumetric mass transfer coefficient. Muroyama et al. (1984) examined the hold-up and mass transfer in an airlift reactor with an internal tube using activated carbon particulates; they observed that the hold-up and mass transfer coefficient increased with gas superficial velocity [4]. Merchuk et al. (1996) investigated the effects of seven types of spargers on the gas hold-up in an airlift reactor with internal flow. They

*Corresponding author
E-mail address: masalehi@guilan.ac.ir

showed that in both uniform and transient bubble regimes, the gas recirculation was higher in spargers with small nozzles than larger nozzles due to the smaller bubbles produced. Therefore, gas spargers with small nozzles resulted in a greater hold-up [5]. The numerous studies on hydrodynamics and mass transfer of airlift reactors have made it clear that increasing the gas superficial velocity results in greater hold-up, liquid circulation velocity, and volumetric mass transfer [6-8].

By adding ethanol to the water-air-calcium alginate system, Freitas et al. (2001) observed that the volumetric mass transfer coefficient increased significantly. The addition of alcohol reduced the mass transfer coefficient k_L due to the increased resistance to mass transfer. On the other hand, the ethanol prevented bubble agglomeration and increased the bubble surface area; therefore, it was suggested that the enhanced overall mass transfer coefficient in the system was due to the increased bubble surface area [9]. Jin et al. (2006) examined and compared the hydrodynamics and mass transfer in two (internal and external) activated sludge airlift reactors. Their results showed that hydrodynamic and mass transfer in the external airlift reactors has better performance in compared with internal airlift reactor [8]. Yang et al. (2009) studied the rheological properties of the activated sludge in membrane airlift reactors and claimed that the viscosity of the activated sludge played an important role in oxygen and mass transfer [10]. Al Taweel et al. (2013) studied the effect of electrolyte on mass transfer in airlift reactors with internal tubes. They used sodium chloride salt at a concentration of 0.01, 0.1 and 1 M and observed that an increase in the aeration velocity and salt concentration yielded a greater K_{La} since the bubbles were less cohesive in high concentrations of salt [11]. Three principal airlift reactor operational parameters, namely gas hold-up, volumetric mass transfer coefficient and liquid circulation velocity were measured in this study; the effect of gas superficial velocity and particulate (activated sludge and carbon) concentration on each of the mentioned parameters were examined and compared in the two reactor types (internal and external).

2. Materials and methods

2.1. Materials

To study the hydrodynamics and mass transfer in the internal airlift reactors and external loop airlift reactor, activated sludge at 1%, 2% and 3% concentrations (prepared from water treatment unit located in Rasht, Iran and activated carbon (MERK™) with 0.00032 and 0.00064 hold-up were used as the solid phase, respectively. Sodium sulfite, Na_2SO_3, (MERK™) was used to absorb oxygen from the liquid and provide an oxygen-free environment. This salt had a high tendency towards bonding with oxygen, thus absorbing it and rendering the water oxygen-free. Therefore, in every flow rate after reaching the steady state,

aeration was stopped and the liquid was deoxygenated using sodium sulfite salt; then, aeration started again for the next flow. These steps were carried out for every flow rate until the end of the experiments.

2.2. Apparatus and experimental method description

A three-phase external loop airlift reactor with a 0.14 down comer-to-riser cross-sectional area ratio was used for the experiments. The bio reactor was made of double glazed glass with a 90 cm height and consisted of three parts, namely the riser (74 cm height and 8 cm diameter), the down comer (55 cm height and 3 cm diameter), and the gas-liquid separator. The temperature was kept constant at the exterior of the reactor using a water bath. The aeration was carried out using an 80W compressor and an antenna sparger with 20 nozzles. The water height was 60 cm without aeration and the volume was 3200 ml. To measure the flow rate of the input air, a rotameter was installed in the line between the compressor and the sparger in a way that it was connected to the sparger from one side and to the compressor from the other. Figure 1 shows the reactor and the equipment.

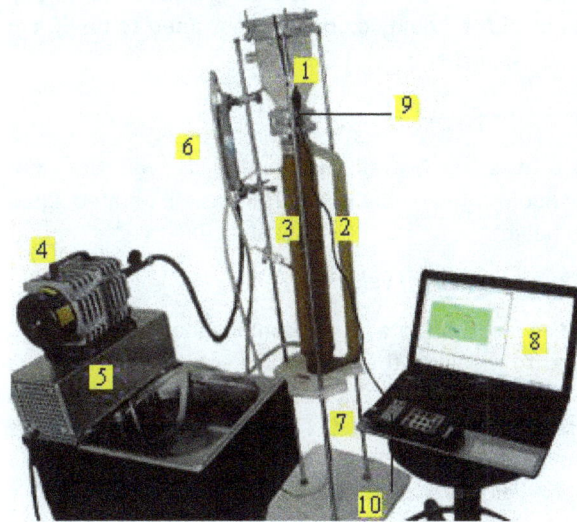

Fig. 1. The external airlift reactor setup: 1) Gas-liquid separator, 2) External loop, 3) Riser, 4) Compressor, 5) Water bath, 6) Sparger, 7) Air sparger, 8) Laptop, 9) Oxygen sensor, 10) Oxygen meter

The same apparatus was used for the internal airlift reactor with a 0.36 down comer-to-riser cross-sectional ratio, with the external part closed and internal 4 and 5 cm diameter tubes were implemented in the reactor 7cm away from the sparger using metal clips. The internal tubes made up two parts inside the reactor, riser and down comer. Both the riser and down comer were inside the reactor in internal airlift reactors.

2.3. Measuring the hydrodynamics and volumetric mass transfer coefficient

2.3.1. Gas superficial velocity

The superficial velocity of gas was calculated by Equation 1, based on the sparger cross-section area [8]

$$U_G = \frac{R_G}{A_G} \tag{1}$$

where A_G (m²) was the sparger cross-sectional area, R_G (m³/s) was the air flow and U_G(m/s) was the gas superficial velocity.

2.3.2. Gas hold-up

Gas hold-up here referred to the overall gas hold-up which was determined by the volumetric expansion method. In fact, the gas hold-up was expressed as the volume increase of the aerated liquid compared to the non-aerated liquid; this was achieved by first measuring the height of the non-aerated liquid column, and then measuring it again after the air bubbles entered the reactor with a specific flow rate. The overall gas hold-up was calculated using Equation 2 [8,12]. The height of the liquid column was assumed to be 60 cm for all the experiments.

$$\varepsilon_g = \frac{H_{LG} - H_L}{H_{LG}} \tag{2}$$

Where ε_g was the overall gas hold-up, H_L was the non-aerated liquid column height, and H_{LG} was the aerated liquid column height.

2.3.3. Liquid circulation velocity

A 0.1 M solution of potassium permanganate dye solution was used to measure the circulation velocity in the down comer of the activated sludge and carbon external loop airlift reactor by injecting 0.4 ml of the permanganate solution to the reactor from the external loop. The circulation velocity was calculated by Equation 3 [7,13].

$$U_{Ld} = \frac{L_d}{t_d} \tag{3}$$

Where U_{LD} was the circulation speed of the liquid in the down comer, L_D was the specified distance in the down comer, and t_d was the required time of permanganate solution to pass down comer length ($L_D = 40$).
The relation between the circulation velocity of the riser and the down comer is presented as Equation 4 [14].

$$U_{Lr}A_r = U_{Ld}A_d \tag{4}$$

Where U_{LR} and U_{LD} were the liquid velocity in the riser and down comer, respectively; A_r and A_d were the cross-sectional area of the riser and down comer, respectively.

2.3.4. Volumetric mass transfer coefficient

The volumetric mass transfer coefficient was measured via the dynamic method. The reactor content was first oxygen depleted and then re-oxygenated. To this end, the liquid phase was deoxygenated using sodium sulfite salt and then aerated for every flow rate [8]. The oxygen concentration was recorded every 5 seconds until reaching the steady state. This was achieved by using an oxygen meter (DO-5510) whose electrode was submerged 7 cm deep in the water column with its sensor connected to the computer using a cable. The volumetric mass transfer coefficient was calculated by the oxygen transfer balance (Equation 5) [15,16].

$$\frac{dC_L}{dt} = k_L a(C_L^* - C_L) \tag{5}$$

The integration of Equation 5 at t = 0 and C_L = 0 will lead to:

$$C_L(t) = C_L^*[1 - \exp\left(\frac{-k_L a}{t}\right)] \tag{6}$$

Where C_L (mg/L) and C_L^* (mg/l) indicated the oxygen concentration in the bulk solution at time t and the oxygen concentration at a steady state, respectively, which were read from the oxygen meter. According to Equation 6, the overall volumetric mass transfer coefficient was obtained from the slope of ln (C_L^*– C_L) in every flow rate [8,17].

3. Results and discussion

3.1. The effect of gas superficial velocity on the gas hold-up

The effect of gas superficial velocity on the hold-up in activated sludge and carbon airlift reactors is illustrated in Figures 2-4. The gas hold-up increased with the superficial velocity of gas. The gas hold-up variation was steeper in lower velocities and gradually became moderate as the velocity increased to the point where, due to turbulence, smaller changes occurred in the gas hold-up. Increasing the activated sludge concentration resulted in a lower hold-up which was attributed to the higher activated sludge viscosity; the effect of the solid phase concentration was relatively negligible in both the internal and external activated carbon airlift reactors.

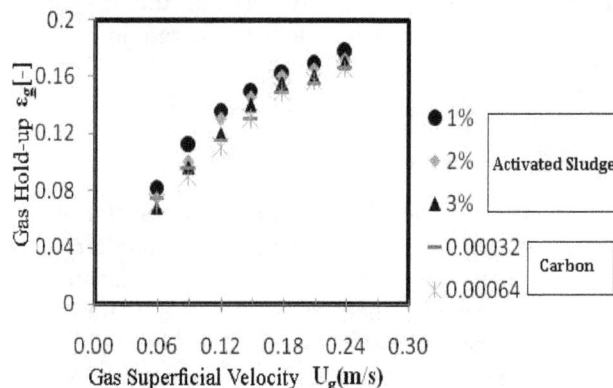

Fig. 2. Comparing the effect of gas superficial velocity on the hold-up in activated sludge and carbon reactors (A_d/A_r=0.14)

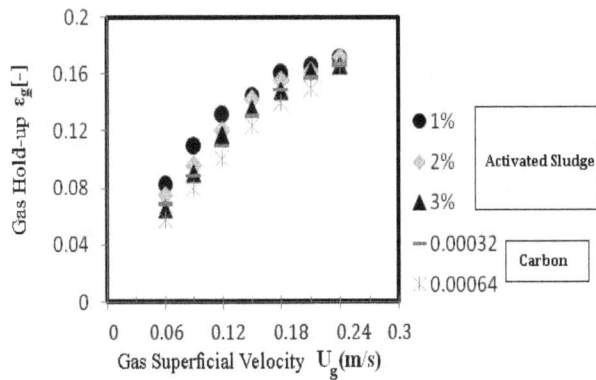

Fig. 3. Comparing the effect of gas superficial velocity on the hold-up in activated sludge and carbon reactors (A_d/A_r=0.36)

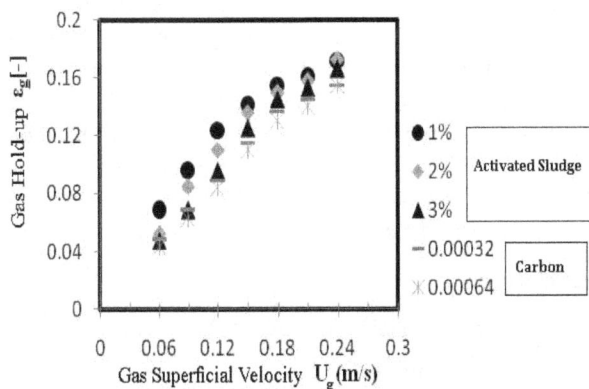

Fig. 4. Comparing the effect of gas superficial velocity on the hold-up in activated sludge and carbon reactors (A_d/A_r=1)

In comparing the activated sludge and carbon airlift reactors using Figures 2-4, we observed that the flow regime switch occurred at velocities higher than 0.15 m/s for the activated sludge reactors and at velocities higher than 0.18 m/s for the activated carbon reactors. The activated sludge liquid had a non-Newtonian behavior; in agreement with the properties of this type of fluids, it generated bigger bubbles and therefore, switched faster to a non-uniform regime [18]. Since the airlift reactors based on water with round carbon particulates had Newtonian behavior, it switched to non-uniform regimes at higher velocities.

3.2. The effect of gas velocity on liquid circulation velocity

Figures 5 and 6 show the effect of the gas superficial velocity on the liquid circulation at the down comer and riser of the activated sludge and carbon external airlift reactors.

The difference in density in the riser and down comer of the airlift reactors provided the driving force for the liquid movement and circulation. By increasing the gas superficial velocity, the liquid circulation velocity increased both at the down comer and riser, while increasing the sludge concentration and solid hold-up resulted in lower velocities. This occurred because the friction between the particles and resistance to flow increased with the sludge concentration and solid hold-up and thus, decreased the liquid velocity. The slope of liquid velocity variation was

steeper at lower gas velocities because at low flows the uniform bubble flow regime governed the system. By increasing the superficial velocity of the gas, the flow became turbulent and a non-uniform flow regime occurred. In Activated carbon and sludge reactors a linear relation was observed between the superficial velocity of gas and the velocity of liquid at $0.06<U_g<0.15$ and $0.06<U_g<0.18$ respectively, in addition a power law relation was observed in higher velocities $U_g \geq .015$ and $U_g \geq .018$

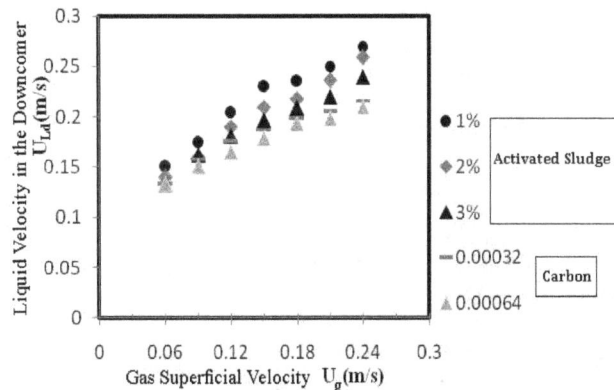

Fig. 5. Comparing the effect of gas superficial velocity on liquid velocity at the downcomer of the activated sludge and carbon external airlift reactor.

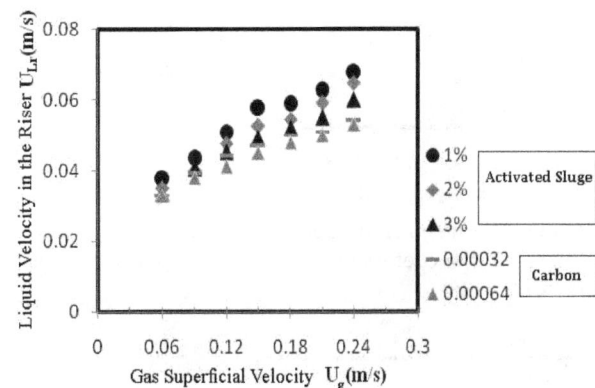

Fig. 6. Comparing the effect of gas superficial velocity on liquid velocity at the riser of the activated sludge and carbon external airlift reactor.

3.3. The effect of gas velocity on the overall mass transfer coefficient

The effect of the superficial velocity of gas on the overall mass transfer coefficient in the activated sludge and carbon airlift reactors is illustrated in Figures 7-9. Increasing the superficial velocity led to a greater gas volume fraction in the liquid and a higher mass transfer rate in all the three reactors. In low velocities there were small bubbles, therefore the K_La increased with a steeper slope while in higher velocities the K_La changed with a slower slope due to the formation of larger bubbles in the riser gas hold-up. For the same gas velocity, as the activated sludge concentration and solid hold-up (activated carbon particles) increased, the

K$_L$a decreased because higher viscosity or solid hold-up promoted bubble agglomeration and a reduction of gas-liquid interface. These results were in agreement with the results presented by Jin et al. (2006) [8].

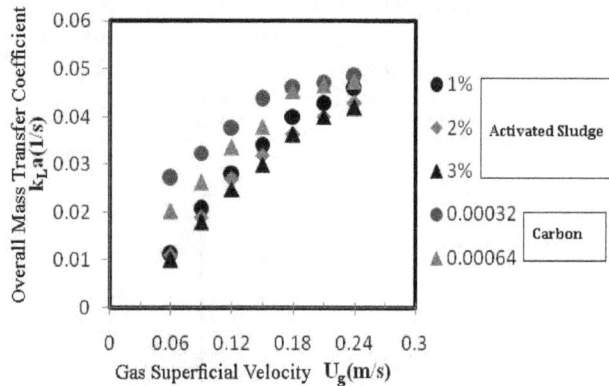

Fig. 7. Comparing the effect of superficial velocity of gas on the overall mass transfer coefficient in activated sludge and carbon reactors (A$_d$/A$_r$=0.14)

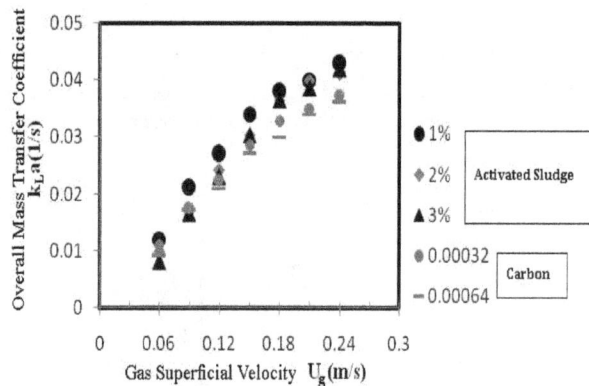

Fig. 8. Comparing the effect of superficial velocity of gas on the overall mass transfer coefficient in activated sludge and carbon ctors (A$_d$/A$_r$=0.36)

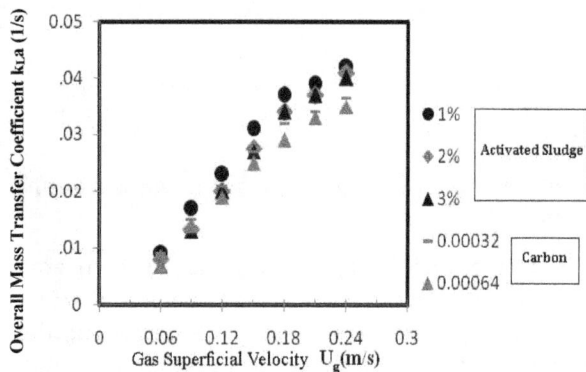

Fig. 9. Comparing the effect of superficial velocity of gas on the overall mass transfer coefficient in activated sludge and carbon reactors (A$_d$/A$_r$=0.36)

The results in the two active sludge and carbon reactors showed a greater K$_L$a difference between the external loop activated carbon airlift reactor and the internal airlift

reactor compared to their difference in the active sludge airlift reactors. In other words, the effect of A$_d$/A$_r$on mass transfer in activated carbon airlift reactors was more than activated sludge reactors; the reason could be the different properties of the fluids.

4. Conclusions

The aim of this study was to examine and compare the effect of the superficial velocity of gas on the hydrodynamics and mass transfer of activated sludge and carbon in internal and external airlift reactors. An increase in the superficial velocity of gas resulted in greater gas hold-up, liquid circulation velocity and volumetric mass transfer coefficient. Meanwhile, increasing the concentration of activated sludge and carbon particulates reduced the hydrodynamics and K$_L$a due to the generation of larger bubbles and a reduction of gas-liquid interface. The results showed that the external activated carbon and sludge airlift reactors performed better than the internal airlift reactors. Furthermore, it was found that the activated sludge reactors had non-Newtonian behavior while the activated carbon reactors had Newtonian behavior.

Nomenclature

U_G:	Gas superficial velocity(m/s)
R_G:	Air flow rate(m^3/s)
A_G:	Sparger cross-sectional area(m^2)
ε_g:	Overall gas holdup
H_L:	Non-aerated liquid column height(m)
H_{LG}:	Aerated liquid column height(m)
U_{LD}:	Liquid velocity in downcomer(m/s)
U_{LR}:	Liquid velocity in riser(m/s)
t_d:	Time(s)
L_D:	Specified distance in downcomer(m)
A_r:	Riser Cross sectional area(m^2)
A_d:	Downcomer cross sectional area(m^2)
$C_L(t)$:	Oxygen concentration in bulk solution at time
C_L^*:	oxygen concentration at steady state (mg/l)
K_La:	Volumetric mass transfer coefficient$(1/s)$

References

[1] Bentifraouine, C., Xuereb, C., Riba, J. P. (1997). An experimental study of the hydrodynamic characteristics of external loop airlift contactors. *Journal of chemical technology and biotechnology, 69*(3), 345-349.

[2] García-Calvo, E., Letón, P. (1996). Prediction of gas hold-up and liquid velocity in airlift reactors using two-phase flow friction coefficients. *Journal of chemical technology and biotechnology, 67*(4), 388-396.

[3] Bakker, W. A. M., Van Can, H. J. L., Tramper, J., De Gooijer, C. D. (1993). Hydrodynamics and mixing in a multiple air-lift loop reactor. *Biotechnology and bioengineering, 42*(8), 994-1001.

[4] Muroyama, K., Mitani, Y., Yasunishi, A. (1985). Hydrodynamic characteristics and gas-liquid mass transfer in a draft tube slurry reactor. *Chemical engineering communications, 34*(1-6), 87-98.

[5] Merchuk, J. C., Gluz, M. (2002). Bioreactors, Air-lift Reactors. Encyclopedia of Bioprocess Technology, Wiley Online Library.

[6] Jin, B., Leeuwen, J. H. V., Doelle, H. W., Yu, Q. (1999). The influence of geometry on hydrodynamic and mass transfer characteristics in an external airlift reactor for the cultivation of filamentous fungi. *World journal of microbiology and biotechnology, 15*(1), 73-79.

[7] Nikakhtari, H., Hill, G. A. (2005). Hydrodynamic and oxygen mass transfer in an external loop airlift bioreactor with a packed bed. *Biochemical engineering journal, 27*(2), 138-145.

[8] Jin, B., Yin, P., Lant, P. (2006). Hydrodynamics and mass transfer coefficient in three-phase air-lift reactors containing activated sludge. *Chemical engineering and processing: Process intensification, 45*(7), 608-617.

[9] Freitas, C., Teixeira, J. A. (2001). Oxygen mass transfer in a high solids loading three-phase internal-loop airlift reactor. *Chemical engineering journal, 84*(1), 57-61.

[10] Yang, F., Bick, A., Shandalov, S., Brenner, A., Oron, G. (2009). Yield stress and rheological characteristics of activated sludge in an airlift membrane bioreactor. *Journal of membrane science, 334*(1), 83-90.

[11] Al Taweel, A. M., Idhbeaa, A. O., Ghanem, A. (2013). Effect of electrolytes on interphase mass transfer in microbubble-sparged airlift reactors. *Chemical engineering science, 100*, 474-485.

[12] Kilonzo, P. M., Margaritis, A., Bergougnou, M. A., Yu, J., Ye, Q. (2007). Effects of geometrical design on hydrodynamic and mass transfer characteristics of a rectangular-column airlift bioreactor. *Biochemical engineering journal, 34*(3), 279-288.

[13] Krichnavaruk, S., Pavasant, P. (2002). Analysis of gas–liquid mass transfer in an airlift contactor with perforated plates. *Chemical engineering journal, 89*(1), 203-211.

[14] Lu, W. J., Hwang, S. J., Chang, C. M. (1995). Liquid velocity and gas holdup in three-phase internal loop airlift reactors with low-density particles. *Chemical engineering science, 50*(8), 1301-1310.

[15] Kadic, E. (2010). Survey of gas-liquid mass transfer in bioreactors.

[16] Clarke, K. G., Correia, L. D. C. (2008). Oxygen transfer in hydrocarbon–aqueous dispersions and its applicability to alkane bioprocesses: A review. *Biochemical engineering journal, 39*(3), 405-429.

[17] Benyahia, F., Jones, L. (1997). Scale effects on hydrodynamic and mass transfer characteristics of external loop airlift reactors. *Journal of chemical technology and biotechnology, 69*(3), 301-308.

[18] Bai, F., Wang, L., Huang, H., Xu, J., Caesar, J., Ridgway, D., Moo-Young, M. (2001). Oxygen mass-transfer performance of low viscosity gas-liquid-solid system in a split-cylinder airlift bioreactor. *Biotechnology letters, 23*(14), 1109-1113.

15

Desorption of reactive red 198 from activated Carbon prepared from walnut shells: Effects of temperature, Sodium carbonate concentration and organic solvent dose

Zohreh Alimohammadi, Habibollah Younesi*, Nader Bahramifar
Department of Environmental Science, Faculty of Natural Resources, Tarbiat Modares University, Noor, Iran

ABSTRACT

This study investigated the effect of temperature, different concentrations of sodium carbonate,and the dose of organic solvent on the desorption of Reactive Red 198 dye from dye-saturated activated carbon using batch and continuous systems. The results of the batch desorption test showed 60% acetone in water as the optimum amount. However, when the concentration of sodium carbonate was raised, the dye desorption percentage increased from 26% to 42% due to economic considerations; 15 mg/L of sodium carbonate was selected to continue the process of desorption. Increasing the desorption temperature can improve the dye desorption efficiency. According to the column test results, dye desorption concentration decreased gradually with the passing of time. The column test results showed that desorption efficiency and the percentage of dye adsorbed decreased; however, it seemed to stabilize after three repeated adsorption/desorption cycles. The repeated adsorption–desorption column tests (3 cycles) showed that the activated carbon which was prepared from walnut shell was a suitable and economical adsorbent for dye removal.

Keywords:
Activated carbon
Batch system
Continuous system
Desorption
Dye

1. Introduction

Dye wastewater discharged from textile and dyestuff industries are one of the most significant sources of pollutants in the environment [1]. Reactive dyes are extensively used in textile industries because of their exclusive properties which include bright color and low energy consumption [2]. It is estimated that around 40% of consumed reactive dyes are discharged into the wastewater of dyeing operations [3]. Reactive dyes are soluble in water, so removing them by flocculation and biodegradation is very difficult. The presence of dye in water can cause many problems including high biochemical oxygen demand (BOD), chemical oxygen demand (COD) and an increase in suspended solids [4]; thus, effective treatment methods such as flocculation, membrane separation processes, oxidation, electrolysis and adsorption are needed to remove them from wastewaters [5]. Among the mentioned

methods, adsorption by adsorbent is one of the most effective methods for removing dye molecules. A successful sorbent material should have a high affinity and capacity for adsorbate molecules as well as regeneration potential. Also, it should be applicable for the majority of reactive dyes and economically affordable [6]. Activated carbon (AC) is an appropriate adsorbent for dye adsorption because of its large surface area, appropriate porosity and hydrophobic properties [7]. ACs can be prepared from a wide range of low-cost carbonaceous material such as agricultural waste. Two processes are commonly used for the preparation of ACs: physical activation that includes material carbonization and chemical activation in which the raw material is impregnated with activating agents such as $ZnCl_2$, H_3PO_4, KOH, K_2CO_3 [8]. In this research, the AC was prepared from walnut shells by chemical activation with KOH (AC –K1). The results of the adsorption of RR198 by activated carbon developed from walnut shells were presented in our

*Corresponding author
E-mail address: hunesi@modares.ac.ir, hunesi@yahoo.com

published report [9]. The results showed that AC –K1 with a surface area of 1439 m²/g has an increased ability to remove RR198 from wastewater. A desorption study can help us to explain the nature of the adsorption and recycling of the dye saturated adsorbent [10]. However, desorption of the AC saturated with dye not only can regenerate the adsorbent but can also return the dye back into the dyeing process. Then, desorption of the dye from AC saturated with dye becomes economically affordable [11]. The objective of this study was to investigate the effect of organic solvent, sodium carbonate, and temperature on desorption of reactive red 198 dye (RR198) from AC made from walnut shells in batch and column systems.

2. Materials and methods

2.1. Materials

Walnut shells collected from a fruit grower in the north of Iran were used as raw material for the preparation of the AC. This agricultural waste was washed with distilled water to remove dust and other inorganic impurities, then it was dried in an oven at 110°C to reduce its moisture content; finally, it was crushed and sieved to a particle size of 0.4–0.8 mm. The reactive red 198 ($C_{27}H_{18}C_lN_7Na_4O_{15}S_5$, molecular weight of 967.5 g/mol) was purchased from the local market. The RR198 characteristic and structure is illustrated in Table 1 and Figure 1. The potassium hydroxide (KOH) was purchased from Scharlo, Spain. The sodium hydroxide (NaOH), hydrochloric acid of 37% (HCl), sodium carbonate (Na_2CO_3) and acetone of 99.5% were purchased from Merck, Germany.

Table.1. Characteristics of Reactive Red 198

Characteristic	Reactive Red 198 (RR-198)
Molecular formula	$C_{27}H_{18}C_lN_7Na_4O_{15}S_5$
Color index name	Reactive Red- 198
λ_{max}	520 nm
Class	Monoazo (-N=N-bond)

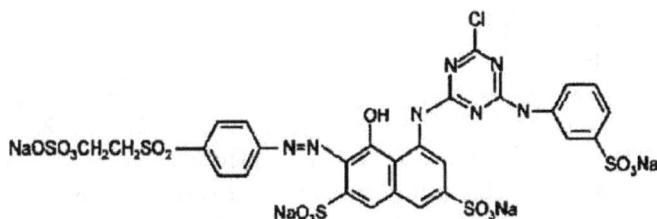

Fig. 1. Structure of reactive red dye

2.2. Preparation of Activated Carbon

Activated carbon was prepared by mixing a certain weight of walnut shell with impregnation ratios of KOH: walnut shell 1 g/g. The impregnated walnut shell was dried at 110°C for 24 hours and pyrolyzed under an argon atmosphere in a

furnace at a temperature of 900°C at an increment rate of 10°C/min for a 1 hour retention time. Then, the samples were removed from the reactor and washed with hot and cold distilled water until the water pH reached about 6-7 in order to remove impurities of the synthesized activated carbon. Finally, the washed samples were dried in an oven at 110°C, weighed and kept in a capped bottle for subsequent use. The sample was named AC –K₁.

2.3. Batch desorption tests

Before the desorption tests, 0.1 g of AC –K1 was added to 250 mL flasks containing 100 mL of RR198 wastewater with a 40 mg/L concentration at a pH=3 and a temperature of 50°C. The flasks were shaken at 120 rpm for 60 min until adsorption reached equilibrium. After filtration, a spectrophotometer device was used (at 520 nm wavelength) to determine the residual dye concentrations in the solutions. The separated activated carbon was added to 100 mL flasks filled with 20 mL of the regenerate with desired concentrations; then the flasks were shaken at 120 rpm for 40 min. After filtration, the dye concentrations in the solutions were measured. Desorption experiments were performed for different temperatures (20, 30, 40, 50 and 60 °C) and certain concentrations of sodium carbonate (5 to 20 mg/L) and organic solvent (30%-80% acetone in water). The desorption efficiency (DE) was determined with the following equation:

$$DE\% = \frac{A_0 - A_d}{A_0} \times 100 \qquad (1)$$

where A_0 and A_d are the amount of dye adsorption and residual dye concentrations in the solutions after desorption, respectively.

2.4. Column tests

Continuous adsorption/desorption experiments were carried out in a double glazed glass column with a 2 cm inner diameter and 20 cm height. The RR198 solution of a certain concentration (40 mg/L) was pumped into a column filled with AC –K1 with a height of 2 cm at a constant flow rate (1 mL/min) using a peristaltic pump (Pump drive PD 5101 Heidolph peristaltic, Schwabach, Germany) until the column was saturated with the dye solution. Also, the top and bottom of the adsorbent was filled with glass wool. Adsorption experiments were carried out at a temperature of 50°C and a pH = 3. Then the saturated column was regenerated with the optimal regenerant solution with 60% acetone in water and 15mg/L sodium carbonate. The column temperature during the tests was 50 °C. After the regeneration of each column, hot water was used to wash the carbon in the column in order to remove the regenerant solution before the next adsorption cycle. The regeneration cycles were repeated three times.

3. Results and discussion

3.1. Dye desorption in batch experiments

3.1.1. Effect of organic solvent dose

The complex structure of RR198 dye makes its removal difficult by organic solvent. Due to the effectiveness of acetone on desorption of activated carbon, acetone was selected for desorption of the dye-saturated activated carbon. Since the dilution of pure solvents with water can increase the desorption efficiency, the effect of organic solvent concentration was investigated in the range of 30%-80% (volume percentage) acetone in water. As shown in Figure 2, the most effective solvent dose for desorption of the RR198 from the activated carbon was 60% acetone in water. Desorption efficiency decreased when a high percentage of acetone in water was applied for the process. This trend may be due to reducing the ability of activated carbon in the low percentage of water condition. Similar observations have been reported for the effect of organic solvents on desorption of red dye on activated carbon in which the maximum desorption efficiency percentage was observed at 60% acetone in water [12].

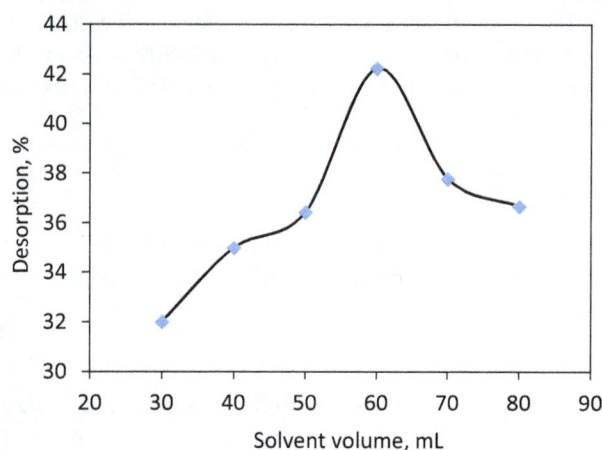

Fig. 2. Effect of solvent dose on the desorption efficiencies of the red dye at 25 °C

3.1.2. Effect of sodium carbonate concentration

The selection of an appropriate base for desorption of the dye from activated carbon was made by considering its dependency with the dye structure. According to this study, we found that pH has a great influence on the dye adsorption onto activated carbon and the High removal percentage is observed at acidic pH condition [9]. The reduced pH level in the acid condition caused an increase of H^+ ion concentration in the solution and therefore, the surface of the adsorbent achieved a positive charge and tended to attract the negatively charged groups of the RR198 structure. Hence, bases appeared to be appropriate solutions for desorption of the RR198 from the saturated carbon activate. The addition of sodium carbonate to aqueous acetone solutions greatly improved the dye

desorption performance. The effect of different sodium carbonate concentrations on desorption in 60% acetone is shown in Figure 3. It can be observed from Figure 3 that the dyes desorption percentage increased from 26% to 42% when the sodium carbonate concentration was raised from 5 to 20 mg/L. This phenomenon can be attributed to the presence of _SO_3Na functional groups in RR198 in which a high concentration of sodium carbonate caused an increase negative charge on the surface of the adsorbent. This was beneficial for desorption of anionic dye molecules. Similar trends have been reported for the effect of NaOH on desorption of reactive red dye on activated carbon [13]. Also, similar observations were obtained for desorption of anionic reactive dyes (Cibacron blue 3GA and Cibacron red 3BA) by mixtures of NaOH in a methanol solution [14].

3.1.3. Effect of temperature

In order to increase the desorption efficiency, batch experiments were performed at 20, 30, 40, 50 and 60°C using 60% acetone in a water solvent and 15 mg/L of sodium carbonate. Since the adsorption of RR198 onto AC is an endothermic process, it is expected that dye desorption falls when temperature increases. But according to Figure 4, the dye desorption percentage increased with an increase in temperature and thus, the maximum dye desorption percentage was observed at 60°C. This trend can be related to the increase of acetone solubility in water with the rising of temperature that led to an increase in desorption efficiency. Given that there is not much difference between the dye percentage at 60 and 50°C because of economic consideration, 50°C was selected as the optimal temperature.

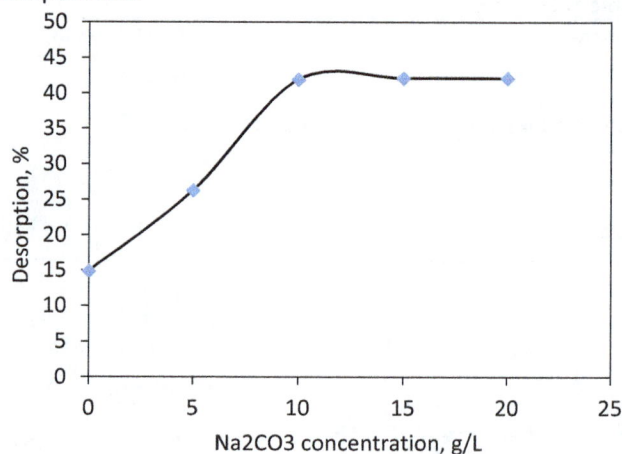

Fig. 3. Effect of sodium carbonate concentration on the desorption efficiencies of the red dye (25 °C, 60% aceton)

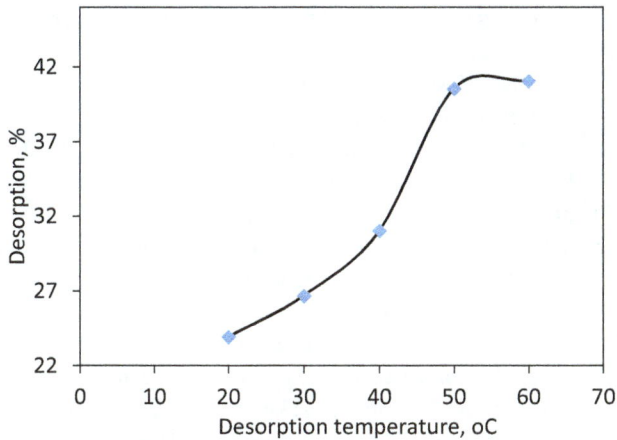

Fig. 4. Effect of temperature on the desorption efficiencies of the red dye (15 mg/L sodium carbonate, 60% aceton)

3.2. Column experiments

Based on the batch test results, the optimum conditions for desorption of dye-saturated activated carbon were found to be 60% acetone in water, 15 mg/L of sodium carbonate and 50°C. As seen in Figure 5, the maximum concentration of desorbed dye (more than 32 mg/L) occurred in the first five minutes; then it gradually decreased as time passed and the total desorption time was 75 minutes. Figure 6 illustrates the regeneration efficiency of red dye for the repeated adsorption/desorption cycles. Figure 6 indicates that the regeneration efficiency decreased from 44.99% to 28.52% and the amount of dye adsorbed decreased from 71.3% to 66.5% after the three repeated adsorption–desorption cycles but after three cycles, adsorbent Has low efficiency in dye adsorption. Additional repeating of the adsorption–desorption column tests with the optimal regenerant showed the ability of AC −K1 to efficiently adsorb and desorb the dye molecules and AC −K1 can be used repeatedly for the removal of dye from waste water. Similar trends have been reported for the desorption of Pb(II) ions on activated tea waste [15].

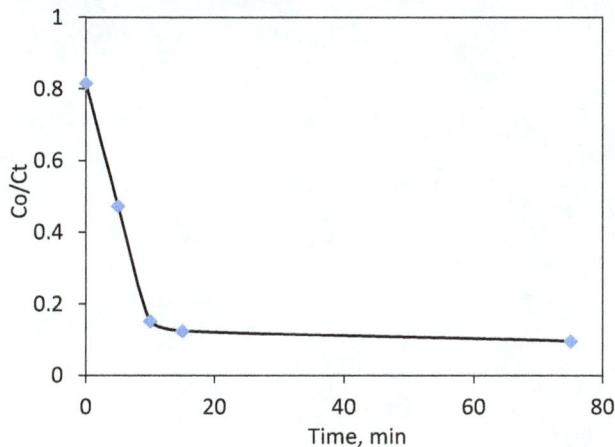

Fig. 5. Desorption curve of column operation (60% acetone, 15 mg/l sodium carbonate, 50 °C)

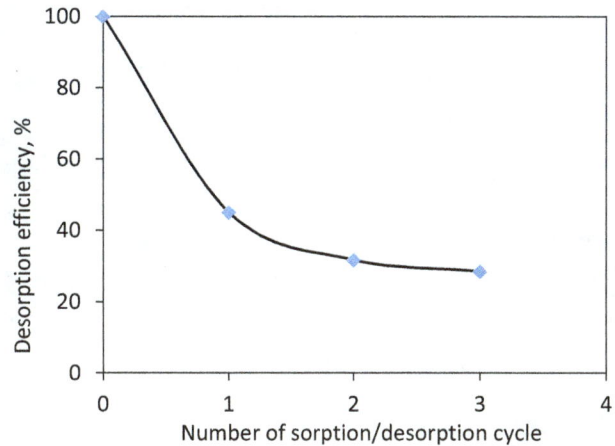

Fig. 6. Adsorption–desorption cycles (60% acetone, 15 mg/l sodium carbonate, 50 °C)

4. Conclusions

The activated carbon was prepared from walnut shell, as a low-cost raw material, using the activating agent KOH. The desorption of RR198 dye from the AC was performed in batch and continuous systems. The batch desorption test results showed that 60% acetone in water and 15 mg/L of sodium carbonate were the optimal regenerants for the red dye. An increase in desorption temperature increased desorption efficiency. The results of the column desorption test demonstrated that the concentration of dye desorbed from the activated carbon decreased as time passed. The percentage of desorption efficiency was decreased after three repeated adsorption/desorption cycles using the optimal regenerant.

Nomenclature

Acronyms	
Obs	Observed values of growth
Mean obs	Mean of observed values
Pred	Predicted values of growth
n	Number of samples
Notations	
A	Asymptotic ln X_t/X_0 as t decreases indefinitely −
A(t)	Precise integral of the adjustment factor h
B	Relative growth rate at time M h^{-1}
C	Asymptotic ln X_t/X_0 as t increases indefinitely −
K	Time at which half maximum growth is achieved h
M	Time at which absolute growth rate is at its maximum h
T	Residence time h

X_0	Initial cell concentration mg L^{-1}
X_{max}	Maximum cell concentration mg L^{-1}
X_t	Cell concentration at time t mg L^{-1}
Y	$\ln(X_t/X_0)$ –
Greek symbols	
λ	Lag phase duration h
μ_{max}	Maximum-specific growth rate h^{-1}
ν	Shape or curvature parameter –

References

[1] Çolak, F., Atar, N., Olgun, A. (2009). Biosorption of acidic dyes from aqueous solution by Paenibacillus macerans: Kinetic, thermodynamic and equilibrium studies. *Chemical engineering journal, 150*(1), 122-130.

[2] Dizge, N., Aydiner, C., Demirbas, E., Kobya, M., Kara, S. (2008). Adsorption of reactive dyes from aqueous solutions by fly ash: Kinetic and equilibrium studies. *Journal of hazardous materials, 150*(3), 737-746.

[3] Santhy, K., Selvapathy, P. (2006). Removal of reactive dyes from wastewater by adsorption on coir pith activated carbon. *Bioresource technology, 97*(11), 1329-1336.

[4] Crini, G., Badot, P. M. (2008). Application of chitosan, a natural aminopolysaccharide, for dye removal from aqueous solutions by adsorption processes using batch studies: a review of recent literature. *Progress in polymer science, 33*(4), 399-447.

[5] Sulak, M. T., Demirbas, E., Kobya, M. (2007). Removal of Astrazon Yellow 7GL from aqueous solutions by adsorption onto wheat bran. *Bioresource technology, 98*(13), 2590-2598.

[6] Karcher, S., Kornmüller, A., Jekel, M. (2002). Anion exchange resins for removal of reactive dyes from textile wastewaters. *Water research, 36*(19), 4717-4724.

[7] Lillo-Ródenas, M. A., Cazorla-Amorós, D., Linares-Solano, A. (2005). Behaviour of activated carbons with different pore size distributions and surface oxygen groups for benzene and toluene adsorption at low concentrations. *Carbon, 43*(8), 1758-1767.

[8] Yagmur, E., Ozmak, M., Aktas, Z. (2008). A novel method for production of activated carbon from waste tea by chemical activation with microwave energy. *Fuel, 87*(15), 3278-3285.

[9] Alimohammadi, Z., Younesi, H., Bahramifar, N. (2016). Batch and Column Adsorption of reactive Red 198 from textile industry effluent by microporous activated carbon developed from walnut shells. *Waste and biomass valorization, 7*(5), 1255-1270.

[10] Salleh, M. A. M., Mahmoud, D. K., Karim, W. A. W. A., Idris, A. (2011). Cationic and anionic dye adsorption by agricultural solid wastes: a comprehensive review. *Desalination, 280*(1), 1-13.

[11] Auta, M., Hameed, B. H. (2011). Preparation of waste tea activated carbon using potassium acetate as an activating agent for adsorption of Acid Blue 25 dye. *Chemical engineering journal, 171*(2), 502-509.

[12] Lu, P. J., Lin, H. C., Yu, W. T., Chern, J. M. (2011). Chemical regeneration of activated carbon used for dye adsorption. *Journal of the Taiwan institute of chemical engineers, 42*(2), 305-311

[13] Cao, J. S., Lin, J. X., Fang, F., Zhang, M. T., Hu, Z. R. (2014). A new absorbent by modifying walnut shell for the removal of anionic dye: kinetic and thermodynamic studies. *Bioresource technology, 163*, 199-205.

[14] Liu, C. H., Wu, J. S., Chiu, H. C., Suen, S. Y., Chu, K. H. (2007). Removal of anionic reactive dyes from water using anion exchange membranes as adsorbers. *Water research, 41*(7), 1491-1500.

[15] Mondal, M. K. (2009). Removal of Pb (II) ions from aqueous solution using activated tea waste: Adsorption on a fixed-bed column. *Journal of environmental management, 90*(11), 3266-3271.

16

Risk assessment of hydrocarbon contaminant transport in vadose zone as it travels to groundwater table

Jamal Alikhani, Jalal Shayegan*, Ali Akbari

Department of Chemical and Petroleum Engineering, Sharif University of Technology, P.O. Box 11155-9465, Azadi Ave., Tehran, Iran

Keywords:

vadose zone

advection-diffusion-reaction equation

contaminant transport

risk assessment

numerical modeling

ABSTRACT

In this paper, a modeling tool for risk assessment analysis of the movement of hydrocarbon contaminants in the vadose zone and mass flux of contamination release into the groundwater table was developed. Also, advection-diffusion-reaction equations in combination with a three-phase equilibrium state between trapped air, soil humidity, and solid particles of unsaturated soil matrix were numerically solved to obtain a one dimensional concentration change in respect to depth of soil and total mass loading rate of hydrocarbons into the groundwater table. The developed model calibrations by means of sensitivity analysis and model validation via data from a site contaminated with BTEX were performed. Subsequently, the introduced model was applied on the collected hydrocarbon concentration data from a contaminated region of a gas refinery plant in Booshehr, Iran. Four different scenarios representing the role of different risk management policies and natural bio-degradation effects were defined to predict the future contaminant profile as well as the risk of the mass flux of contaminant components seeping into the groundwater table. The comparison between different scenarios showed that bio-degradation plays an important role in the contaminant attenuation rate; where in the scenarios including bio-degradation, the contaminant flux into the ground water table lasted for 50 years with the maximum release rate of around 20 gr per year while in the scenarios without including bio-degradation, 300 years of contaminant release into groundwater table with the maximum rate of 100 gr per year is obtained. Risk assessment analysis strongly suggests a need for bioremediation enhancement in the contaminated zones to reduce the contaminant influx to groundwater.

1. Introduction

Groundwater is located beneath the ground surface in soil pore spaces and in the fractures of rock formations. Pollutants released into the soil work their way down into the groundwater, polluting it and triggering certain problems. These underground contaminants, one of the most common of these problems, have beset the use of groundwater worldwide. Therefore, any risk assessment tools could play an essential role in identifying which sites are more likely to infect the groundwater [1]. Hydrocarbon contaminants that permeate to the lower levels, transit from the vadose zone to reach the groundwater table. In their pathway, it is important to consider soil types and pollutant properties in the model, which contain an immense range of properties [2, 3]. Volatile organic compounds (VOCs) are mostly non-aqeous phase liquids (NAPLs) and have leaked randomly into the vadose zone from different sources, i.e., in this study from the underground facilities of a gas refinery plant. NAPL contaminants from various sources which percolate in the vadose zone can remain there for decades and act as a long term contaminated source [4, 5] presented one of the first analytical models for contaminant transport in the vadose zone and it was later developed by [6]. In addition,

*Corresponding author

E-mail address: shayegan@sharif.edu

many extensions of their assumptions as well as some new assumptions have been provided by other researchers [7]; [8, 4] evaluated analytical and numerical mathematical models for the transport of VOCs in the unsaturated zone, especially those that have been recently developed, and provided a comparison of the results. They concluded that these models were mostly used for NAPL sources rather than VOCs. They also mentioned the lack of laboratory and field data for validation. Published one or three dimensions numerical models and commercial codes have also been evaluated by [9] and the results indicated that the most of these studies investigated pollution transport in the unsaturated zone in the aqueous phase only. The three dimensional models need more input data; nevertheless, [3] concluded that in some common states a one dimensional (1D) model gives an acceptable result, especially when the contaminated area is shallow or large and has high water contents or a high biodegradation constant. However, the input parameters of a one dimensional model are less than that of a three dimensional model. Even though the assumptions used in a 1D model overestimate the risk, this aspect is not considered as a negative point. [9] also mentioned that in the majority of these models, the contaminant transport modeling was independent of each of the other compounds when there was a mixture of compounds as contaminants. Contaminants located in the voids within the soil, in addition to being dissolved in the pore water, are adsorbed into the soil particles [5, 7] or replaces the air trapped in the soil [10] and the degradation process can occur in all phases [11, 12]. A 1D contaminant transport model in the vadose zone that takes into account the effect of immobile water content has been investigated by [13]. They solved a dual porosity model whereby moisture is split into two parts, θ_m and θ_{im}, indicating mobile and immobile soil moisture, respectively. Finally, they were unable to achieve a clear result regarding whether immobile humidity should be considered or ignored. Their findings only noted that the immobile moisture affects the contaminant transport. A study of one dimensional hydraulic transport of contaminants in the vadose zone in both cases of steady state and transient flow were examined by [14]. They introduced the dimensionless Damköhler number for sorption and degradation kinetics for characterization and comparison of soil contamination profile. They concluded that a steady state flow conditions along with assuming a constant Darcy velocity in the soil does not cause significant problems, with the exception of special cases such as extreme infiltration rate. In this study, a one dimensional model of the hydrocarbon contaminant's movement in the vadose zone is described. This model was used in four different scenarios representing different risk management policies over the measured soil hydrocarbon content of a gas refinery plant. To obtain the field data of the hydrocarbon contaminant

concentration in the unsaturated zone, a soil sampling and laboratory analysis of the samples were conducted.

2. Materials and Methods

The field sampling and analyzing was done in a gas refinery plant located in the SPGC area, Booshehr, Iran. Different locations of the plant were determined to be contaminated by the release of hydrocarbon contaminants from underground pipes. To begin with, a map of the suspected locations was prepared. The contaminated areas were dug and those with a higher contamination level were drilled with an excavation machine. During the drilling, soil samples were taken from various depths (roughly at one meter intervals) and were kept in a sealed container and placed in a cold box; they were then transferred to the refrigerator to be kept until the lab tests could be performed. The soil samples were refluxed with methylene chloride for 30 minutes. The extracted solution was then injected to the gas chromatography–mass spectrometry (GC-MS) to determine the components spectrum of the contaminated soil (Agilent Technologies 6890N GC- Agilent Mass selective detector 5973N). A HP-5MS column, 30 m in length and 0.25 mm in diameter, was used. The instrument condition was: a 250 °C injection temperature, 100 °C for 2 min; then the temperature was increased by 10 °C/min for 20 min. Helium was used as a carrier gas in 1 ml/min and run in a split-less mode. A mass detector was run at 230 °C in an ion source and at 150 °C for the mass analyzer. The total petroleum hydrocarbon (TPH) analysis was also performed on the collected sample soil according to US EPA Methods 413.2 and 418.1 and ASTM Method D3921. The TPH was determined by an IR at 2940 cm^{-1} (InfraCal TOG/TPH Analyzer -Wilkins Enterprise). Most of the compounds found in the contaminated soil were hydrocarbons that have a boiling temperature between 50-250 °C. According to the European Union definition, they are considered VOCs [4].

3. Model Description

In this section, a one dimensional vertical transport contaminant is described in three phases. The contaminant movement within and between the liquid, gas, and solid phase is considered as: (1) a solute dissolved in water, (2) a gas in the vapor phase, and (3) as an adsorbed compound in the solid phase. A conceptual sketch of the most effective mechanisms that are considered in the simulation is shown in Fig. 1. The model is based on the following assumption sets:

(i)   Soil is homogenous with constant properties,

(ii)   Steady-state flow pattern is considered; by means of Darcy velocity is constant with time $(\frac{dq}{dt} = 0)$,

(iii)   Linear, reversible, and isotherm equilibrium partitioning is held between phases,

(iv)   Non-aqueous phase liquid transport is not considered,

(v)   Advection and degradation are considered in the liquid phase, diffusion in the gas phase and adsorption in the solid phase, and,

(vi)   It is assumed that the non-aqueous phase is available in the immobile parts of the soil. This means that only the aqueous phase is transported and NAPL is not moved by gravity or other driving forces such as leakage of a new amount of pollutant infiltration. This assumption is more reasonable in the low NAPL concentration.

3.1. Governing Equations

Based on the available data and the aim of our study, a one dimensional model was used to predict a contaminant concentrations profile and groundwater infiltration. In this model, only the vertical transport of dissolved contaminants is modeled in three phases of gas, liquid, and solid which are indicated with C_g, C_l and C_s, respectively. The concentration in the liquid phase is obtained by solving the following equation that accounts for advection and considering first-order degradation:

$$\frac{\partial C_l}{\partial t} = -v\frac{\partial C_l}{\partial z} - \lambda C_l \tag{1}$$

where, C_l (M L^{-3}) shows the solute concentration in the water phase, λ (T^{-1}) is the first-order degradation rate, and v (L T^{-1}) is the seepage velocity that is given by:

$$v = \frac{q}{\theta_w} \tag{2}$$

where, q (L T^{-1}) stands for the water infiltration rate and θ_w (-) is the volumetric water content [15, 8, 3].

The gas phase concentration of contaminant is determined by the following diffusion equation:

$$\frac{\partial C_g}{\partial t} = D_P\frac{\partial^2 C_g}{\partial z^2} \tag{3}$$

where, C_g (M L^{-3}) is the contaminant concentration in the gas phase and D_p is the gaseous diffusion coefficient in the air-filled pore spaces as given by:

$$D_P = D_{air}\frac{\theta_a^{7/3}}{\emptyset^2} \tag{4}$$

where, D_{air} (L^2 T^{-1}) is the diffusion coefficient in the air, θ_a (-) is the volumetric air content, and $\emptyset$ (-) is the porosity. Eq. (4) is an expression of the Millington Equation [16] for the gaseous diffusion coefficient in the air-filled pore spaces [17, 15, 8].

The total mass of contaminant in the soil mass M_T (M M^{-1}) then can be expressed as:

$$M_T = [\theta_w C_l + \theta_a C_g + \rho_b C_s] \tag{5}$$

where, C_s (M M^{-1}) is the contaminant concentration in the solid phase [15]. Equilibrium partitioning between the concentrations of different phases can be obtained as:

$$C_1 = \frac{M_T}{R'} \tag{6}$$

$$C_g = K_H C_l \tag{7}$$

$$C_s = \frac{K_d M_T}{R'} \tag{8}$$

Where:

$$R' = R\theta_w + K_H\theta_a \tag{9}$$

$$R = 1 + \frac{\rho_b}{\theta_w}K_d \tag{10}$$

$$K_d = K_{oc}f_{oc} \tag{11}$$

In the above equations, R (-) is the retardation coefficient, K_d (L^3 M^{-1}) is the distribution coefficient, K_H (-) is the dimensionless Henry's constant, ρ_b (M L^{-3}) is the bulk density, K_{oc} (L^3 M^{-1}) is the partition coefficient between the contaminant and natural organic matter, and f_{oc} (-) is the fraction organic carbon content of the soil [15, 3].

3.2. Solution Procedure

The partial differential equations (1) and (3) are solved by the Crank-Nicholson method in the finite difference approach. A zero concentration gradient for gas concentration was assumed for the lower boundary condition at the groundwater table. Also, contaminant concentration in the atmosphere above the soil surface was assumed to be zero. According to Eq. (6), if C_l in each time and location is higher than the maximum capacity of water solubility ($C_{l\_soluble}$) then C_l was made equal to $C_{l\_soluble}$ to calculate C_g. Mass loading of contamination $\dot{m}_{loading}$ (M T^{-1}) to groundwater was determined according to the following equation:

$$\dot{m}_{loading} = C_l|_{@ \text{ groundwater table}} \cdot A \cdot q \tag{12}$$

Where $C_l|_{@ \text{ groundwater table}}$ is the water phase concentration at the groundwater table, A (L^2) is the contaminated zone area, and q (L T^{-1}) is the infiltration rate. Where $C_l|_{@ \text{ groundwater table}}$ is the water phase concentration at the groundwater table, A (L^2) is the contaminated zone area, and q (L T^{-1}) is the infiltration rate.

4. Model Application

4.1 Model Validation

The model was validated using the BTEX released data by [3]. In their case study, free phase fuel oil had been leaked from a fuel oil tank for an unknown period. The pipe was 0.5 m below the ground surface and the contaminant source was estimated to have an area of 5×5 m², with residual phase contamination extending to a depth of 3 m below ground surface. The source zone was considered immobile. The contaminant source contained approximately 320 kg fuel oil. The soil samples of the source zone contained approximately 0.1% BTEX (Benzene, Toluene, Ethylbenzene, and Xylenes). The mass fraction of BTEX in the pore gas samples was reported to be around 1–3%; the average source zone concentration in the gas phase was estimated to be 11.25 mg/m³ corresponding to a water phase concentration of 75 µg/l. The properties of benzene as a toxic component in the leaked fuel oil were used in the model. The properties of the contaminated site soil and the contaminant's components are tabulated in Table 1. Fig. 2 shows the distribution of the contaminant as a function of vertical distance from the source to the water table. The distribution of the contaminant is calculated at a

steady state condition and was compared to the field measured data. An acceptable agreement between measured and modeled data was observed.

4.2. Model Sensitivity Analysis

In more general terms, uncertainty and sensitivity analysis investigate the robustness of a study when the study includes some form of computational modeling. Predictions are highly dependent on the quantity and quality of space and time data. In this case a sensitivity analysis study was performed to evaluate the impact of various input parameters on soil contaminant level and loading to groundwater. A qualitative description of each parameter's sensitivity to the calculated groundwater impact and soil concentration profile is exhibited in the Tables 2 and 3.

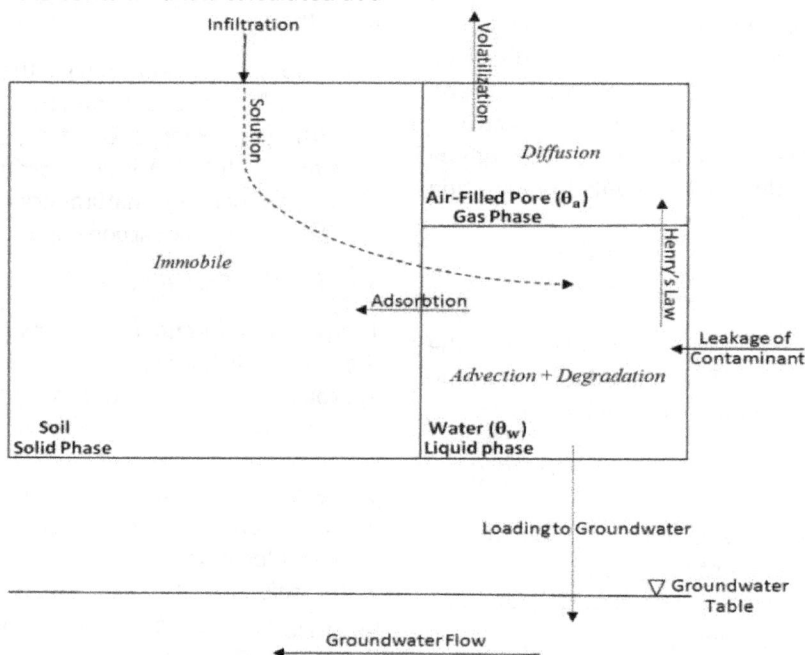

Fig. 1. Conceptual sketch of the most effective mechanisms considered in the simulation. In solid phase as an immobile phase: adsorption and solution adsorbed contaminant into the water pass through the soil, in liquid phase: advection, degradation and desorption into the gas phase and in outrance loading to groundwater, in gas phase: diffusion and volatilization to the atmosphere is considered.

Table 1. Model parameters data and chemical properties for benzene used for model validation [3]

q (mm/year)	Area (m²)	WS*1 (mg/lit)	$C_{w,source}$*2 $(\mu g/lit)$	H*3 (m)	λ (day⁻¹)	Time (year)
200	25	1732	75	18 [b]	0.01	200
f_{oc} (-)	θ_t (-)	θ_w (-)	ρ_b (g/ml)	D_{air} (m²/day)	K_H (-)	K_{oc} (ml/g)
0.02	0.45	0.15	1.7	0.76	0.15	22

*1: water solubility of benzene

*2: average source zone concentration in the water phase

*3: groundwater table below land surface

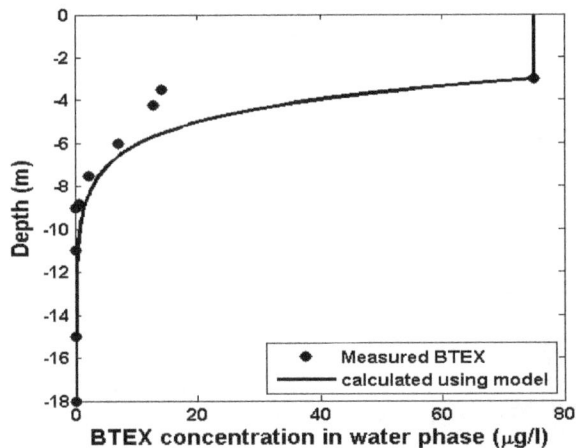

Fig. 2. The comparison of contaminant concentration in the unsaturated zone below the source is calculated using 1D numerical model (in solid line). The results are based on a constant source concentration in the water phase of 75 µg/l, which is the average of the measured concentrations within the source zone. Also shown are field measurements of total pore water BTEX concentrations below the source shown as (•).

4.3. Model Application in contaminated site

Fig. 3 shows the total contamination profile for one of the most contaminated drilled boreholes mainly due to the underground leakage. The current contamination level was used as an initial value in the model; then, prediction was made for spreading of the plume and loading to groundwater in the future years for four different scenarios. Not surprisingly, results obtained from GC-MS analysis showed that the soil contained numerous chemical compounds. Clearly, estimation of contaminant transport for each of these compounds was nearly impossible. To overcome to this difficulty, the contaminated soil was extracted by water and the total hydrocarbon content of solution was found 49.5 mg/lit. Consequently, all of the hydrocarbon compounds were considered as a single component with the above-mentioned solubility. This is helpful for a reliable risk assessment of plume and contaminant influx down to the groundwater. Soil properties such as bulk density, porosity, humidity and annual precipitation were taken from geotechnical studies performed in the area. The contaminant and soil properties are summarized in Table 4.

4.3.1. Different scenarios for predicting contaminant plume and loading to groundwater

On the way down to risk assessment goals, the model was applied in different cases by defining of different scenarios with two optimistic and pessimistic extremities based on the rate of bio-degradation and future rate of leakage. Comparing results obtained by different scenarios can be considered to make a suitable decision for site remediation and prevent further leakages. For each scenario the model was run for different upcoming years until the remained contaminant in the soil reached to the end. Four different scenarios are considered as follows:

4.3.1.1. First scenario-most optimistic state view

In this scenario it was assumed that there will be no contaminant leakage in the future, i.e., the source point behaves like a non-permanent source. Bio-degradation was also considered as an effective process. The bio-degradation rate for total hydrocarbons was assumed 0.01 day^{-1}. Results are shown for first scenario in Figs. 3 and 7.

4.3.1.2 Second Scenario- optimistic state

In this scenario it was assumed that there will be a continuous leakage of the contaminant for next 15 years (permanent source). Furthermore, in these 15 years it is assumed that there would be an increase of contaminant content by 5% each year in the higher part of the source, the first 6 meters of the source height. Similar to the first scenario, bio-degradation rate of 0.01 day^{-1} was considered as effective process in this scenario. Results are shown in Figs. 4 and 7.

4.3.1.3 Third scenario-pessimistic state

Unlike the first and second scenarios, in this scenario it was assumed that biodegradation is not significant. Similar to the first scenario, the leakage has already been stopped and the source point behaves like a non-permanent source. Results are shown for this scenario in Figs. 5 and 8.

Table 2. Parameter sensitivity of model for soil concentration profile.

Parameters	Area	f_{oc}	q	θ_t	ρ_b	θ_w	D_{air}	K_H	K_{oc}	λ
High		×	×						×	×
Moderate				×	×	×				
Low	ineffective						×	×		

Table 3. Parameter sensitivity of model for mass loading to groundwater.

Parameters	Area	f_{oc}	q	θ_t	ρ_b	θ_w	D_{air}	K_H	K_{oc}	λ
High	×	×	×						×	×
Moderate						×				
Low				×	×		×	×		

4.3.1.4 Fourth scenario-most pessimistic state

In this scenario, source condition for the next 15 years is similar to the second scenario, but bio-degradation is not significant. Results are shown in Figs. 6 and 8.

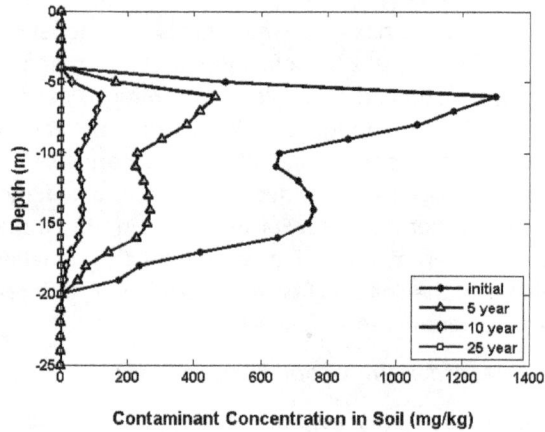

Fig. 3. Total soil contaminant profiles in different years-scenario 1. Also shown are profile of the total contamination in TOG (mg/Kg) in May 2010 located at phase 2&3 SPGC as an initial value.

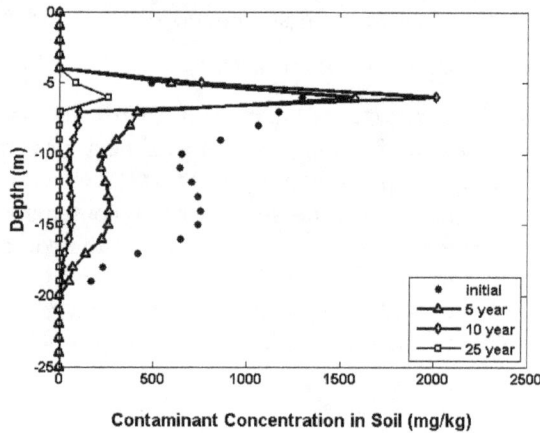

Fig. 4. Total soil contaminant profiles in different years-scenario 2.

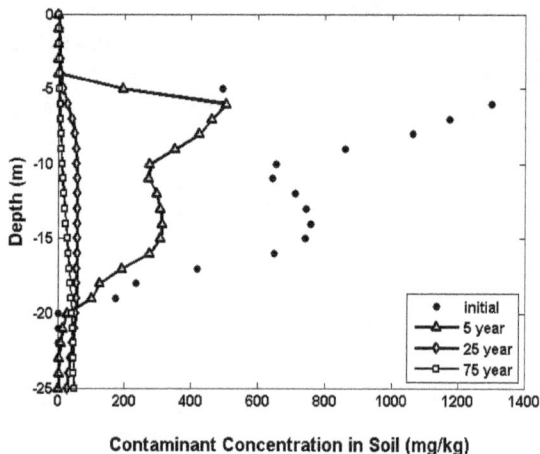

Fig. 5. Total soil contaminant profiles in different years-scenario 3.

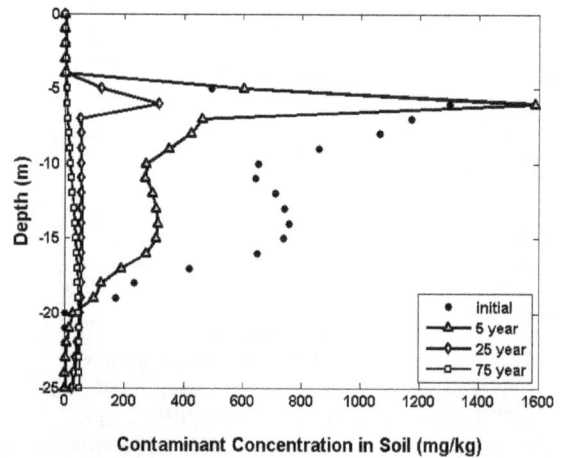

Fig. 6. Total soil contaminant profiles in different years-scenario 4.

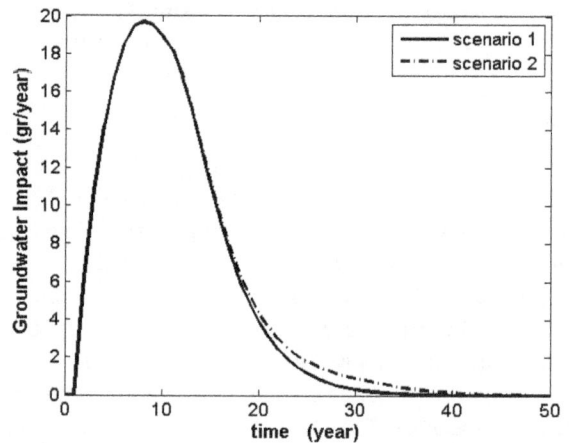

Fig. 7. Comparison of different mass loading to groundwater for scenarios 1 and 2.

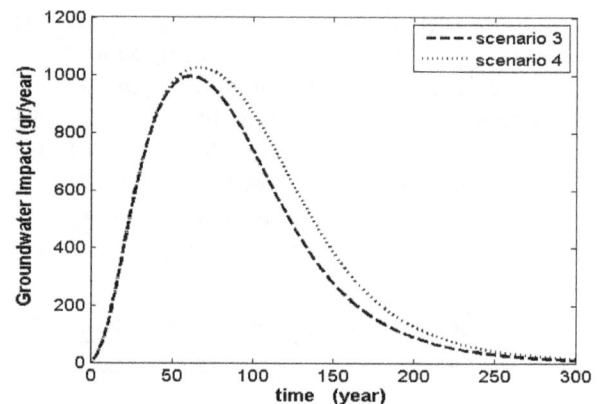

Fig. 8. Comparison of different mass loading to groundwater for scenarios 3 and 4.

5. Discussion

Numerous hydrocarbon compounds were identified in the GC-MS analysis. To avoid out of control complexity and to have a convenient model, all the hydrocarbons were considered as a lump component presented by TPH (mg/kg). Unlike the analytical solution models, it is possible to use the numerical model for heterogeneous soils. The model presented here is fortified in such a way that if the porosity profile, humidity, or infiltration of surface waters in different season is available, the model can predict the future situation. In the risk assessment of contaminated sites, according to the variation in the contaminant properties and polluted soil, the lack of accurate data used in modeling is considered normal. For this study, a reasonable range of each parameter respondent to local conditions and pollutants materials is considered [18]; and then, in this range, based on the sensitivity analysis, the minimum or maximum value is selected. The focus in this study was based on biodegradation. In the pre-defined scenarios, one of the most important factors considered in each scenario is the presence or absence of the bio-degradation mechanism as an effective process. First order bio-degradation is considered with the roughly low hydrocarbon decomposition rate of 0.01 day^{-1}. However, even this low rate bio-degradation indicates a very significant role in natural attenuation while the results of different scenarios are being compared. The results are shown in Figs. 3 and 4 for the first and second scenarios, respectively that included bio-degradation which shows faster attenuation in the contaminant profile in comparison with the results that are shown in Figs. 5 and 6 for third and fourth scenarios with no bio-degradation, respectively. Total period of mass flux of contaminant into groundwater table in Fig. 7 shows 40 and 50 years for the first and second scenarios respectively when the bio-

degradation process was considered as an active process; on the other hand for same conditions but without bio-degradation process the mass flux period of roughly 300 years as it is shown in Fig. 8 is obtained for the third and fourth scenarios showing that bio-degradation has a significant role on the contaminant attenuation. The contamination leakage may persist for the next 15 years (equivalent to the remaining effective life of the refinery) and 5% of the amount will be added each year; this is an idea that occurred in the second and fourth scenarios that represented poor environmental conservation policies in the plant. However in this case study, due to high contaminant leakage rate, additional contaminant leakage in upper layers has a less effective impact than bio-degradation on the soil concentration profile or the extent of contamination in the groundwater. This effect is more obvious when comparing Fig. 4 with Fig. 3 and Fig. 5 with Fig. 6. In the scenarios in which bio-degradation has been considered, the decomposition rate was assumed as a first order reaction. To be more accurate, a biological reaction rate follows the Monod kinetic model. However, when the substrate concentration is low, such as the concentration of this contaminant in the soil, the Monod equation can be approximated by first order reaction [19]. Using the presented model is plausible even when the biological processes are fortified. These kinds of fortification, especially done as in-situ processes, e.q., nutrient addition and humidity adjustment, increase the rate of degradation that can be considered in the model.

Table 4. Model parameters applied for contaminated site in our case study

q (m/year)	Area (m^2)	Aqs^{*2} (mg/l)	Δz (m)	H^{*1} (m)	Δt (year)	Time (year)
0.22[a]	100[b]	49.5	1	25[b]	1	variable
f_{oc} (-)	θ_t (-)	θ_w (-)	ρ_b (g/ml)	D_{air} (m^2/day)	K_H (-)	K_{oc} (ml/g)
0.003[d]	0.37[c]	0.23[c]	1.78[c]	0.8235[d]	0.10736[d]	163[d]

*1: groundwater table below land surface   *2: average solubility of pollutants in water
References: a: South Pars special economic zone database, b: site visit & field sampling, c: geotechnical studies performed in the area, d: [18]

6. Conclusions

In this study, a one-dimensional multiphase vertical model is presented with the aim of being applicable to risk assessment of contaminated sites. After sensitive analysis and validation, the model was applied for a contaminated zone of a gas refinery plant. Different scenarios were considered to evaluate the associated risks. Based on the simulation results the following conclusions were made:

1. In this particular contaminated zone, leakage of contaminant has occurred on the nearby water surface and the groundwater contamination will occur in the near future.

2. Considering different scenarios, maximum groundwater impact was calculated as nearly 1000 g/year for the pessimistic view with no bio-degradation in effect and 20 g/year for the optimistic view including bio-degradation, and the extant time of the contaminant in an optimistic view was 50 years and for the pessimistic view was around 250 years.

3. Bio-degradation rate is found as a critical process for the contaminant attenuation in the unsaturated zone and reduction of contaminant loading to groundwater.

4. Bio-degradation considerably decreases the spread of contaminant down to groundwater, so by enhancing the bio-degradation rate with bio-stimulation of the contaminated soil, the chance of groundwater contamination could be reduced.

Acknowledgements

The authors would like to acknowledge the financial support of the R&D division of South Pars Gas Complex Company.

References

[1] Cushman, D. J., Driver, K. S., Ball, S. D. (2001). Risk assessment for environmental contamination: an overview of the fundamentals and application of risk assessment at contaminated sites. *Canadian Journal of civil engineering, 28*(S1), 155-162.

[2] Falta, R. W., Javandel, I., Pruess, K., Witherspoon, P. A. (1989). Density-driven flow of gas in the unsaturated zone due to the evaporation of volatile organic compounds. *Water resources research, 25*(10), 2159-2169.

[3] Troldborg, M., Binning, P. J., Nielsen, S., Kjeldsen, P., Christensen, A. G. (2009). Unsaturated zone leaching models for assessing risk to groundwater of contaminated sites. *Journal of contaminant hydrology, 105*(1), 28-37.

[4] Rivett, M. O., Wealthall, G. P., Dearden, R. A., McAlary, T. A. (2011). Review of unsaturated-zone transport and attenuation of volatile organic compound (VOC) plumes leached from shallow source zones. *Journal of contaminant hydrology, 123*(3), 130-156.

[5] Jury, W. A., Spencer, W. F., Farmer, W. (1983). Behavior assessment model for trace organics in soil: I. Model description. *Journal of environmental quality, 12*(4), 558-564.

[6] Shoemaker, C. A., Culver, T. B., Lion, L. W., Peterson, M. G. (1990). Analytical models of the impact of two-phase sorption on subsurface transport of volatile chemicals. *Water resources research, 26*(4), 745-758.

[7] Mendoza, C. A., & Frind, E. O. (1990). Advective-dispersive transport of dense organic vapors in the unsaturated zone: 1. Model development. *Water resources research, 26*(3), 379-387.

[8] Shan, C., Stephens, D. B. (1995). An analytical solution for vertical transport of volatile chemicals in the vadose zone. *Journal of contaminant hydrology, 18*(4), 259-277.

[9] Karapanagioti, H. K., Gaganis, P., Burganos, V. N. (2003). Modeling attenuation of volatile organic mixtures in the unsaturated zone: codes and usage. *Environmental modelling & software, 18*(4), 329-337.

[10] Gioia, F., Murena, F., Santoro, A. (1998). Transient evaporation of multicomponent liquid mixtures of organic volatiles through a covering porous layer. *Journal of hazardous materials, 59*(2), 131-144.

[11] Aelion, C. M., Bradley, P. M. (1991). Aerobic biodegradation potential of subsurface microorganisms from a jet fuel-contaminated aquifer. *Applied and environmental microbiology, 57*(1), 57-63.

[12] Höhener, P., Duwig, C., Pasteris, G., Kaufmann, K., Dakhel, N., Harms, H. (2003). Biodegradation of petroleum hydrocarbon vapors: laboratory studies on rates and kinetics in unsaturated alluvial sand. *Journal of contaminant hydrology, 66*(1), 93-115.

[13] Kartha, S. A., Srivastava, R. (2008). Effect of immobile water content on contaminant transport in unsaturated zone. *Journal of hydro-environment research, 1*(3), 206-215.

[14] Kuntz, D., Grathwohl, P. (2009). Comparison of steady-state and transient flow conditions on reactive transport of contaminants in the vadose soil zone. *Journal of hydrology, 369*(3), 225-233.

[15] Ravi, V., & Johnson, J. A. A One-Dimensional Finite Difference Vadose Zone Leaching Model.

[16] Millington, R. J. (1959). Gas diffusion in porous media. *Science, 130*(3367), 100-102.

[17] Falta, R. W., Pruess, K., Javandel, I., Witherspoon, P. A. (1992). Numerical modeling of steam injection for the removal of nonaqueous phase liquids from the subsurface: 1. Numerical formulation. *Water resources research, 28*(2), 433-449.

[18] Mackay, D., Shiu, W. Y., Ma, K. C., Lee, S. C. (2006). *Handbook of physical-chemical properties and environmental fate for organic chemicals.* CRC press.

[19] Bekins, B. A., Warren, E., Godsy, E. M. (1998). A comparison of zero-order, first-order, and monod biotransformation models. *Groundwater, 36*(2), 261-268.

17

Biokinetic coefficients determination of a MBR for Styrene and Ethylbenzene as substrate based on the Andrews model

Seyed Mojtaba Seyedi[1], Hossein Hazrati[2,3,*], Jalal Shayegan[1]

[1] Department of Chemical and Petroleum Engineering, Sharif University of Technology, Tehran, Iran
[2] Department of Chemical Engineering, Sahand University of Technology, Tabriz, Iran
[3] Environmental Center of Sahand University of Technology, Tabriz, Iran

Keywords:
Biokinetic coefficient
Hydraulic retention time
Solid retention time
Volatile organic compounds

ABSTRACT

In this study, a lab-scale membrane bioreactor (MBR) was operated for a period of more than 10 months to determine the biokinetic coefficients of the system under the hydraulic retention times (HRT) of 20, 15 and 10 hrs and sludge retention times (SRT) of 5-20 days. The results revealed that the biological removal efficiency of styrene and ethylbenzene at a solid retention time of 20 day and a hydraulic retention time of 15 hr was higher compared to a SRT of 10 day and at the same HRT. The results also showed that the yield (Y), the endogenous decay coefficient (k_d), the maximum specific growth rate (μ_{max}), and the saturation constant (K_s) for styrene and ethylbenzene as substrate were 0.60 and 0.60 mg/mg, 0.25 and 0.25 day^{-1}, 0.188 and 0.363 h^{-1}, and 0.146 and 2.82 mg /l, respectively. Furthermore, ethylbenzene was more appropriate as a source of carbon to activated sludge in the membrane bioreactor than the styrene which had a lower μ_{max} than ethylbenzene.

1. Introduction

Many different chemical industries in Iran produce wastewater that contains volatile organic compounds (VOCs). These materials are man-made and/or naturally occurring highly reactive hydrocarbons [1]. These contaminants not only have destructive effects on the environment but also raise health concerns for workers. Wastewater treatment processes have been established to respond appropriately and relieve various anxieties concerning public health. These methods include physical techniques such as activated carbon adsorption [2], chemical procedures such as ozonation [3], and biological methods such as conventional activated sludge process [4], rotation biological contactor (RBC) processes [5], and stabilized biofilm [6]. Among the biological methods, conventional activated sludge has been employed in many industries as well as in the petrochemical industry. It was reported that biodegradation, stripping, and absorption are three main mechanisms that facilitate volatile organic

compound removal in the conventional activated sludge systems [1]. Due to low mixed liquor suspended solid (MLSS) concentration in conventional activated sludge plants (CASPs), they are one of the most significant VOC emission sources. A membrane bioreactor involves an activated sludge process in which the sedimentation unit is replaced with a membrane. The MBR process can achieve a higher MLSS concentration compared to CASP; therefore, these systems reduce VOC emissions. The main disadvantage of a MBR is the reduction of output flux due to the clogging of the membrane. Previous works show that membrane fouling is related to operating parameters such as hydraulic retention time, sludge retention time, and sludge specifications [6,7,8] HRT and SRT are the main parameters of biological processes such as CASP and MBR for wastewater treatment [9,10]. Further, defining the degradation kinetics of VOCs by bacterial populations is one of the principle steps to forecast and optimize the activated sludge processes on an industrial scale. Mathematical models have been developed to evaluate the

*Corresponding author
E-mail address: h.hazrati@sut.ac.ir

biodegradation rates of organic contaminants that include substrate utilization, bacterial growth and decay, and the utilization of electron acceptors. Hence, the employment of an appropriate kinetic model is necessary. For instance, the models derived from Monod simulation are employed for population growth studies in the case of the microbial growth kinetics [9]. Generally, the Monod kinetic model is used in reports with a pure culture, limited substrate, and non-inhibitory biomass growth [10,11]. However, modifications of the Monod model, which includes an inhibition term, are used when substances are above a certain concentration. In addition, these models have been used to investigate the effects of substrate inhibition on biomass growth at high substrate concentrations [12, 13,14]. Up to now, there have been only a few investigations concerning styrene and ethylbenzene removal via MBRs. The literature survey showed that there is a lack of information related to the biokinetic coefficients of membrane bioreactors. Furthermore, no attempt has been made to determine the biokinetic coefficients of styrene and ethylbenzene in the MBR. The main goal of this study was to determine the biokinetic coefficients of membrane bioreactors for styrene, ethylbenzene and optimum HRT, which would enable process engineers to determine the minimum reactor volume and recognize the process control through reactor simulation. In order to achieve this aim, a lab scale membrane bioreactor was operated at various HRTs and SRTs.

1.1. Biokinetic coefficients determination

Basic equations that describe the interaction between the growth of microorganisms and the utilization of the growth limiting substrate in activated sludge processes are based on the Monod model; Monod was the first to initially suggest the idea of microbial growth kinetics controlled by an empirical model (Eq. 1).

$$\mu_i = \frac{\mu_{max_i} S_i}{K_{s_i} + S_i} \tag{1}$$

where the specific growth rate of biomass is μ_i (h^{-1}), μ_{max_i} is the maximum specific growth rate of biomass (h^{-1}), S_i is the substrate concentration (mg/L), and K_{s_i} is the substrate half-saturation constant (i.e. substrate concentration at half μ_{max_i}). The Monod model presented the concept of a growth limiting substrate through the parameters μ_{max} and Ks [14]. However, this model becomes unsatisfactory when a substrate prevents its own biodegradation. Therefore, a modified version of the Monod model, which is named the Andrews model, was employed to deliver an improved fitting for the data achieved from the sole substrate tests. In this case, the Andrews model, shown as Eq. (2), was used for substrate inhibition [15]. This model was justified and employed in this study based on the satisfactory/reliable results from previous reports [15,16], the toxic nature of VOCs (in this

case styrene and ethylbenzene), and the possibility of substrate inhibition.

$$\mu_i = \frac{\mu_{max_i} S_i}{K_{s_i} + S_i + S_i^2 / K_I} \tag{2}$$

Again, in Eq. (2), μ_{max_i} is the maximum specific growth rate (h^{-1}), K_{s_i} is the half velocity constant and K_I is the substrate inhibition constant, which quantified the influence of a toxic compound on its biodegradation. The three kinetic parameters, μ_{max_i}, K_{s_i} and K_I could be estimated by fitting Eq. (2) to experimentally obtained specific growth rates as a function of substrate concentration. It is important to note that the Andrews model is nonlinear. The existence of mixtures of chemicals in industrial and municipal wastewaters is a significant issue in biodegradation or bioremediation developments. The complexity of the degradation models sharply increased based on the growing number of substrates caused by the interaction between them. Furthermore, the kinetic parameters for a single substrate were not able to illustrate the phenomena observed during the mixture biodegradation. Uncompetitive inhibition, non-competitive inhibition and competitive inhibition are some interactions that can occur when multiple substrates are present [17]. Hence, several models have been established in order to define the specific growth rate through the degradation of multiple interacting substrates. One of the most common types of these models is obtained through the summation of specific growth rates. For instance, during competitive inhibition, substrates compete for binding sites to be metabolized by the mix culture; in this environment, a sum kinetics model incorporating purely competitive substrate kinetics is useful and is shown in Eq. 3 [18,22].

$$\mu_{tot} = \mu_1 + \mu_2 = \frac{\mu_{max_1} S_1}{K_{s_1} + S_1 + (\frac{K_{s_1}}{K_{s_2}})S_2} + \frac{\mu_{max_2} S_2}{K_{s_2} + S_2 + (\frac{K_{s_2}}{K_{s_1}})S_1} \tag{3}$$

Nevertheless, there is a model that accounts for substrate interactions without directly specifying the type of interaction [21]. This model is formulated by incorporating an interaction parameter $I_{i,j}$ as an unknown and is shown in Eq. 4.

$$\mu_{tot} = \mu_1 + \mu_2 = \frac{\mu_{max_1} S_1}{K_{s_1} + S_1 + I_{2,1} S_2} + \frac{\mu_{max_2} S_2}{K_{s_2} + S_2 + I_{1,2} S_1} \tag{4}$$

This model is known as sum kinetics with interaction parameters or SKIP. $I_{i,j}$ Specifies the degree to which substrate i affects the biodegradation of substrate j. According to the model, the stronger inhibition has a direct relationship with the large value of j [19]. The value of the interaction parameter is calculated by fitting the SKIP model to a binary of mixture data sets. Therefore, the specific growth rate of the biomass from the utilization of substrate i can be expressed by:

$$\mu_i = \frac{\mu_{max_i} S_i}{K_{s_i} + S_i + I_{j,i} S_j + I_{k,i} S_k + \cdots} \tag{5}$$

This work employed three substrates as the carbon source: 1) styrene 2) ethylbenzene and 3) ethanol. The biological removal of ethylbenzene and styrene are important due to their low biodegradability compared to ethanol; the value of the interaction parameters for styrene and ethylbenzene are presented in Table 1.

Table 1. The value of interaction parameters for styrene and ethylbenzene

Interaction parameter	$I_{1,2}$	$I_{1,3}$	$I_{2,1}$	$I_{2,3}$
Value	0.4	0.08	1.64	0.12

Not: Styrene: 1, Ethylbenzene: 2 and Ethanol: 3

The equations describing the performance of the reactor are the mass balance equations of both the biomass and substrate. The biomass balance can be expressed by:

$$V \frac{dX_i}{dt} = \mu_i X_i V - k_{d_i} X_i V - Q_w X_i \tag{6}$$

where V= reactor volume (m³); X_i= biomass concentration that is produced from the utilization of substrate i in the reactor (mg/l); k_{d_i}= biomass decay coefficient for fraction i (1/d); and Q_w =wastage flow rate (m³/d); and t = time (d). At steady-state conditions, $\frac{dX_i}{dt}$ = 0; hence, Eq. 6 becomes:

$$\mu_i - k_{d_i} = \frac{Q_w}{V} \tag{7}$$

Since the SRT is defined as:

$$SRT = \theta_c = \frac{VX}{Q_w X} = \frac{V}{Q_w} \tag{8}$$

Therefore, Eq. 7 becomes:

$$\mu_i = k_{d_i} + \frac{1}{\theta_c} \tag{9}$$

Substituting the value of μ_i from Eq. 9 into Eq. 5 yields the following equation that describes the steady-state condition of the substrate concentration in the reactor:

$$S_i = \frac{\left(k_{d_i} + \frac{1}{\theta_c}\right)\left(K_{s_i} + I_{j,i} S_j + I_{k,i} S_k + \cdots\right)}{\mu_{max_i} - \left(k_{d_i} + \frac{1}{\theta_c}\right)} \tag{10}$$

On the other hand, the substrate balance can be expressed as:

$$V \frac{dS_i}{dt} = QS_{0_i} - \mu_i \frac{XV}{Y_i} - S_i(Q - Q_w) \\ - Q_w S_i \tag{11}$$

where Y_i is the maximum cell yield for substrate i; S_{0_i}= initial concentration of substrate i (mg/L); and S_i is concentration of substrate i in the reactor. At the steady state, $\frac{dS_i}{dt} = 0$, therefore :

$$\frac{Q}{V} = \left(S_{0_i} - S_i\right) = \frac{\mu_i X}{Y_i} \tag{12}$$

Substituting Eq. 9 into Eq. 12 results in:

$$\frac{Q(S_{0_i} - S_i)}{VX} = \frac{1}{Y_i}\left(\frac{1}{\theta_c}\right) + \frac{k_{d_i}}{Y_i} \tag{13}$$

Eq. 12 is plotted as $\frac{Q(S_{0_i}-S_i)}{VX}$ versus $\frac{1}{\theta_c}$; the biokinetic coefficients k_{d_i} and Y_i can be determined from the slope and the Y-intercept of the equation. To determine the biokinetic coefficients, μ_{max_i} and K_{s_i}, Eq. 10 can be rearranged to become:

$$\frac{\theta_c}{1 + k_{d_i}\theta_c} = \frac{\left(K_{s_i} + I_{j,i} S_j + I_{k,i} S_k + \cdots\right)}{\mu_{max_i}}\left(\frac{1}{S_i}\right) + \frac{1}{\mu_{max_i}} \tag{14}$$

Substituting the value of k_{d_i} in Eq. (14) and plotting $\frac{\theta_c}{1+k_{d_i}\theta_c}$ versus $\frac{1}{S_i}$, the biokinetic coefficients, μ_{max_i} and K_{s_i} can also be examined from the slope and the Y-intercept of the equation. Similarly, it could be applied to other substrates; videlicet μ_{max_j}, μ_{max_k} K_{s_j} and K_{s_k} can also be determined by applying the above method.

2. Materials and methods

2.1. Experimental setup

The dimensions of the membrane bioreactor for this setup were 60×22×6.5 cm (Figure 1). The effective volume in the reactor was 7 L. The membrane used in this study was a Micro-Filtration (MF) type with an effective area of 0.1 m², a pore nominal diameter of 0.4 μm, and an A4 sheet size. The membrane was produced by the SINAP Company and was made of Poly-Ethylene (PE). The aeration process in the MBR was done for two purposes: first, to supply the oxygen needed for biological processes; and secondly, to clean the membrane surface and reduce the fouling rate. To achieve the second goal, a poly (methyl methacrylate) (PMMA) plate was used as a baffle to keep the air bubbles near the membrane surface so that they can make proper tensions with it and wash the sediments out of the surface. The aerobic sludge used in the MBR basin was supplied from the activated sludge of the Tabriz Petrochemical Company and then adapted with synthetic feed for one month.

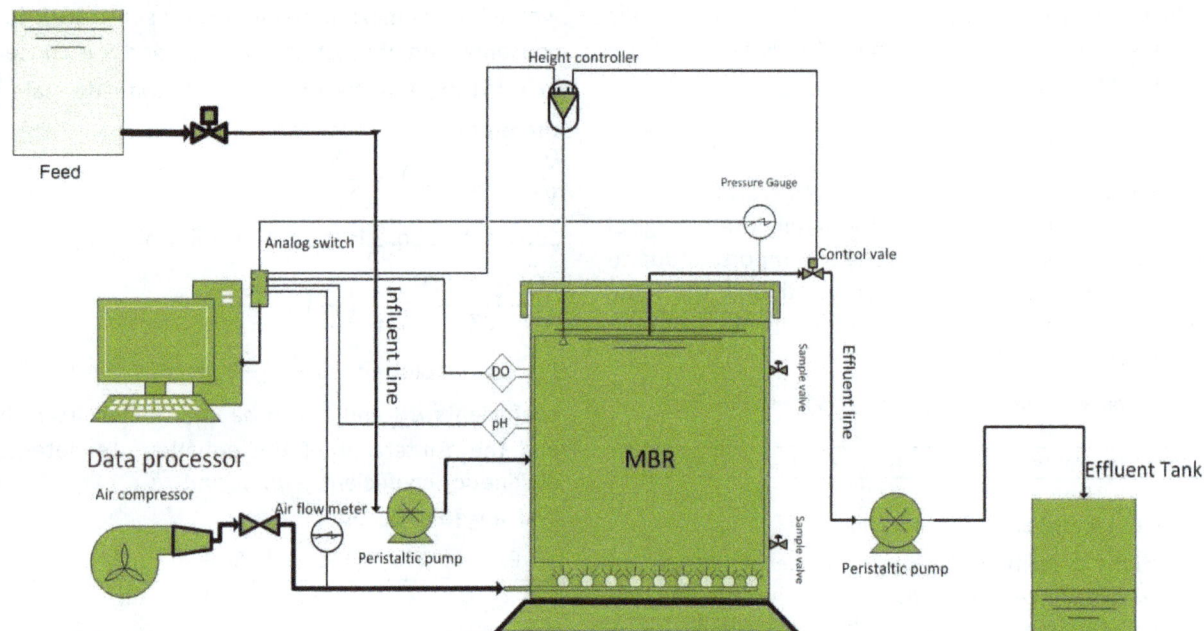

Fig. 1. Schematic of lab-scale experimental setup

2.2. Influent wastewater

The synthetic wastewater used in this research was formulated to simulate petrochemical industrial wastewater in terms of chemical oxygen demand (COD), styrene, and ethylbenzene concentrations which were 1200, 50-100 and 50-100 mg/L respectively. Ethanol was used as a carbon source which created a COD concentration of about 1200 mg/L. The synthetic wastewater compositions used in the present study are described in Table 2.

Table 2. The synthetic wastewater compositions

Components	Concentration (mg/L)
Ethanol	370-400
Styrene (STR)	50-100
Ethylbenzene (EB)	5-100
NH_4Cl	560
K_2HPO_4	35
KH_2PO_4	45
$MgSO_4.7H_2O$	13
$CaCl_2.2H_2O$	7
$FeCl_3$	5
$ZnSO_4$	2
$NaHCO_3$	500
EDTA ($C_{10}H_{16}N_2O_8$)	7

2.3. Analytical methods and operation parameters

The styrene and ethylbenzene concentrations were analyzed using a gas chromatograph (GC). The GC (Young Lin, ACME-6100) was set with a Flame Ionizing Detector (FID) and an attached silica capillary column (DB-5, 0.53 mm

I.D., 30 m length, 1 mm film thickness) that was designed to be well suited for the analysis of volatile components. The carrier gas was helium flowing at 15 mL/min. The oven temperature was maintained at 70 °C for a 1 min duration and then raised to 140 °C. The temperatures of the injector and the detector were fixed at 200 and 240 °C, respectively. The styrene and ethylbenzene concentrations in the liquid phase were estimated using the head-space method [20]. The gas flow rate from the bioreactors headspace was measured using a flow meter. The MLSS, MLVSS, and COD were estimated according to standard methods [21].

3. Results and discussion

3.1. Experimental results for MBR

3.1.1. Styrene, ethylbenzene and COD removal efficiency in HRT=20 hr and SRT= 20 d

The styrene, ethylbenzene, and COD removal efficiency is presented in Figure 2. As it can be seen, in the steady state condition for a HRT of 20 hrs, the COD removal efficiency in the reactor was around 98 percent; the styrene and ethylbenzene removal was more than 99 percent. In addition, the concentration of ethylbenzene and styrene in the exit air from the reactor was measured daily. Figure 3 shows that in an HRT of 20 hrs, the concentrations of ethylbenzene and styrene in the reactor exit air in a steady-state condition was 0.7 ppm (equal 1.16% stripping removal) and 1ppm (equal 1.65% stripping removal), respectively. This fact indicated that the mechanism of removal in the reactor was not a consequence of the volatility of styrene and ethylbenzene. Also, the absorption of a pollutant by a microbial culture can only be considered as an important mechanism whenever the partition coefficient of octanol - water (log K_{ow}) was more than 4 [20], while this coefficient for styrene and ethylbenzene was about 3.15 and 2.85, respectively [7,21]. Moreover, previous studies revealed that the styrene absorption by sludge as a removal mechanism was insignificant [23]. Therefore, the removal mechanism in the reactor was mainly through a biodegradation mechanism.

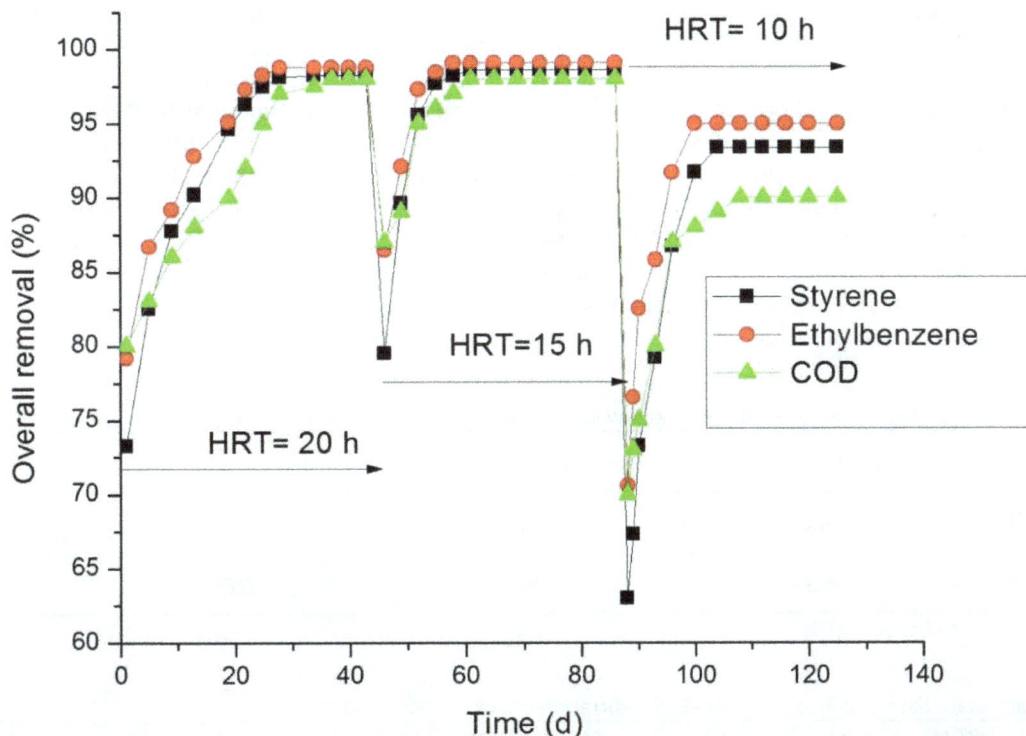

Fig. 2. Variations of COD, styrene and ethylbenzene removal during the operation of the MBR (HRT of 20h (days 0-43th), 15h (days 44- 85th) and 10h (days 86-125th)

3.1.2. Styrene, Ethylbenzene and COD Removal Efficiency in HRT=15 & 10 h and SRT =20 d

In an HRT of 15 hr, except during the first few days (days 44 to 47th) in which the removal efficiency in the reactor was extremely reduced, the removal efficiency of COD, styrene and ethylbenzene increased and after reaching a steady state was 98, 99.9 and 99.9%, respectively. Because of an unexpected boost in the amount of organic load that entered the system on the 44th day, the microorganisms were under shock and for this reason, a declining trend was observed. Following this stage, the microorganisms adapted themselves to the new conditions which gradually increased the efficiency of the system and eventually reached a steady state condition. After a change in the retention time from 20 to 15 hours, the stripping removal efficiency of styrene and ethylbenzene in the reactor slightly decreased. This could be attributed to styrene and ethylbenzene concentrations in the exit air which decreased from 1 and 0.7 ppm to 0.8 and 0.5 ppm, respectively. In a previous study, it was also reported that HRT reduction decreased the removal efficiency through volatility [10]. After the change in the retention time from 20 to 15 hours, the biological removal efficiency of styrene and ethylbenzene in the reactor increased. Thus, when the retention time was reduced, two parameters affected the removal efficiency. Firstly, organic loading increased slightly and as a consequence, the amount of MLSS grew in the reactor. The other parameter was the drop in contact time between the contaminants and the sludge. It was obvious that an increase in MLSS had a positive effect and a reduction in the contact time had a negative influence on the biological removal efficiency. Nevertheless, since the MLSS concentration in the reactor increased, the negative effect of the contact time was neutralized and the removal efficiency increased. When the HRT declined to 10 hrs, the organic load rate in the system increased; on the other hand, the contact time between the activated sludge and wastewater decreased significantly compared to the previous states (e.g. in HRTs of 20 and 15 hrs). Therefore, the removal efficiency in the reactor was reduced significantly. Under this circumstance, the removal efficiency of COD, styrene and ethylbenzene was 90, 99.9, and 99.9 percent, respectively, but the biological removal for styrene and ethylbenzene was 93 and 94%. Further, the styrene and ethylbenzene concentration in the exit air of the reactor was 4 and 3 ppm, respectively. It should be noted that the concentration of exit gases also increased because of the fall of the MLSS value. Therefore, in an HRT of 10 hr, the biological removal efficiency in the reactor was reduced significantly compared to an HRT of 15 and 20 hr.

3.1.3. Styrene, ethylbenzene biological removal efficiency in HRT=15 h and SRT = 10 day

The biological removal efficiency of styrene and ethylbenzene at a SRT of 10 and a HRT of 15 hrs was measured in a steady state. The biological removal of styrene and ethylbenzene was about 94.6% and 98.7% while the SRT of 20d was about 99% and 99%, respectively. Moreover, the concentration of ethylbenzene and styrene in the reactor exit air was measured daily under this condition. The results showed that in the SRT of 10 day, the concentrations of ethylbenzene and styrene in the reactor exit air was 3.2 and 4.9 ppm, respectively. The compression of biological removal efficiency and VOCs concentration for

styrene and ethylbenzene in two SRTs (10 and 20) are presented in Table 3.

3.2. Determination of biokinetic coefficients

The determination of the biokinetic coefficients at an MLSS concentration of 4000 mg/L was initiated using an SRT of 20 days. During the investigation, the SRT was varied between 20 and 5 days. Table 4 shows the steady-state data obtained at an MLSS concentration of 4000 mg/L, while Figures 3 and 4 show the determination of the coefficients using Eqs. 13 and 14. The values of the biokinetic coefficients were found to be as follows: 1. for styrene substrate: $Y = 0.599$ mg/mg, $k_d = 0.25$ day^{-1}, $\mu_{max} = 0.188$ h^{-1}, and $K_s = 1.457$ mg/L; and 2. for ethylbenzene: $Y = 0.599$ mg/mg, $k_d = 0.25$ day^{-1}, $\mu_{max} = 0.364$ h^{-1}, and $K_s = 2.821$ mg /L.

Table 3. Comparison of biological removal efficiency and VOCs concentration in the exit air for two SRTs

SRT (d)	HRT (hr)	Biological removal of STR	Biological removal of EB	STR concentration in air	EB concentration in air
10	15	93.6	95.7	4.9	3.2
20	15	98.5	99.1	0.8	0.5

Note: Styrene (STR), Ethylbenzene (EB)

Table 4. The steady-state data obtained at an MLSS concentration of about 4000 mg/L

SRT (d)	HRT (h)	X (mg/L)	Q (L/d)	S_0 (mg/L) for STR & EB	S (mg/L) for STR	S (mg/L) for EB
20	20	4000	0.20	100	0.12	0.10
15	20	5000	0.20	100	0.13	0.11
10	15	5100	0.27	80	0.14	0.12
5	18	3500	0.22	80	0.19	0.15
3	15	3400	0.27	80	0.25	0.21

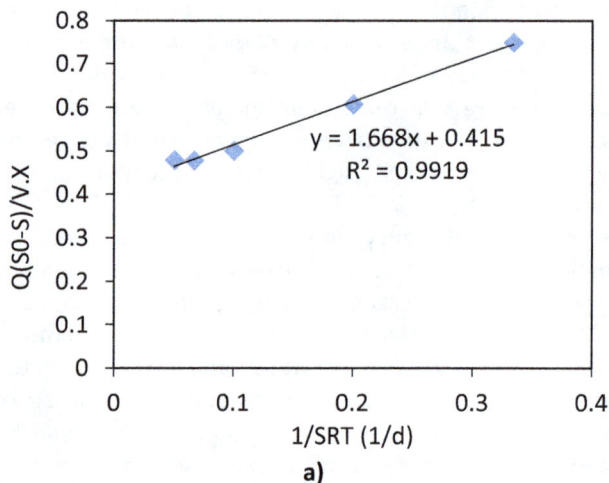

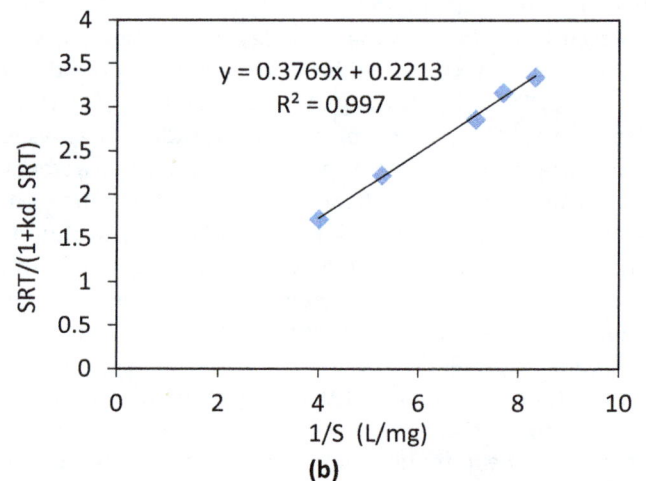

Fig. 3. Determination of (a) Y and kd (b) μ_{max} and Ks for styrene substrate

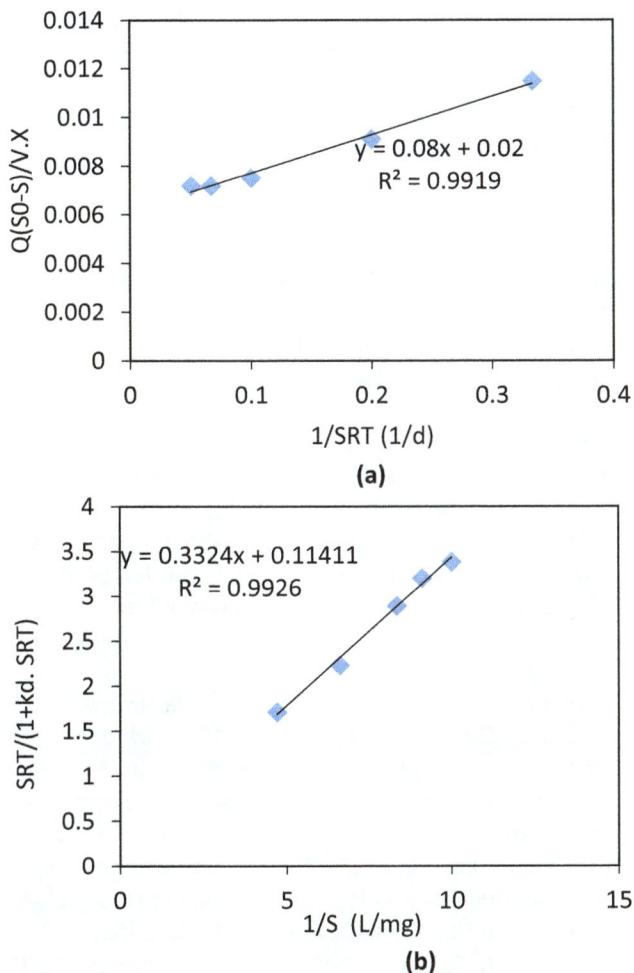

Fig.4. Determination of (a) Y and kd (b) μ_{max} and Ks for ethylbenzene substrate

The results showed that the yield (Y) and the endogenous decay coefficient (k$_d$) was the same for styrene and ethylbenzene as the substrate, but the maximum specific growth rate (μ_{max}) for ethylbenzene was more than styrene. As can be seen in Table 5, in comparison to prior

research that considered the special pure culture to evaluate the kinetics of biodegradation, the biokinetic coefficients obtained for both styrene and ethylbenzene in this study were different from the previously gained values [21,22]. This clearly showed that the type of substrate and bacterial consortium can have a significant effect on the determination of the biokinetic coefficients. The μ_{max} value showed the capability of the microbial culture in MBR to use the special pollutant as a source of carbon and energy. Although some microorganisms showed an excessive ability to biodegradation, the other culture cannot appropriately use these components as a source of energy. Therefore, the abundance of the microorganisms led to a competition between the bacterium cultures for the common substrate [23,24]. Furthermore, it can be seen in this case that the ethylbenzene was more appropriate as a source of carbon to activated sludge in the MBR than the styrene, which had a lower μ_{max} than the ethylbenzene. In addition, the different values of μ_{max} demonstrated different pathways in order to completely catabolize the selected components using the microbial species picked to attack and catabolized the carbon sources [22].

4. Conclusions

The operation of a MBR system for the biological removal of volatile organic compounds such as styrene and ethylbenzene demonstrated the 15-hour time as an optimum HRT value at a SRT of 20 day. The operation of a lab scale MBR confirmed that it can be a feasible procedure to reduce VOC emissions from petrochemical wastewater. The results showed that the yield (Y) and the endogenous decay coefficient (k$_d$) was the same for styrene and ethylbenzene as for the substrate, but the maximum specific growth rate (μ_{max}) for ethylbenzene was more than the styrene. Further, the values of the biokinetic coefficients, except that of k$_d$, were within the normal range reported for these components.

Table 5. Comparison between kinetic parameters estimated for the biodegradation of styrene and ethylbenzene in different studies

Strain	Substrate	μ_{max} (1/h)	Y(mg/mg)	K$_S$ (mg/L)	k$_d$ (d^{-1})	pH	References
Activated sludge in MBR	Styrene	0.188	0.6	0.146	0.25	7	This study
Exophiala jeanselmei	Styrene	0.630	-	0.1	3.3	5.7	[27]
Activated sludge in MBR	Ethylbenzene	0.363	0.6	2.82	0.25	7	This study
Pseudomonas putida F1	Ethylbenzene	0.260	-	1.5	20	-	[28]

References

[1] Harper, D. B. (1977). Microbial metabolism of aromatic nitriles. Enzymology of C–N cleavage by Nocardia sp. (Rhodochrous group) NCIB 11216. *Biochemical journal*, *165*(2), 309-319.

[2] Lin, C. K., Tsai, T. Y., Liu, J. C., Chen, M. C. (2001). Enhanced biodegradation of petrochemical wastewater

using ozonation and BAC advanced treatment system. *Water research, 35*(3), 699-704.

[3] Pendashteh, A. R., Fakhru'l-Razi, A., Chuah, T. G., Radiah, A. D., Madaeni, S. S., Zurina, Z. A. (2010). Biological treatment of produced water in a sequencing batch reactor by a consortium of isolated halophilic microorganisms. *Environmental technology, 31*(11), 1229-1239.

[4] Alemzadeh, I., Vossoughi, M. (2001). Biodegradation of toluene by an attached biofilm in a rotating biological contactor. *Process biochemistry, 36*(8), 707-711.

[5] Hsien, T. Y., Lin, Y. H. (2005). Biodegradation of phenolic wastewater in a fixed biofilm reactor. *Biochemical engineering journal, 27*(2), 95-103.

[6] Deowan, S. A., Galiano, F., Hoinkis, J., Johnson, D., Altinkaya, S. A., Gabriele, B., Figoli, A. (2016). Novel low-fouling membrane bioreactor (MBR) for industrial wastewater treatment. *Journal of membrane science, 510*, 524-532.

[7] Alkmim, A. R., da Costa, P. R., Moser, P. B., França Neta, L. S., Santiago, V. M., Cerqueira, A. C., Amaral, M. C. (2016). Long-term evaluation of different strategies of cationic polyelectrolyte dosage to control fouling in a membrane bioreactor treating refinery effluent. *Environmental technology, 37*(8), 1026-1035.

[8] Nguyen, L. N., Hai, F. I., Nghiem, L. D., Kang, J., Price, W. E., Park, C., Yamamoto, K. (2014). Enhancement of removal of trace organic contaminants by powdered activated carbon dosing into membrane bioreactors. *Journal of the Taiwan institute of chemical engineers, 45*(2), 571-578.

[9] Liu, Y. (2007). Overview of some theoretical approaches for derivation of the Monod equation. *Applied microbiology and biotechnology, 73*(6), 1241-1250.

[10] De Lucas, A., Rodriguez, L., Villasenor, J., Fernandez, F. J. (2005). Biodegradation kinetics of stored wastewater substrates by a mixed microbial culture. *Biochemical engineering journal, 26*(2), 191-197.

[11] Kumaran, P., Paruchuri, Y. L. (1997). Kinetics of phenol biotransformation. *Water research, 31*(1), 11-22.

[12] Babaee, R., Bonakdarpour, B., Nasernejad, B., Fallah, N. (2010). Kinetics of styrene biodegradation in synthetic wastewaters using an industrial activated sludge. *Journal of hazardous materials, 184*(1), 111-117.

[13] Gąszczak, A., Bartelmus, G., Greń, I. (2012). Kinetics of styrene biodegradation by Pseudomonas sp. E-93486. *Applied microbiology and biotechnology, 93*(2), 565-573.

[14] Monod, J. (1949). The growth of bacterial cultures. *Annual Reviews in Microbiology, 3*(1), 371-394.

[15] Andrews, J. F. (1968). A mathematical model for the continuous culture of microorganisms utilizing inhibitory substrates. *Biotechnology and bioengineering, 10*(6), 707-723.

[16] Littlejohns, J. V., Daugulis, A. J. (2008). Kinetics and interactions of BTEX compounds during degradation by a bacterial consortium. *Process biochemistry, 43*(10), 1068-1076.

[17] Segel, I. H. (1975). *Enzyme kinetics* (Vol. 360). Wiley, New York.

[18] Yoon, H., Klinzing, G., Blanch, H. W. (1977). Competition for mixed substrates by microbial populations. *Biotechnology and bioengineering, 19*(8), 1193-1210.

[19] Hazrati, H., Shayegan, J., Seyedi, S. M. (2015). Biodegradation kinetics and interactions of styrene and ethylbenzene as single and dual substrates for a mixed bacterial culture. *Journal of environmental health science and engineering, 13*(1), 72.

[20] Eckenfelder, W. W. (1989). *Industrial water pollution control*. McGraw-Hill.

[21] Cox, H. H. J., Moerman, R. E., Van Baalen, S., Van Heiningen, W. N. M., Doddema, H. J., Harder, W. (1997). Performance of a styrene-degrading biofilter containing the yeast Exophiala jeanselmei. *Biotechnology and bioengineering, 53*(3), 259-266.

[22] Trigueros, D. E., Módenes, A. N., Kroumov, A. D., Espinoza-Quiñones, F. R. (2010). Modeling of biodegradation process of BTEX compounds: Kinetic parameters estimation by using particle swarm global optimizer. *Process biochemistry, 45*(8), 1355-1361.

[23] Jung, I. G., Park, C. H. (2005). Characteristics of styrene degradation by Rhodococcus pyridinovorans isolated from a biofilter. *Chemosphere, 61*(4), 451-456.

[24] Trigueros, D. E., Módenes, A. N., Kroumov, A. D., Espinoza-Quiñones, F. R. (2010). Modeling of biodegradation process of BTEX compounds: Kinetic parameters estimation by using Particle Swarm Global Optimizer. *Process biochemistry, 45*(8), 1355-1361.

18

Evaluation of adsorption efficiency of activated carbon/chitosan composite for removal of Cr (VI) and Cd (II) from single and bi-solute dilute solution

Hakimeh Sharififard[1,*], Mahboubeh Nabavinia[2], Mansooreh Soleimani[2]

[1] Department of Chemical Engineering, Yasouj University, Yasouj, Iran,

[2]Department of Chemical Engineering, Amirkabir University of Technology, Tehran, Iran

ABSTRACT

The aim of this study was to evaluate the adsorption capacity of a novel activated carbon coated with chitosan for the removal of Cr (VI) and Cd (II) ions from single and bi-solute dilute aqueous solutions. In addition, the adsorption abilities of activated carbon (AC), chitosan (CH) and chitosan / activated carbon composite (CHAC) were compared. The adsorption studies were performed in a batch system, and the effects of various operating parameters such as solution pH, particle size and the dose of adsorbent were considered for the removal of Cr (VI) and Cd (II) via the Taguchi method. The equilibrium experimental data were well fitted to the Langmuir isotherm for single and bi-solute solutions. The adsorption capacities of the AC and CH adsorbents were improved by means of the synthesized CHAC composite. As expected, the competitive adsorption of metal ions on the CHAC surface led to a reduction in the adsorption capacity from 90.9 mg/g to 41.94 mg/g for Cr (VI) ions and 52.63 mg/g to 30.21 mg/g for Cd (II) ions, respectively. The adsorption potential of CHAC for Cr (VI) was greater than Cd (II) for the different metal solution–adsorbent systems. The kinetic studies indicated that the adsorption process was best described by pseudo-second-order kinetics for single and bi-solute solutions.

Keywords:
Activated carbon
Cadmium
Chitosan
Chromium
Composite

1. Introduction

Today, water contamination with different pollutants is a major environmental problem. Heavy metals and dyes are major water pollutants [1-12]. Heavy metals are widely distributed in the environment and are harmful due to their toxicity. They are non-biodegradable and their concentration is accentuated through bioaccumulation via the food chain in living organisms. Their accumulation causes different diseases and dysfunctions. Cadmium and chromium are toxic metals found in several industrial discharges and effluents. Cadmium is released into the environment through the combustion of fossil fuels, metal production (zinc, iron and steel production), cement production, electroplating, and the manufacturing of batteries and pigments. The presence of cadmium (II) in water, even at very low concentrations, is extremely harmful to the aquatic environment; in humans, it causes muscular cramps, chronic pulmonary problems, renal degradation, proteinuria, skeletal deformity, and testicular atrophy [13]. The World Health Organization (WHO) guideline for Cd in drinking water is set at a maximum concentration of 0.003 mg/L [14]. Chromium found in the environment mainly exists in two oxidation states: Cr (III) and Cr (VI). Cr (VI) contamination is more concerning, as it is highly toxic, mutagenic and carcinogenic to living organisms; Cr (III) is generally less

*Corresponding author

E-mail address: hakimeh.sharifi@gmail.com

toxic and believed to be essential in glucose metabolism in mammals [15]. Chromium usually presents in the effluents of electroplating, tanning, mining, and fertilizer industries. The maximum chromium levels permitted in wastewater are 5 mg/L for Cr (III) and 0.05 mg/L for Cr (VI) [16]. Different technologies such as ion exchange, electrocoagulation process, and emulsion liquid membrane have been studied for heavy metal removal from industrial wastewaters [17-20]. Among these technologies, adsorption is more advantageous due to its high separation efficiency, low cost, and easy operation. Recently, the use of natural materials as the adsorbent for heavy metals removal from wastewater has gained considerable importance in the adsorption process [21-22]. Chitin, poly-N-acetylglucosamine, is a natural polymer found in several sources such as crustaceans and fungal cell walls. Chitosan, poly D-glucosamine, is the deacetylated form of chitin that has a large variety of applications in areas such as biochemistry, pharmacy, medicine, and agriculture as well as wastewater treatment [23]. The cationic character of chitosan is unique in that it is the only pseudo-natural cationic polymer. Chitosan has high nitrogen content in the form of amine groups. The amine and hydroxyl groups of chitosan have the capacity to absorb metals through several mechanisms including chemical interactions (such as chelation) and electrostatic interactions (for example, ion exchange or the formation of an ion pair). The kind of interaction depends on the metal type, its chemistry, and the solution pH [24]. In recent years, chitosan and its derivatives such as cross-linked chitosan, chitosan beads and composite chitosan have been studied as an adsorbent for the removal of dyes and heavy metal ions from aqueous solutions [24-30]. The main disadvantages of chitosan are related to its weak mechanical strength and its dissolution in acidic solution. To improve the performance of chitosan as an adsorbent, the modification of its physical and chemical properties is necessary. The immobilization of chitosan on common substances is a new method for its modification. If a proper and commercial material is used as a support for immobilizing the chitosan, its mechanical and chemical properties will be improved; also, much lower quantities of chitosan will be needed to build a new adsorbent. Recently, chitosan composites have been developed to adsorb heavy metals and dyes from wastewater. Different kinds of substances have been used to form a composite with chitosan such as sand, perlite, clay, and PVC beads [28-30]. Activated carbon adsorbents can adsorb many types of pollutants from industrial wastes because of their unique porous structure, high specific surface area, and high mechanical and chemical resistance, [31-

32]. Also, activated carbon can be used as a support for immobilizing the chitosan on its surface [33]. This research work analyzed the synergistic effects of chitosan and activated carbon on heavy metal removal from a multicomponent solution in the form of a chitosan/activated carbon composite. The adsorption ability of this new composite was investigated for Cr (VI) and Cd (II) removal from single and bi-solute solutions under different adsorption conditions. Also, the equilibrium and kinetics of the adsorption process were analyzed in order to understand the adsorption mechanism.

2. Material and methods

2.1. Materials

Chitosan flakes (CH) with a minimum of 80% degrees of deacetylation and a medium molecular weight of (750,000) were purchased from the Fluka Company. The Cd $(NO_3)_2$ and Cr $(NO_3)_3.9H_2O$ supplied by the Merck Company and the deionized water were used for the preparation of the metal solutions. The initial pH of the solution was adjusted with HCl and NaOH solutions (5 M) using a pH meter (Metrohm, 780). Commercially activated carbon pellets (AC), oxalic acid (0.2 M) and NaOH (0.7 M) were used for immobilizing the chitosan onto the activated carbon.

2.2. Synthesis of composite adsorbent

2.2.1. Stability of chitosan

Several researchers [28,34-35] have shown that chitosan is soluble in organic acids and to a certain extent, in mineral acids; however, its solubility depends on molecular weight and the degree of deacetylation of the chitosan. In our previous works, the stability of CH was investigated and the results showed that the applied CH was stable after 44 h of contact time in the acidic solution (pH = 1, 2) and remained in its original state as flakes. After this period of time, the CH began to gel and dissolve in the acidic media. Our previous results showed that the stability of CHAC in an acidic solution is 66 h [35-36].

2.2.2. Synthesis of chitosan/activated carbon composite (CHAC)

The chitosan/activated carbon composite (CHAC) was synthesized using a synthesis method described in our previous work [35-36]. Briefly, 20 g of AC was poured into 0.2 M of oxalic acid for 4 h. After washing with deionized water, filtration was done and the acid treated AC was dried in an oven. Then, ten grams of CH was added to 1 L of 0.2 M oxalic acid solution under continuous stirring at 40 °C to form a viscous gel. About 20 g of the acid treated AC was added slowly to the CH gel and stirred for 12 h at 40 °C. The CHAC composite was prepared by the dropwise addition of the AC gel mixture into a 0.7 M NaOH precipitation bath. The beads were filtered and washed several times with deionized water to a neutral pH and dried in an air oven (50 °C).

2.3. Adsorbents Characterization

The adsorbents under study (CH, AC, and CHAC) were characterized with standard methods for different properties.

2.3.1. Bulk density and pHzpc

To measure bulk density, a 10 mL cylinder was filled to a specified volume with dried adsorbent. The cylinder was weighed. The bulk density was then calculated as follows [37]:

Bulk density= [weight of dry material　　(1)
(g)/volume of packed dry material (cm³)]

In order to measure the pHzpc of the adsorbent, 50 mL of 0.1 mol/L NaCl solution was put into glass containers and the initial pH of these solutions was adjusted in the range of 1-12 using 0.1 mol/L HCl or NaOH solutions (Merck Co.). Then, 0.1 g of adsorbent was added to each container. The containers were agitated for 48 h at 25 °C. At the end of this time period, filtration was done and the final pH of the residual solutions was measured. The point in which the final pH equalled the initial pH of the NaCl solution was defined as pHzpc [35].

2.3.2. Specific surface area, surface chemistry, and morphology

The specific surface area (S_{BET}) and total pore volume of the adsorbents were determined from the adsorption–desorption isotherm of N_2 at 77 K. A Quantachrom NOVA 1000 surface area analyzer was applied to determine these parameters via the 3-point BET (Brunaeur–Emmet–Teller) method. The surface chemistry of CH, AC, and CHAC were determined using Fourier transform infrared radiation (FTIR). The analysis was performed on PerkinElmer, Spectroum GX, under 4 cm^{-1} resolution,(sample/KBr =1/100) within the range of a 400–4000 cm^{-1} wavenumber. The element analysis of CH and AC was determined by elemental analysis with a CHN analyzer (Perkin-Elmer 2400). The morphology of the CH, AC and CHAC composites were determined by direct measurement on the scanning electron micrographs (SEM, Vega II Tescan, MV2300).

2.4. Effect of removal parameters

In the removal experiments, CHAC was sieved and separated into three different particle sizes. All the dried samples of CH, AC, and CHAC were kept in a desiccator. The removal of Cd (II) from the single aqueous solution using CHAC was performed in our previous work [36]. In this study, the effects of three

operating parameters including solution pH, particle size, and dose of CHAC were investigated in three levels for the removal of Cr (VI) from the single solution. The Taguchi method was used for the experimental design. The degree of freedom related to the three sets of three-level adsorption parameters was six and in accordance with the Taguchi method; the standard orthogonal array L9, with three columns and nine rows, could be used for these experiments [38]. The experimental layout for these parameters using the L9 orthogonal array is listed in Table1.

Table 1. Arrangement of parameters in L9 orthogonal array

Parameters	pH	Adsorbent particle size	Adsorbent dose(g/L)
1	1	0.210	2
2	1	0.425	4
3	1	1.000	6
4	3	0.210	4
5	3	0.425	6
6	3	1.000	2
7	6	0.210	6
8	6	0.425	2
9	6	1.000	4

The removal experiments for Cr (VI) were conducted in a batch manner by mixing a measured weight of CHAC with a known particle size in 50 mL of metal solution (initial concentration 50 mg/L) with a known initial pH in 100 mL Erlenmeyer flasks. These flasks were shaken at 25 ºC using an orbital shaker at 250 rpm for 2 h. After this time, the adsorbent was separated by means of Whatman filter paper (No. 42), and the final solution concentration was analyzed using an inductively coupled plasma (ICP) analyzer (Varian 735ES) in the aqueous phase. The removal percentage of Cr (VI) removed by CHAC was determined from the following equation:

$$Removal\ percentage = \frac{(C_0 - C_f)}{C_0} \times 100 \qquad (2)$$

In this equation, C_0 and C_f are the initial and final metal ion concentration (mg/L), respectively. All the above-mentioned removal experiments were repeated two times, and the average values were reported.

2.5. Kinetic studies

Adsorption kinetic experiments for single and bi-solute solutions of Cr (VI) and Cd (II) ions were performed at a pH = 6.0 , a temperature of 25 ∘C , an adsorbent dose of 2 g/L, and an agitating rate of 250 rpm. The operation conditions were selected based on our experimental works. The effect of the initial pH, adsorbent dose and adsorbent particle size was investigated for the removal of Cr (VI) and Cd (II) using AC and CH. The results indicated that the optimum operating conditions were an initial of pH=6, the particle size of the adsorbent =0.425 mm, and an adsorbent dose = 6 g/L for both metals removal. For the single solution, a certain dosage of AC, CH and CHAC was placed into the flasks containing 50 mL (50 mg/L) metal ion solution. For the bi-solute solution, a certain dosage of CHAC was placed in the flasks containing 50 mL metals solution (initial concentration of metals 50 mg/L: concentration ratio of Cr (VI) and Cd (II) was 1). The contents of the flasks were agitated on an orbital shaker for prescribed periods of time (5-150 min). After these time periods, the adsorbent was separated by Wattman filter paper and the solution concentration was analyzed using an inductively coupled plasma (ICP) analyzer. The metal uptake capacity, q, for each metal ion was calculated as:

$$q = \frac{V}{m}(C_o - C_t) \qquad (3)$$

where q_t, C_t, V and m were the amount of solute adsorbed per unit weight of adsorbent (mg/g) at t (min), final metal ions concentration (mg/L), solution volume (L) and dry weight of adsorbent (g), respectively. The kinetics experiments were repeated two times and the average values were reported.

2.5.1. Kinetic models

In order to investigate the mechanism of adsorption, the kinetic models were exploited to test the experimental data. In this work, three general kinetic models including pseudo-first-order, pseudo-second-order and intra-particle diffusion (Waber–Morris model) were used to test kinetic data. These models are presented in Table 2.

Table 2. Kinetics models

Kinetic model	Equation
Pseudo-first-order[1]	$\log(q_e - q) = \log q_e - \dfrac{k_1 t}{2.303}$
Pseudo-second-order[2]	$\dfrac{t}{q} = \dfrac{1}{k_2 q_e^2} + \dfrac{1}{q_e} t$
Intra-particle diffusion[3]	$q_t = k_{id} t^{0.5} + C$

[1.] k_1(min^{-1}) is pseudo-first-order rate constant, q and q_e are the amounts of solute adsorbed per unit weight of adsorbent (mg/g) at t (min) and equilibrium time, respectively.
[2.] k_2(g/mg min) is pseudo-second-order rate constant.
[3.] k_{id} (mg/g.min$^{1/2}$) is intraparticle diffusion rate constant

2.6. Equilibrium studies

For the assessment of adsorption equilibrium and to determine the adsorption capacity, the adsorption equilibrium for the single and bi-solute solution of Cr (VI) and Cd (II) ions were performed at a pH = 6.0, an adsorbent dose of 2 g/L, a temperature of 25°C, an agitating rate of 250 rpm and an equilibrium time of 24 h. The initial concentrations of Cr (VI) and Cd (II) for the single metal isotherm were 10–50 mg/L. For the bi-

solute system, one metal concentration was varied from 10 mg/L to 50 mg/L and the concentration of the other metal was kept constant (20 mg/L). After equilibrium time, the metal solution was filtered and the residual concentration of the metal ions was determined. The amount of Cr (VI) and Cd (II) adsorbed by adsorbents, q_e, was determined by Equation (3). These adsorption tests were repeated two times.

2.6.1. Adsorption isotherm

The three most widely used adsorption isotherms are the Langmuir, Freundlich and Dubinin –Radushkevich (D-R) isotherms as expressed in Table 3. In this study, the experimental data was analyzed in terms of these isotherms. The E (kJ/mol) parameter from the D-R isotherm gave information in regard to whether the adsorption mechanism was an ion-exchange or physical adsorption. If the magnitude of E was between 8 and 16 kJ/mol, the adsorption process was followed by an ion-exchange (chemical nature); if the values of E < 8 kJ/mol, the adsorption process had a physical nature [39].

Table 3. Isotherm models for equilibrium data

Isotherm	Model
Langmiur[1]	$\dfrac{1}{q_e} = \dfrac{1}{q_{max}.K_L} \cdot \dfrac{1}{C_e} + \dfrac{1}{q_{max}}$
Freundlich[2]	$\log q_e = \dfrac{1}{n}\log C_e + \log K_f$
Dubinin –Radushkevich[3]	$\ln q_e = \ln q_m - \beta \varepsilon^2$
	$\varepsilon = RT\ln\left(1 + \dfrac{1}{C_e}\right)$
	$E = \dfrac{1}{(2\beta)^{1/2}}$

[1] q_{max} is maximum adsorption capacity of adsorbent (mg/g) and K_L is the Langmuir constant related to energy of adsorption (L/mg).
[2] K_f is the Freundlich constant related to adsorption capacity of adsorbent (mg$^{1-(1/n)}$L$^{1/n}$/g) and $1/n$ is the Freundlich –exponent related to adsorption intensity.
[3] q_m is the theoretical saturation capacity (mol/g), β is a constant related to the mean free energy of adsorption per mole of the adsorbate (mol^2/J^2), E is mean free energy (kJ/mol) of adsorption per molecule of the adsorbate.

3. Results and discussion

3.1. Adsorbents Characterization

The physical and chemical properties of AC, CH, and CHAC are listed in Table 4. The surface area and total pore volume values were arranged in the following sequence: AC > CHAC > CH. The lower surface area and pore volume of CHAC was due to CH molecules blocking the pores of AC.

The improvement of the CH properties may occur because AC acts as a support for it.

Table 4. Characteristics of AC, CH and CHAC

Parameter	AC	CH	CHAC
Bulk density (g/cm³)	0.60	0.21	0.27
BET surface area (m²/g)	922.33	16.37	362.30
Total pore volume (cm³/g)	0.57	0.02	0.23
Average pore diameter (°A)	24.70	44.90	12.68
pH$_{ZPC}$	6.9	6.4	6.5
Element analysis of CH and AC		CH	AC
C		40.67%	87.29%
H		7.27%	0.62%
N		7.61%	0.35%

The result of the elemental analysis indicated that the nitrogen content of CH was about 8%, this parameter was related to the presence of amine groups in the CH structure [35]. The FTIR spectra of AC, CH, and CHAC were presented in Figure 1 of our previous work [36] and the description of these FTIR spectra was expressed in our work [36]. In brief, these spectra confirmed the presence of amine and amide groups in the chitosan structure. Also, the presence of carbonyl, lactonic and carboxyl, and carboxylic acid groups in the structure of the activated carbon has been determined. The peaks at 1548 cm^{-1}-1602 cm^{-1} in the FTIR spectra of CHAC are the specific peaks for N–H scissoring from the amines and amides. These findings confirm that CH was successfully immobilized on AC [35-36]. Figure 1 shows the SEM microscopies of the AC, CH, and CHAC adsorbents. These images revealed that CH coated the surface of AC and reduced the pores of AC.

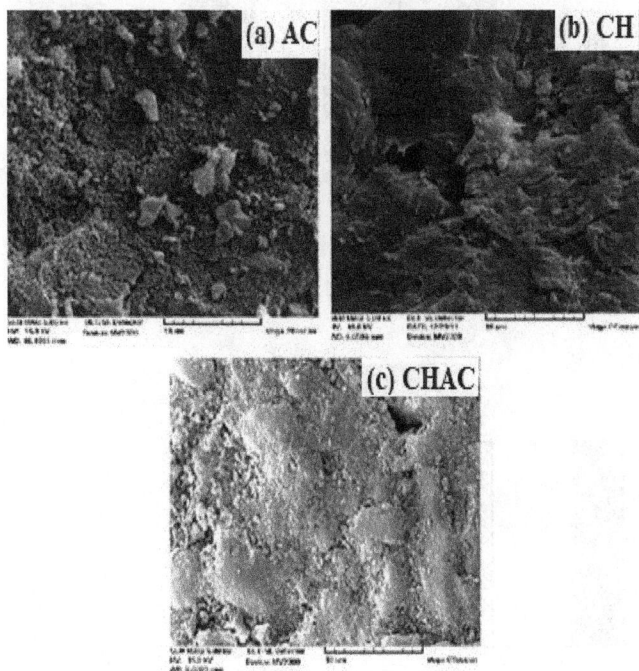

Fig. 1. SEM micrograph of: (a) AC, (b) CH, (c) CHAC

3.2. Effect of removal parameters

To analyse the results in the Taguchi method, the percentage of Cr (VI) and Cd (II) removal was selected as the performance characteristic of the process, and the optimum operating conditions were determined based on this parameter. The observed results for the effect of pH solution, adsorbent particle size and adsorbent dose on Cr (VI) and Cd (II) removal from a single component solution using CHAC are presented in Figure 2. The results indicated that the optimum operating conditions were a pH=6, the particle size of the adsorbent =0.425 mm, and the adsorbent dose = 6 g/L. Under these conditions, 100% of Cr (VI) and Cd (II) were removed from the dilute solution by CHAC after only 2 h. The results for cadmium removal were also presented in our previous article [36]. The pH-dependent metal removal can be mainly related to the functional groups of adsorbent and also to the metal chemistry. The pH$_{ZPC}$ of CHAC was 6.5. At a pH < pH$_{ZPC}$, the surface charge of the adsorbent was net positive and could uptake anions by the electrostatic attraction mechanism; at a pH > pH$_{ZPC}$, the surface charge of the adsorbent was net negative [35]. In the acid media, $HCrO_4^-$, $Cr_2O_7^{2-}$ and H_2CrO_4 were the predominant species of chromium. In basic solutions, chromium exists in the form of CrO_4^{2-}. If the pH ranged from 2 to 6, $HCrO_4^-$ and $Cr_2O_7^{2-}$ were predominant in equilibrium; but at a pH< 1, the main species was H_2CrO_4 [40]. Cadmium existed as a free Cd^{2+} species along the whole acid pH range. The metal complexes precipitated at a pH ≥ 7 and therefore, the separation may not be due to adsorption [41]. In an acidic pH (1-4), the surface charge of CHAC was positive and could attract chromium anions by electrostatic attraction. At a pH > 4, the surface of the adsorbent had a higher negative charge. But the maximum removal percent was achieved at a pH of 6. These results can be attributed to the complexation of Cr (VI) with the surface functional groups of CHAC. According to the experimental results, the strong pH-dependent adsorption indicated that the adsorption was dominated by the surface complexation model for chromium. For cadmium adsorption, in a low pH, the surface charge of CHAC was positive and could not attract Cd^{2+} cations; however, at high pH values, the surface of the adsorbent had a higher negative charge which resulted in a higher cadmium attraction. The obtained results showed that these three parameters affected the performance characteristic of the removal process, but the percentage contribution was different for each parameter. Therefore, the percentage contribution of these parameters should be evaluated. The ANOVA table (Analysis of variance) was used for this objective in the Taguchi method. In the ANOVA table, the F- ratio could be used to determine which parameters had a significant effect on the performance characteristic. The calculated values of the F-ratio were compared with the F values predicted by the statistical F distribution (in Fischer tables) [42] at 95% confidence levels for CHAC. According to the rule, when the F calculated in the ANOVA table was bigger than the standard F, the parameter was significantly influenced by the response variable at the respective confidence level. The results of the ANOVA are shown in Tables 5 and 6 for Cd (II) and Cr (VI) removal by CHAC, respectively [36].

Table 5. Results of ANOVA table for Cd (II) removal by CHAC [36]

Removal parameters	Sums of squares	Var (V)	F-ratio	Percent of contribution
pH	11119.36	5559.68	222.61	97.10
Adsorbent dose	108.96	54.48	2.18	0.51
Adsorbent particle size	121.44	60.72	2.43	0.62
Other/Error	49.94	24.97		1.75

Table 6. Results of ANOVA table for Cr (VI) removal by CHAC [36]

Removal parameters	Sums of squares	Var (V)	F-ratio	Percent of
pH	6324.89	3162.44	85.80	95.29
Adsorbent dose	80.24	40.12	1.08	0.09
Adsorbent particle size	80.90	40.45	1.09	0.10
Other/Error	73.71	36.85		4.49

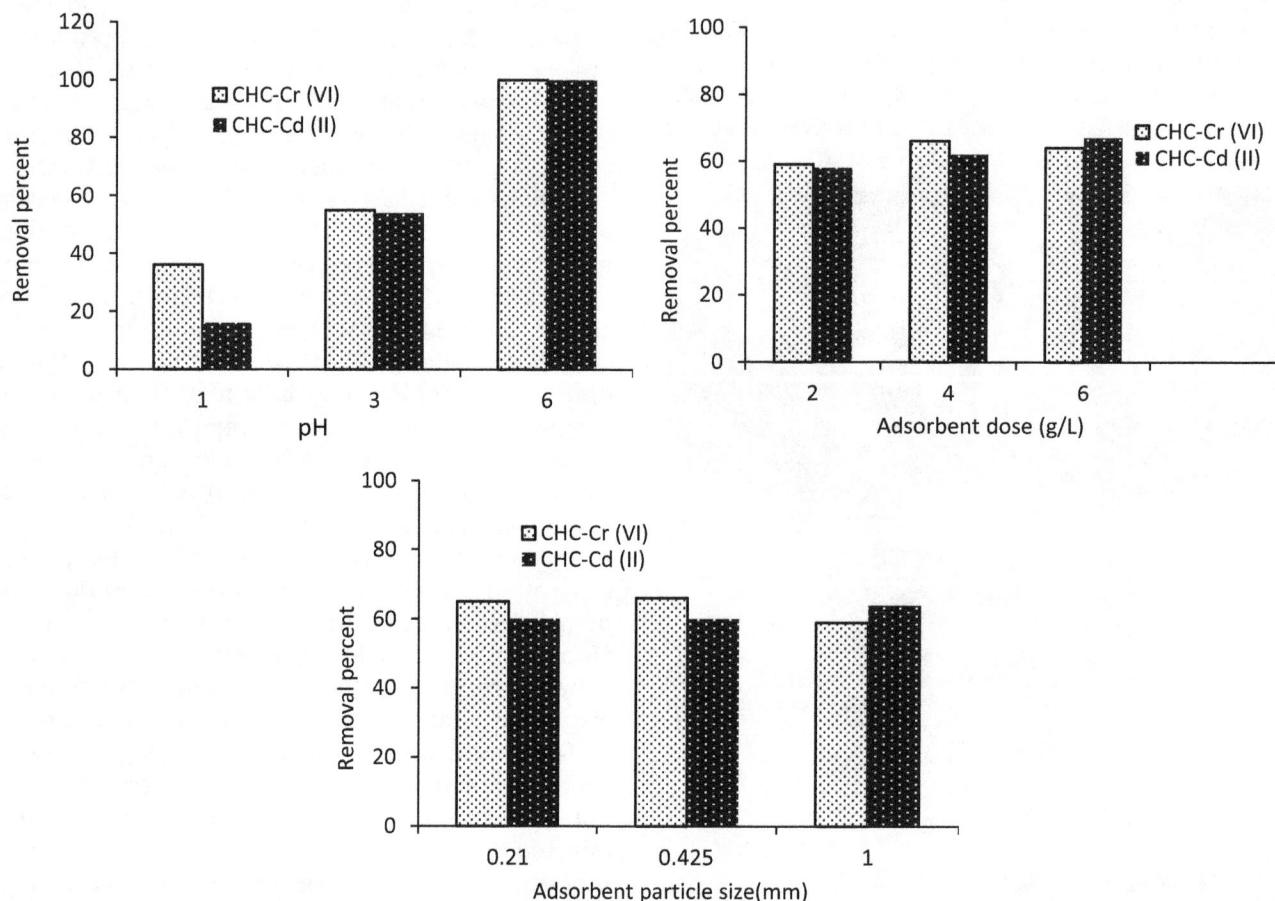

Fig. 2. Influence of (a) pH, (b) adsorbent particle size, (c) adsorbent dose on Cr (VI) and Cd (II) removal by CHAC

These tables indicate that the pH solution was the most significant removal parameter. According to these tables, the maximum percentages of error in these results were 1.754 for Cd (II) and 4.497 for Cr (VI), which was not of significance. The Fisher tables indicated a 95% confidence for Cd (II) and Cr (VI) removal by CHAC,

$F_{0.05,2,2}$ = 19. According to these values, the F calculated in the results in the ANOVA tables was bigger than the standard F for the pH solution. Therefore, this parameter had a significant effect on the performance characteristic for adsorption using CHAC. The F calculated in the ANOVA tables for the adsorbent particle size and the adsorbent dose were smaller than that of the standard F; therefore, it can be concluded that these parameters did not have a main effect on the metals removal.

3.3. Comparison of adsorption performance of AC, CH and CHAC

3.3.1. Adsorption kinetics

Figure 3 shows the effect of contact time on the Cd (II) and Cr (VI) removal from a single component solution using AC, CH, and CHAC. The results demonstrated that the removal rate was faster for the first 5 min. As shown in this figure, the removal process reached equilibrium at 100 min for all adsorbents. The experimental data was tested in terms of pseudo-first-order, pseudo-second-order, and intra-particle diffusion kinetic models. The constants and regression coefficient (R^2) of these models are given in Table 7.

Table 7. Kinetic models parameters for single solute adsorption of Cd (II) and Cr (VI) by AC, CH and CHAC

		Pseudo-first-order			Pseudo-second-order			Intraparticle diffusion		
	$q_{e, exp}$ (mg/g)	k_1 (min^{-1})	q_e (mg/g)	R^2	k_2 (g/mg min)	q_e (mg/g)	R^2	k_{id} (mg/g min$^{1/2}$)	C	R^2
AC-Cd (II)	7.87	0.12	2.95	0.94	0.06	8.03	0.99	0.55	4.80	0.83
CH-Cd (II)	7.67	0.09	5.48	0.94	0.02	7.99	0.99	0.77	2.73	0.97
CHAC-Cd (II)	10.30	0.11	2.76	0.97	0.08	10.60	0.99	0.38	8.10	0.97
AC-Cr (VI)	12.30	0.04	1.35	0.98	0.09	12.50	0.99	0.13	11.1	0.98
CH-Cr (VI)	11.80	0.05	3.90	0.98	0.03	12.04	0.99	0.35	8.49	0.93
CHAC-Cr (VI)	12.50	0.68	0.26	0.96	0.48	12.63	1	0.03	12.2	0.93

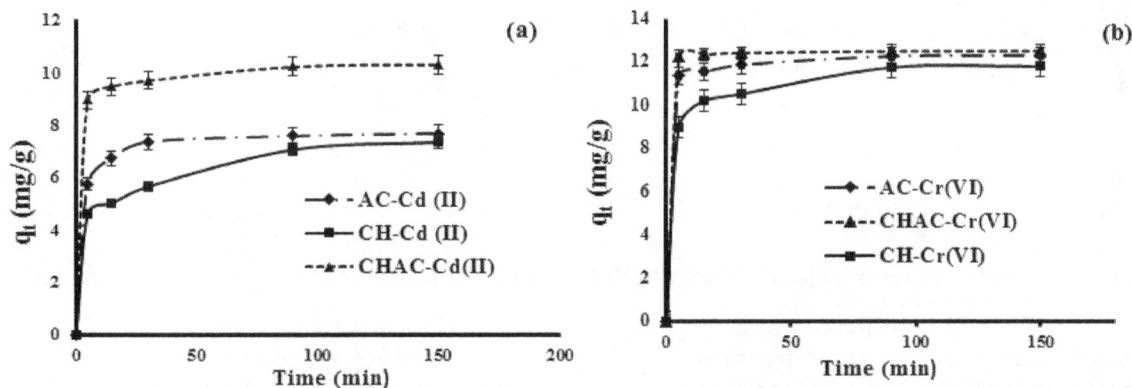

Fig. 3. Influence of contact time on removal efficiency from single solution using AC, CH and CHAN for (a): Cd (II), (b): Cr (VI)

For all adsorbent-metal systems, the pseudo-second order model had a high correlation coefficient ($R^2 \geq$ 0.99). In addition, the calculated q_e from pseudo-second-order model was very similar to experimental q_e for Cd (II) and Cr (VI) adsorption by AC, CH, and CHAC. Therefore, the pseudo-second-order model was the best model to predict the kinetic behavior of Cd (II) and Cr (VI) adsorption. This suggested that the adsorption mechanism was a chemical adsorption for Cd (II) and Cr (VI) adsorption on the three adsorbents. The k_2 values for the adsorption by CHAC were higher than adsorption by AC and CH; this indicated that the metals adsorption by CHAC was faster. The overall reaction kinetics for the adsorption of Cd (II) and Cr (VI) was a pseudo-second-order process. However, this could not

be emphasized in regard to the rate-limiting step. It is always important to predict the rate-limiting step in an adsorption process to understand the mechanism of adsorption. Generally, the dsorption process involves three types of mechanisms:

- The liquid-film diffusion (external diffusion) involves the movement of adsorbate molecules from the bulk of the solution towards the external surface of the adsorbent.
- The adsorption of the adsorbate molecules on the surface of the porous adsorbent. This step is assumed to be very rapid and the binding process can be physical or chemical.

- In the intra-particle diffusion (internal diffusion), the adsorbate molecules move in the interior of the adsorbent particles.

The intra-particle diffusion model suggested by Weber and Morries was used for the probability of intra-particle and liquid-film diffusion mechanisms. The plot of q_t against $t^{0.5}$ can be employed to test the linearity of the experimental values. If the plot was linear and passed through the origin, the internal diffusion was the slowest (rate controlling) step in the adsorption process. But, if the plot was non-linear or linear but did not pass though the origin, it suggested that the adsorption process may be controlled by film diffusion and internal diffusion together. Based on these data, the intra particle diffusion model had high correlation coefficient values ($R^2 \geq 0.97$) for Cd (II) adsorption by CH and CHAC and Cr (VI) adsorption by AC. But the plots of q_t versus $t^{0.5}$ did not pass through the origin and have the intercept; therefore, the intra particle diffusion may

not be the rate controlling step. This indicated that the film/internal diffusion or/and chemical reaction controlled the adsorption rate of Cr (VI) and Cd (II) onto the AC, CH, CHAC surface.

3.3.2. Adsorption Equilibrium

The equilibrium results for Cd (II) and Cr (VI) adsorption from a single component solution (Fig. 4) by AC, CH and CHAC were fitted for the Langmuir, Freundlich and Dubinin-Raduskevich (D-R) isotherms. The values of the constants were calculated and are listed in Table 8. From the data presented in Table 8, it can be said that the Langmiur isotherm best fitted with the equilibrium data of Cd (II) and Cr (VI) adsorption on all adsorbents; this suggested that the present adsorption process was probably controlled by a monolayer adsorption process rather than multiple adsorption ones.

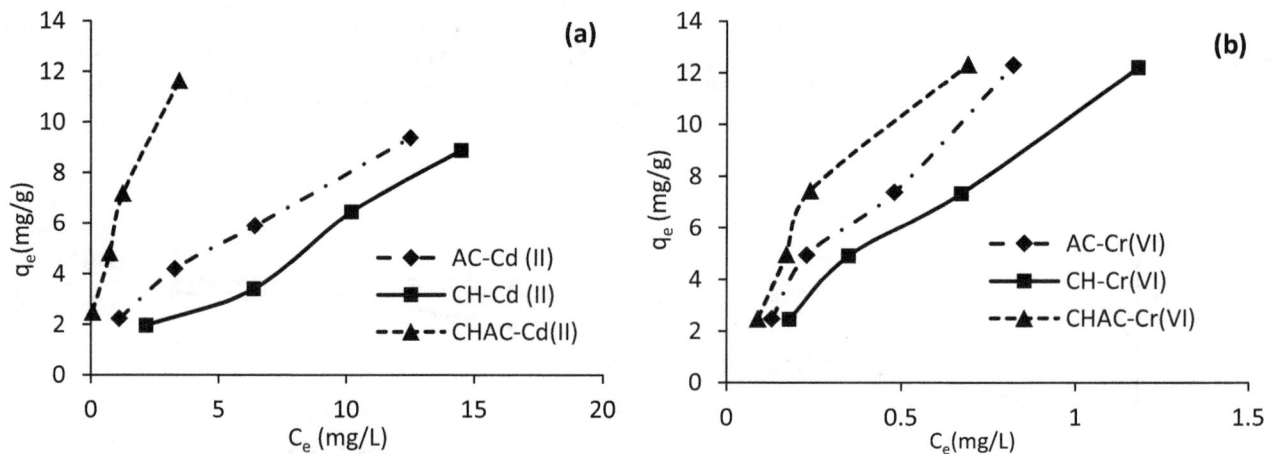

Fig. 4. Adsorption isotherms for (a): Cd (II) and (b): Cr (VI) adsorption from single solution by AC, CH and CHAC

According to the D-R isotherm parameters, the mean free energies E (kJ/mol) of adsorption per molecule of the adsorbate were 10.05 kJ/mol, 12.9 kJ/mol and 13.36 kJ/mol for Cd (II) adsorption and 13.1 kJ/mol, 9.41 kJ/mol and 12.5 kJ/mol for Cr (VI) adsorption using AC,

CH and CHAC, respectively; this corresponded to the ion-exchange process and electrostatic attraction (chemical nature). The maximum adsorption capacities of these adsorbents for Cd (II) and Cr (VI) are presented in Table 8.

Table 8. Equilibrium isotherms parameters for single solute adsorption of Cd (II) and Cr (VI) by AC, CH and CHAC

Adsorption system	Langmuir model			Freundlich model			D-R model			
	q_{max} (mg/g)	K_L (L/mg)	R^2	N	K_f ($mg^{1-(1/n)} L^{1/n} /g$)	R^2	β (mol^2/J^2)	q_m (mol/g)	R^2	E (kJ/mol)
AC-Cd (II)	10.30	0.24	0.98	1.70	2.07	0.94	4.5e-9	6.84	0.89	10.05
CH-Cd (II)	10	0.25	0.99	1.35	1.01	0.92	3e-9	5.19	0.88	12.90
CHAC-Cd (II)	50.50	0.11	0.98	1.34	4.93	0.95	2.6e-9	15.95	0.95	13.90
AC-Cr (VI)	52.63	0.39	0.99	1.21	14.40	0.97	2.9e-9	10.07	0.96	13.10
CH-Cr (VI)	41.60	0.35	0.99	1.11	10.60	0.97	5.7e-9	12.40	0.95	9.41
CHAC-Cr(VI)	90.90	0.43	0.98	1.56	17.80	0.90	3.2e-9	16.36	0.95	12.50

The maximum adsorption capacity of CHAC was higher than AC and CH for both metals adsorption and these

results indicated that by immobilizing CH on AC, the adsorption capacities of these adsorbents improved. In addition, by employing this process, much lower quantities of AC and CH (they are expensive adsorbents) would be needed in the adsorption process which in turn changed the removal process into a cost-effective and environmentally friendly process. The maximum adsorption capacity of other adsorbents for Cd (II) and Cr (VI) are summarized in Table 9. As it can

be seen from Table 9, the maximum adsorption capacities of AC, CH and CHAC were higher than several other adsorbents. The comparison of q_{max} values indicated that these adsorbents exhibited reasonable capacities for Cd (II) and Cr (VI) adsorption from aqueous solutions.

Table 9. Adsorption capacity of different adsorbents for Cd (II) and Cr (VI) adsorption

Adsorbent	Liquid phase concentration (mg/L)	q_m (mg/g)		Ref.
		Cd (II)	Cr (VI)	
Orange peel–Fe$_2$O$_3$	16	71.43	-	[8]
Chitosan	50-400	-	97.4	[26]
Urea-modified wheat straw	-	39	-	[41]
Olive stones	-	5.46	14.34	[43]
Turkish fly ash	0.2-6	0.29	-	[44]
Bentonite-iron oxide composite	10-700	63.29	-	[45]
Coffee grounds	10-700	15.65	-	[46]
Bamboo charcoal	20-100	12.08	-	[47]
PANI/CFs	1-56	-	18.1	[48]
Waste oil palm frond powder	-	-	90.09	[49]
Raw dolomite	5-50	-	10.1	[50]
Activated carbon prepared from peanut shell	10-100	-	16.26	[51]
Sweet potato peels	50-500	18	-	[52]
Fe$_3$O$_4$-SO$_3$H MNP	10-200	80.90	-	[53]
Bamboo charcoal grafted by Cu^{2+}-N-aminopropylsilane complexes	2-12	-	17.93	[54]
Treated waste newspaper	5-70	-	59.88	[55]
Activated carbon prepared from longan seed	50-500	-	169	[56]
AC	10-50	10.3	51.6	This work
CH	10-50	10	41.6	This work
CHAC	10-50	52.63	90.9	This work

3.4. Adsorption of Cd (II) and Cr (VI) from bi-solute solutions

3.4.1. Adsorption kinetic

Figure 5 shows the kinetic adsorption profiles of Cd (II) and Cr (VI) on CHAC from bi-solute and single systems.

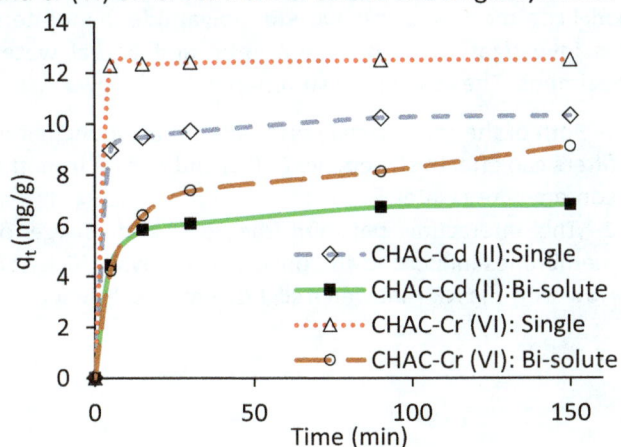

Fig. 5. Influence of contact time on Cd (II) and Cr (VI) removal from single and bi-solute solution by CHAC

The CHAC showed increasing Cd (II) and Cr (VI) adsorption with time, the same as that in the single systems. The results illustrated that in the binary system, the adsorption capacity of CHAC decreased 65% and 50% for Cr (VI) and Cd (II), respectively. This trend was a result of the competitive adsorption of Cd (II) and Cr (VI) ions. These results demonstrated that Cr (VI) was more sensitive to competitive adsorption than the Cd (II). These kinetic data were analyzed in terms of the pseudo-first-order, pseudo-second-order and intra-particle diffusion equations and the parameters of these equations are given in Table 10. Based on these results, the pseudo-second-order was the best model to predict the kinetic behavior of Cd (II) and Cr (VI) removal from bi-solute solutions with CHAC. In addition, the calculated q_e from the pseudo-second-order model was very similar to the experimental q_e for Cd (II) and Cr (VI) removal by CHAC. This suggested that the main adsorption mechanism was a chemical adsorption for these metals on CHAC. Also, the R^2 value of intra-particle model was low (0.8) for adsorption kinetics data of bi-solute systems. These results showed that the film diffusion was important for the adsorption of ions metals from bi-solute solution. The k_2 values for the adsorption of two metals from single solutions were higher than that of the adsorption from bi-solute solutions; this indicated that the competitive adsorption led to a decrease in the adsorption rate.

3.4.2. Adsorption isotherms

The adsorption equilibrium isotherms obtained for Cd(II) and Cr(VI) ions by CHAC in the single and bi-solute system are shown in Figure 6. It was found that the adsorption of each metal decreased in the presence of another ion because of the competition of Cd (II) and Cr (VI) ions for adsorption on the active sites of the CHAC surface. These equilibrium data were analyzed with the Langmuir, Freundlich, and Dubinin–Radushkevich (D-R) isotherms; the constants of these models for bi-solute systems are summarized in Table 11. The results showed that the Langmuir isotherm best fitted with the equilibrium data of both metals adsorption on CHAC in the bi-solute system. The adsorption capacity of CHAC for Cr (VI) removal decreased from 90.9 mg/g in the single system to 41.94 mg/g in the bi-solute system; for Cd (II), it

decreased from 52.63 mg/g in the single system to 30.21 mg/g in the bi-solute system. In general, a mixture of different adsorbates may exhibit three possible types of behaviour: synergism (the effect of the mixture is greater than that of each of the individual adsorbates in the mixture); antagonism (the effect of the mixture is less than that of each of the individual adsorbates in the mixture); and non-interaction (the mixture has no effect on the adsorption of each of the adsorbates in the mixture). The $q_{max,2}/q_{max,1}$ ratios ($q_{max,2}$ and $q_{max,1}$ were the maximum adsorption capacity in the bi-solute and the single system from the Langmuir model, respectively) for the sorption of one metal in the presence of another metal were 0.46 for Cr (VI) and 0.57 for Cd (II) removal by CHAC, respectively. The ratios were all < 1, indicating that the adsorption of these metals was depressed by the presence of other metal ions in the bi-solute solution; hence, the effect of the mixtures seemed to be antagonistic. The $q_{max,2}/q_{max,1}$ for the Cr (VI) adsorption was smaller than the Cd (II) adsorption and this showed that Cr (VI) was very sensitive to the presence of Cd (II). It can be seen that the adsorption capacity followed the order as Cr (VI) > Cd (II) which was consistent with the order in single systems.

Table 10. Kinetic models parameters for bi-solute adsorption of Cd (II) and Cr (VI) by CHAC

Kinetic model	$q_{e,exp}$ (mg/g)	Pseudo-first-order			Pseudo-second-order			Intra-particle diffusion		
		k_1 (min^{-1})	q_e (mg/g)	R^2	k_2 (g/mg min)	q_e (mg/g)	R^2	k_{id} (mg/g min$^{1/2}$)	C	R^2
CHAC-Cd (II) Bi-solute	6.8	0.11	4.18	0.	0.05	6.89	0.99	0.64	3.38	0.85
CHAC-Cr (VI) Bi-	10.1	0.09	51.28	0.	0.02	9.9	0.99	1.14	2.31	0.85

Table 11. Equilibrium isotherms parameters for bi-solute adsorption of Cd (II) and Cr (VI) by CHAC

Adsorption system	Langmuir model			Freundlich model			D-R model			
	q_{max} (mg/g)	K_L (L/mg)	R^2	n	K_f (mg$^{1-(1/n)}$L$^{1/n}$/g)	R^2	β (mol^2/J^2)	q_m (mol/g)	R^2	E (kJ/mol)
CHAC- Cd(II) Bi-solute	30.21	0.03	0.98	1.56	1.15	0.96	4.9e^{-9}	7.31	0.94	10.1
CHAC-Cr(VI) Bi-solute	41.94	0.05	0.98	1.41	2.87	0.9	5.1e^{-9}	9.94	0.95	9.9

4. Conclusions

In this paper, commercially activated carbon (AC), chitosan (CH) and chitosan/activated carbon composite (CHAC) were used as adsorbents for Cr (VI) and Cd (II) removal from single and bi-solute solutions. The Taguchi analysis of the experimental data showed that the solution pH had a significant effect on the removal process and the optimum operating conditions were as follows: pH = 6, particle size of the adsorbent =0.425 mm and adsorbent dose = 6 g/L. Under optimum operation conditions, 100% of the Cr (VI) and Cd (II) were separated from the single-dilute solution with CHAC after 2 h. The Langmuir isotherm provided the best fit with the equilibrium data of Cr (VI) and Cd (II) removal on AC, CH in the single system and CHAC in the single and the bi-solute system. The CHAC had a higher adsorption capacity than that of AC and CH for Cr (VI) and Cd (II). The results showed that the adsorption capacity of CHAC for each metal decreased in the presence of another ion because of the competition of the Cr (VI) and Cd (II) ions; also, Cr (VI) was highly sensitive to this competitive effect. It was observed that the adsorption kinetics of Cr (VI) and Cd (II) on

AC, CH in single systems and CHAC in bi-solute systems could be suitably analyzed with the pseudo-second-order model.commercial spiral wound polyamide nanofilters was investigated for both synthetic and actual water specimens. The results are summarized as follows:

1- Both of the commercial polyamide spiral wound nano-filters can effectively remove Cr (VI) and nitrate from the contaminated water.

2- The interaction between the ions and charge of membranes influenced the pollutant removal efficiency. However, the NF-I showed a slightly better efficiency

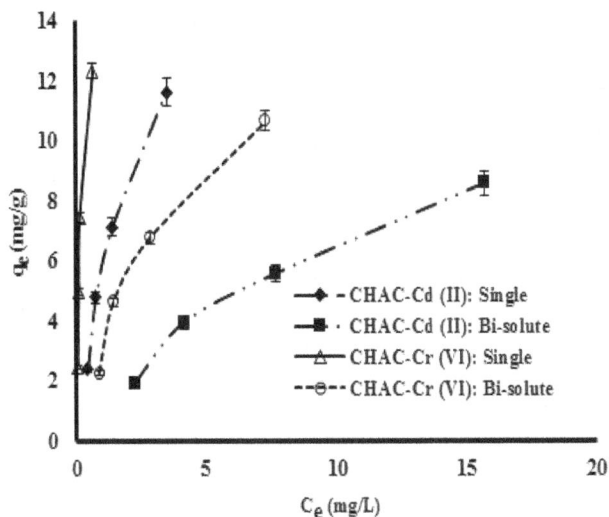

Fig. 6. Adsorption isotherms of Cd (II) and Cr (VI) adsorption from single and bi-solute solution by CHAC

References

[1] Gupta, V. K., Jain, R., Mittal, A., Saleh, T. A., Nayak, A., Agarwal, S., Sikarwar, S. (2012). Photo-catalytic degradation of toxic dye amaranth on TiO_2/UV in aqueous suspensions. *Materials science and engineering: C, 32*(1), 12-17.

[2] Mittal, A., Mittal, J., Malviya, A., Gupta, V. K. (2009). Adsorptive removal of hazardous anionic dye "Congo red" from wastewater using waste materials and recovery by desorption. *Journal of colloid and interface science, 340*(1), 16-26.

[3] Mittal, A., Mittal, J., Malviya, A., Kaur, D., Gupta, V. K. (2010). Decoloration treatment of a hazardous triarylmethane dye, Light Green SF (Yellowish) by waste material adsorbents. *Journal of colloid and interface science, 342*(2), 518-527.

[4] Gupta, V. K., Agarwal, S., Saleh, T. A. (2011). Synthesis and characterization of alumina-coated carbon nanotubes and their application for lead removal. *Journal of hazardous materials, 185*(1), 17-23.

[5] Mittal, A., Mittal, J., Malviya, A., Gupta, V. K. (2010). Removal and recovery of Chrysoidine Y from aqueous solutions by waste materials. *Journal of colloid and interface science, 344*(2), 497-507.

[6] Gupta, V. K., Ali, I., Saleh, T. A., Nayak, A., Agarwal, S. (2012). Chemical treatment technologies for waste-water recycling—an overview. *Royal society of chemistry advances, 2*(16), 6380-6388.

[7] Gupta, V. K., Jain, R., Nayak, A., Agarwal, S., Shrivastava, M. (2011). Removal of the hazardous dye—tartrazine by photodegradation on titanium dioxide surface. *Materials science and engineering: C, 31*(5), 1062-1067.

[8] Gupta, V. K., Nayak, A. (2012). Cadmium removal and recovery from aqueous solutions by novel adsorbents prepared from orange peel and Fe_2O_3 nanoparticles. *Chemical engineering journal, 180*, 81-90.

[9] Saravanan, R., Sacari, E., Gracia, F., Khan, M. M., Mosquera, E., Gupta, V. K. (2016). Conducting PANI stimulated ZnO system for visible light photocatalytic degradation of coloured dyes. *Journal of molecular liquids, 221*, 1029-1033.

[10] Devaraj, M., Saravanan, R., Deivasigamani, R., Gupta, V. K., Gracia, F., Jayadevan, S. (2016). Fabrication of novel shape Cu and Cu/Cu_2O nanoparticles modified electrode for the determination of dopamine and paracetamol. *Journal of molecular liquids, 221*, 930-941.

[11] Gupta, V. K., Kumar, R., Nayak, A., Saleh, T. A., Barakat, M. A. (2013). Adsorptive removal of dyes from aqueous solution onto carbon nanotubes: a review. *Advances in colloid and interface science, 193*, 24-34.

[12] Saleh, T. A., Gupta, V. K. (2012). Column with CNT/magnesium oxide composite for lead (II) removal from water. *Environmental science and pollution research, 19*(4), 1224-1228.

[13] Zaini, M. A. A., Okayama, R., Machida, M. (2009). Adsorption of aqueous metal ions on cattle-manure-compost based activated carbons. *Journal of hazardous materials, 170*(2), 1119-1124.

[14] World Health Organization. (2010). *World health statistics 2010*. World health organization.

[15] Losi, M. E., Amrhein, C., Frankenberger Jr, W. T. (1994). Environmental biochemistry of chromium. In *Reviews of environmental contamination and toxicology* (pp. 91-121). Springer New York.

[16] Levankumar, L., Muthukumaran, V., Gobinath, M. B. (2009). Batch adsorption and kinetics of chromium (VI) removal from aqueous solutions by Ocimum americanum L. seed pods. *Journal of hazardous materials, 161*(2), 709-713.

[17] Vasudevan, S., Lakshmi, J., Sozhan, G. (2011). Effects of alternating and direct current in electrocoagulation process on the removal of cadmium from water. *Journal of hazardous materials, 192*(1), 26-34.

[18] Mekatel, H., Amokrane, S., Benturki, A., Nibou, D. (2012). Treatment of polluted aqueous solutions by Ni2+, Pb2+, Zn2+, Cr+ 6, Cd+ 2 and Co+ 2 Ions by ion exchange process using faujasite zeolite. *Procedia Engineering, 33*, 52-57.

[19] Elkady, M. F., Abu-Saied, M. A., Rahman, A. A., Soliman, E. A., Elzatahry, A. A., Yossef, M. E., Eldin, M. M. (2011). Nano-sulphonated poly (glycidyl methacrylate) cations exchanger for cadmium ions removal: effects of operating parameters. *Desalination, 279*(1), 152-162.

[20] Gherasim, C. V., Bourceanu, G., Olariu, R. I., Arsene, C. (2011). A novel polymer inclusion membrane applied in chromium (VI) separation from aqueous solutions. *Journal of hazardous materials, 197*, 244-253.

[21] Areco, M. M., Hanela, S., Duran, J., dos Santos Afonso, M. (2012). Biosorption of Cu (II), Zn (II), Cd (II) and Pb (II) by dead biomasses of green alga Ulva lactuca and the development of a sustainable matrix for adsorption implementation. *Journal of hazardous materials, 213,* 123-132.

[22] Tang, Y., Chen, L., Wei, X., Yao, Q., Li, T. (2013). Removal of lead ions from aqueous solution by the dried aquatic plant, Lemna perpusilla Torr. *Journal of hazardous materials, 244,* 603-612

[23] Rinaudo, M. (2006). Chitin and chitosan: properties and applications. *Progress in polymer science, 31*(7), 603-632.

[24] Erosa, M. D., Medina, T. S., Mendoza, R. N., Rodriguez, M. A., Guibal, E. (2001). Cadmium sorption on chitosan sorbents: kinetic and equilibrium studies. *Hydrometallurgy, 61*(3), 157-167.

[25] Konaganti, V. K., Kota, R., Patil, S., Madras, G. (2010). Adsorption of anionic dyes on chitosan grafted poly (alkyl methacrylate) s. *Chemical engineering journal, 158*(3), 393-401.

[26] Santana Cadaval Jr, T. R., Camara, A. S., Dotto, G. L., Pinto, L. A. D. A. (2013). Adsorption of Cr (VI) by chitosan with different deacetylation degrees. *Desalination and water treatment, 51*(40-42), 7690-7699.

[27] Repo, E., Warchol, J. K., Kurniawan, T. A., Sillanpää, M. E. (2010). Adsorption of Co (II) and Ni (II) by EDTA-and/or DTPA-modified chitosan: kinetic and equilibrium modeling. *Chemical engineering journal, 161*(1), 73-82.

[28] Ngah, W. W., Teong, L. C., Hanafiah, M. A. K. M. (2011). Adsorption of dyes and heavy metal ions by chitosan composites: A review. *Carbohydrate polymers, 83*(4), 1446-1456.

[29] Wang, Y., Qi, Y., Li, Y., Wu, J., Ma, X., Yu, C., Ji, L. (2013). Preparation and characterization of a novel nano-absorbent based on multi-cyanoguanidine modified magnetic chitosan and its highly effective recovery for Hg (II) in aqueous phase. *Journal of hazardous materials, 260,* 9-15.

[30] Swayampakula, K., Boddu, V. M., Nadavala, S. K., Abburi, K. (2009). Competitive adsorption of Cu (II), Co (II) and Ni (II) from their binary and tertiary aqueous solutions using chitosan-coated perlite beads as biosorbent. *Journal of hazardous materials, 170*(2), 680-689.

[31] Khani, H., Rofouei, M. K., Arab, P., Gupta, V. K., Vafaei, Z. (2010). Multi-walled carbon nanotubes-ionic liquid-carbon paste electrode as a super selectivity sensor: application to potentiometric monitoring of mercury ion (II). *Journal of hazardous materials, 183*(1), 402-409.

[32] Khani, H., Rofouei, M. K., Arab, P., Gupta, V. K., Vafaei, Z. (2010). Multi-walled carbon nanotubes-ionic liquid-carbon paste electrode as a super selectivity sensor:

application to potentiometric monitoring of mercury ion (II). *Journal of hazardous materials, 183*(1), 402-409.

[33] Bansal, R. C., Goyal, M. (2005). *Activated carbon adsorption.* CRC press.

[34] Guibal, E. (2004). Interactions of metal ions with chitosan-based sorbents: a review. *Separation and purification technology, 38*(1), 43-74.

[35] Sharififard, H., Zokaee Ashtiani, F., Soleimani, M. (2013). Adsorption of palladium and platinum from aqueous solutions by chitosan and activated carbon coated with chitosan. *Asia-Pacific journal of chemical engineering, 8*(3), 384-395.

[36] Heydari, S., Sharififard, H., Nabavinia, M., Kiani, H., Parvizi, M. (2013). Adsorption of chromium ions from aqueous solution by carbon adsorbent. *International journal of environmental, chemical, ecological, geological and geophysical engineering, 7*(12), 649-652.

[37] Conshohocken, W., 2000. ASTM international standard test method for apparent density of activated carbon, Designation: D 2854-96, U.S.A.

[38] Montgomery, D. C., 2006. Design and analysis of experiments, third ed., Wiley, New York.

[39] Namasivayam, C., Sangeetha, D. (2006). Removal and recovery of vanadium (V) by adsorption onto ZnCl2 activated carbon: Kinetics and isotherms. *Adsorption, 12*(2), 103-117.

[40] Sahu, S. K., Verma, V. K., Bagchi, D., Kumar, V., Pandey, B. D. (2008). Recovery of chromium (VI) from electroplating effluent by solvent extraction with tri-n-butyl phosphate. *Indian journal of chemical technology, 15*(4), 397-402.

[41] Farooq, U., Khan, M. A., Athar, M., Kozinski, J. A. (2011). Effect of modification of environmentally friendly biosorbent wheat (Triticum aestivum) on the biosorptive removal of cadmium (II) ions from aqueous solution. *Chemical engineering journal, 171*(2), 400-410.

[42] Fisher, R. A. (1925). *Statistical methods for research workers.* Genesis publishing Pvt Ltd.

[43] Rouibah, K., Meniai, A. H., Deffous, L., Lehocine, M. B. (2010). Chromium VI and cadmium II removal from aqueous solutions by olive stones. *Desalination and water treatment, 16*(1-3), 393-401.

[44] Bayat, B. (2002). Comparative study of adsorption properties of Turkish fly ashes: I. The case of nickel (II), copper (II) and zinc (II). *Journal of hazardous materials, 95*(3), 251-273.

[45] Orolínováá, Z., Mockovčiaková, A., Škvarla, J. (2010). Sorption of cadmium (II) from aqueous solution by magnetic clay composite. *Desalination and water treatment, 24*(1-3), 284-292.

[46] Azouaou, N., Sadaoui, Z., Djaafri, A., Mokaddem, H. (2010). Adsorption of cadmium from aqueous solution onto untreated coffee grounds: Equilibrium, kinetics

and thermodynamics. *Journal of hazardous materials, 184*(1), 126-134.

[47] Wang, F. Y., Wang, H., Ma, J. W. (2010). Adsorption of cadmium (II) ions from aqueous solution by a new low-cost adsorbent—Bamboo charcoal. *Journal of hazardous materials, 177*(1), 300-306.

[48] Qiu, B., Xu, C., Sun, D., Wei, H., Zhang, X., Guo, J., Wei, S. (2014). Polyaniline coating on carbon fiber fabrics for improved hexavalent chromium removal. *Roral society of chemidtry advances, 4*(56), 29855-29865.

[49] Shahadat, M., Rafatullah, M., Teng, T. T. (2015). Characterization and sorption behavior of natural adsorbent for exclusion of chromium ions from industrial effluents. *Desalination and water treatment, 53*(5), 1395-1403.

[50] Albadarin, A. B., Mangwandi, C., Ala'a, H., Walker, G. M., Allen, S. J., Ahmad, M. N. (2012). Kinetic and thermodynamics of chromium ions adsorption onto low-cost dolomite adsorbent. *Chemical engineering journal, 179*, 193-202.

[51] Al-Othman, Z. A., Ali, R., Naushad, M. (2012). Hexavalent chromium removal from aqueous medium by activated carbon prepared from peanut shell: adsorption kinetics, equilibrium and thermodynamic studies. *Chemical engineering journal, 184*, 238-247.

[52] Asuquo, E. D., Martin, A. D. (2016). Sorption of cadmium (II) ion from aqueous solution onto sweet potato (Ipomoea batatas L.) peel adsorbent: characterisation, kinetic and isotherm studies. *Journal of environmental chemical engineering, 4*(4), 4207-4228.

[53] Chen, K., He, J., Li, Y., Cai, X., Zhang, K., Liu, T., Liu, J. (2017). Removal of cadmium and lead ions from water by sulfonated magnetic nanoparticle adsorbents. *Journal of colloid and interface science, 494*, 307-316.

[54] Wu, Y., Ming, Z., Yang, S., Fan, Y., Fang, P., Sha, H., Cha, L. (2017). Adsorption of hexavalent chromium onto Bamboo Charcoal grafted by Cu^{2+}-N-aminopropylsilane complexes: Optimization, kinetic, and isotherm studies. *Journal of industrial and engineering chemistry, 46*, 222-233.

[55] Dehghani, M. H., Sanaei, D., Ali, I., Bhatnagar, A. (2016). Removal of chromium (VI) from aqueous solution using treated waste newspaper as a low-cost adsorbent: Kinetic modeling and isotherm studies. *Journal of molecular liquids, 215*, 671-679.

[56] Yang, J., Yu, M., Chen, W. (2015). Adsorption of hexavalent chromium from aqueous solution by activated carbon prepared from longan seed: Kinetics, equilibrium and thermodynamics. *Journal of industrial and engineering chemistry, 21*, 414-422

19

Decolorization of ionic dyes from synthesized textile wastewater by nanofiltration using response surface methodology

Najmeh Askari[1], Mehrdad Farhadian[1*], Amir Razmjou[2]

[1] Department of Chemical Engineering, Faculty of Engineering, University of Isfahan, Isfahan, Iran

[2] Department of Biotechnology, Faculty of Advanced Science and Technology, University of Isfahan, Isfahan, Iran

Keywords:
Environment
Wastewater Treatment
Water Reuse, Nanofiltration
Ionic dye
Textile Wastewater

ABSTRACT

Decolorization of aqueous solutions containing ionic dyes (Reactive Blue 19 and Acid Black 172) by a TFC commercial polyamide nanofilter (NF) in a spiral wound configuration was studied. The effect of operating parameters including feed concentration (60-180 mg/l), pressure (0.5-1.1 MPa) and pH (6-10) on dye removal efficiency was evaluated. The response surface method (RSM) was utilized for the experimental design and statistical analysis to identify the impact of each factor. The results showed that an increase in the dye concentration and pH can significantly enhance the removal efficiency from 88% and 87% up to 95% and 93% for Reactive and Acid dye, respectively. Results showed that dye removal efficiency increased by an increase in pressure from 0.5 to 0.8 MPa, while further increase in pressure decreased the removal efficiency. The maximum dye removal efficiencies which were predicted at the optimum conditions by Design Expert software were 97 % and 94 % for Reactive Blue 19 and Acid Black 172, respectively. According to the results of this study, NF processes can be used at a significantly lower pressure and fouling issue for reuse applications as an alternative to the widely used RO process.

1. Introduction

The textile industry is known as one of the largest consumers of fresh water and color. According to Lucas et al. [1-3] between 25 and 250 m³ water is consumed per each ton of product. Therefore, these industries produce a high volume of wastewater including complex structures such as wetting agents, dyes, fixing agents, softeners and many other additives [4, 5]. Color, high pH, high COD and low biodegradability are characteristics of textile wastewater [6-8] The annual production of synthetic dyes globally, exceeds over 700000 tons and around 10% to 15% of the used dyes in the dyeing process appear in the sewage [9,10] There are around 10000 different dyes which should be monitored in waste water streams [11]. Released textile wastewater containing dyes are the main sources of toxicity in environment, which imposes a significant risk to the living organisms. There are a few reports that have associated the textile chemicals to causing cancer and other mutagenic diseases [12]. Consideration and appropriate treatment should be taken into account to ensure that the final discharge will not cause any harm to society and the environment. A variety of methods such as, chemical precipitation, adsorption, ion exchange process, electro-coagulation, membrane systems, and even biological treatments have been used for dye removal from industrial effluent. Among them, chemical precipitation and reduction processes not only need separation stage and produce high amounts of sludge, but also require significant treatment chemicals [13]. Biologically assisted approaches have also been reported as an inefficient treatment due to the complex and stable structure of synthetic dyes [14]. The pressure driven membrane processes (MF, UF, NF and RO) have been found to be

---

*Corresponding author
E-mail address: m.farhadian@eng.ui.ac.ir

efficient and environmentally friendly. Although RO is known as the most efficient separation technique in terms of permeate quality and dye rejection, the required cost and high pressure as well as fouling phenomenon has limited its application. In comparison with RO, NF can be used in textile wastewater treatment to remove dyes at a significantly lower pressure and thus lesser fouling [15, 16]. Hassani et al. investigated dye removal rate of four different feed types of dyes (Acidic, Disperse, Reactive and Direct) by utilizing a spiral wound nanofiltration membrane (MWCO of 90 KDalton). Their results showed that increasing the dye concentration can lead to higher dye removal efficiency (98 %). Also, at different pressures the removal efficiency for Acidic and Reactive dyes reaches to 99.7 %. However, the effect of pH on color removal efficiency was not considered [17]. Sanchuan et al., found that the trans-membrane pressure and dye concentration can significantly affect the dye retention and water permeability. According to their results, the dye retention, water permeability and salt rejection rate of an aqueous solution containing 2000 mg/l Congo red and 10000 mg/l NaCl were 99.8%, 7.0 l/m^2.h.bar, and lower than 2.0%, respectively. It should be pointed out here that they did not study the optimization of operational parameters [18]. Sahinkaya et al., used a combination of activated sludge with NF to treat denim textile wastewater and to reach reuse standards quality. The COD removal efficiency in the activated sludge reactor was 91±2% and 84±4% based on the total and soluble feed COD, respectively. They managed to achieve 75% color removal efficiency through the adsorption of color on biomass or precipitation within the reactor. The effective parameters and optimum conditions had not been investigated [19]. In an investigation performed by Liu et al. reverse osmosis (RO) and nanofiltration (NF) were evaluated and compared for a textile effluent treatment in terms of COD removal, salinity reduction and permeate flux. However, color removal efficiency and the impressive parameters had not been studied [20]. Acid and Reactive dyes can escape from conventional treatments, because they show resistance against microbial, chemical and photolytic degradation [21]. As mentioned before, NF can be used efficiently for treating textile effluent at low pressure and less fouling, which significantly minimizes dyes escaping. The aim of the present work is to optimize the operational parameters of a NF process for removal of Reactive Blue 19 (RB 19) and Acid black 172 (AB 172) from an aqueous solution. The relationship between dye removal efficiency and three main parameters including pH, initial dye concentration and pressure, were evaluated by applying response surface methodology.

2. Materials

Reactive Blue 19 (with molecular formula of $C_{22}H_{16}N_2Na_2O_{11}S_3$, MW=626 gmol^{-1}, λ_{max}=592nm) and Acid Black 172 (with molecular formula of $C_{20}H_{12}N_3NaO_7S$ and MW=461gmol^{-1}, λ_{max}=572nm) were supplied by Alvan Sabet Co. and were used without further purification. The structures of Reactive Blue 19 and Acid Black 172 are presented in Fig. 1. To adjust pH, HCl (37%), NaOH were purchased from Merck Company of Germany.

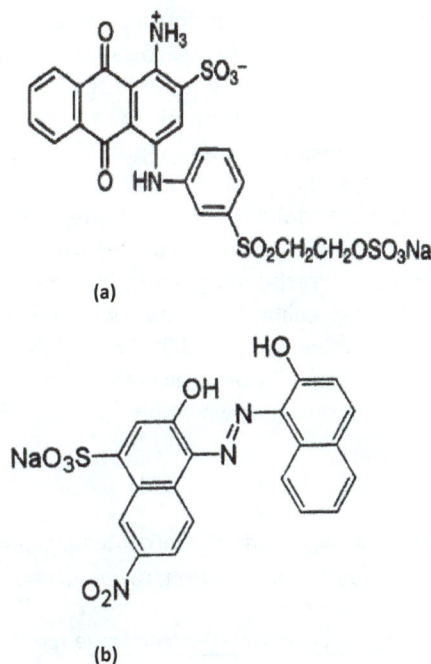

Fig. 1. Reactive Blue 19 structure (a) and Acid Black 172 (b).

2.1. Experimental set-up

In this study as schematically presented in Fig. 2, a continuous co-current NF set up was used. A commercial polyamide thin film composite membrane in a spiral wound configuration was used for the NF process. The membrane specifications are presented in Table 1. The pumps used in this system are diaphragm-type. The pumps output flow and pressure are 0.8 l/min and 6 bar, respectively.

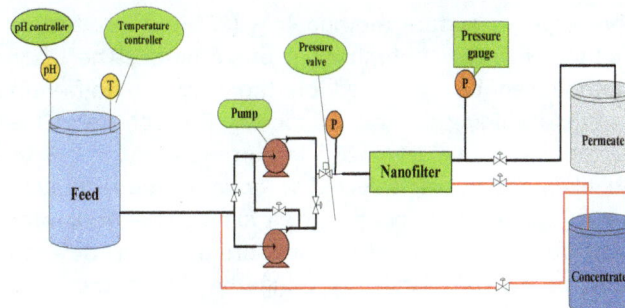

Fig. 2. Schematic represents the NF experimental apparatus.

2.2. Experimental procedure

The synthesized wastewater was prepared by mixing the dye powders (Reactive Blue 19, Acid Black 172) in three concentrations of 60, 120 and 180 mg/L in distilled water. The pH of solution was adjusted by 0.1M HCl and 0.1M NaOH. The temperature of the solution was kept constant at room temperature (25 C^0) with recovery percentage of 75±3% of the feed volume. The recovery percentage was kept constant by varying the feed and permeate fluxes. All measurements were performed according to the American Public Health Association water and wastewater examination methods [22]. The permeated dye concentrations were obtained by analyzing the absorbance at maximum wavelength of each dye (592 and 572nm for RB19 and AB172) using a V-570 spectrophotometer following the standard method No. 2120C (Spectrophotometric method – single wavelength method). The corresponding concentration was then calculated from a calibration curve with ten points (R^2 was 0.98 for reactive dye and 0.95 for acid dye). The dye removal percentage was calculated by using NF membrane rejection as shown below:

$$DR(\%) = \left[1 - \frac{C_P}{C_f}\right] \times 100 \tag{1}$$

Where DR is the dye removal percentage, C_p and C_f are permeate and feed concentration, respectively.

Table1. Commercial polyamide TFC membrane specifications.

Provider	CSM Company, Korea
Skin layer	Polyamide
Maximum tolerable pressure	20 bar
pH range	2-11
Isoelectric point	4.5
Surface electrical charge	Negative
Active surface (m^2)	0.35

2.3. Response surface methodology

The response surface methodology (RSM) is an effective method for the optimization of responses.The RSM method can be employed on the basis of different designs including Central Composite, Box-Behnken, One Factor, D-Optimal, etc. [23]. In current study, the Box-Behnken design was selected to optimize the responses for a three level factors design. A full factorial design for the three parameters of pH, pressure and concentration requires 27 runs while by using the design the total number of experiments was reduced to 15. The

confidence level (CL) for randomly conducted experiments was 95% to avoid possible errors due to the systematic bias. In this study, the aim was to obtain maximum removal percentage of Reactive Blue 19 and Acid Black 172, which were considered as the responses. The contour plots and the analysis of variance (ANOVA) evaluation were used to analyze the results. The contributory factors and their selected levels are presented in Table 2.

Table 2. Factors and selected parameters.

Factors	Level 1	Level 2	Level 3
Dye	60.0±2.0	120±3.0	180.0±5.0
pH	6.00±0.10	0.80±0.1	10.00±0.1
Pressure (MPa)	0.50±0.10	0.80±0.1	1.100±0.1

3. Results and discussion

The experimental design and the responses of the experiments for both dyes are presented in Table 3 and 4. Results are the average values obtained by 3 parallel experiments. The ANOVA table for the dye removal efficiency of Reactive Blue 19 and Acid Black 172 is shown in Table 5 and 6, respectively.

Table 3. Box-Behnken design results for Reactive Blue 19.

Exp. No.	Dye Conc.	pH	Pressure (MPa)	Dye Removal
1	120.0±2	8.00±0.1	0.8±o.1	93.8
2	60.00±3	8.00±0.1	1.1±o.1	90.5
3	180.0±5	10.0±0.1	0.8±0.1	95.7
4	180.0±2	6.00±0.1	0.8±o.1	91.4
5	60.00±2	8.00±0.1	0.5±0.1	89.6
6	60.00±5	10.0±0.1	0.8±0.1	93.4
7	120.0±3	6.00±0.1	1.1±0.1	89.3
8	120.0±3	10.0±0.1	0.5±0.1	91.0
9	180.0±5	8.00±0.1	1.1±0.1	92.8
10	120.0±5	8.00±0.1	0.8±0.1	92.9
11	120.0±2	8.00±0.1	0.8±0.1	93.5
12	120.0±3	10.0±0.1	1.1±0.1	92.7
13	60.00±3	6.00±0.1	0.8±0.1	89.6
14	120.0±3	6.00±0.1	0.5±0.1	88.9
15	180.0±3	8.00±0.1	0.5±0.1	91.7

3.1. Effect of pH

The results in Figs. 3a and 4a indicate that an increase in pH from 6 to 10 has a positive effect on the color removal efficiency. This might be related to the two factors of electrostatic repulsive force and membrane swelling. From Table 1, the isoelectric point for applied commercial NF membranes is at pH of 4.5 above which the membrane surface is negatively charged. With an increase in pH, the electrical repulsive force between the membrane surface and the dye molecules increases and thus the dye removal efficiency rises to 95% and 93% for Reactive Blue and Acid Black, respectively. According to Mahmoodi et al. [24], the membranes will swell at higher pH values and cause the pores to shrink. Reduction in pore size will directly increase the membrane rejection and removal efficiency. However, as the experiment was conducted at constant flux and no significant pressure increment was observed when pH increased, the contribution of pore shrinkage on the removal efficiency enhancement is negligible.

Table 4. Box-Behnken design results for Acid Black 172.

Experiment No.	Dye Concentration ($mg\ L^{-1}$)	pH	Pressure (MPa)	Dye Removal Percentage
1	120.0±3	8.00±0.1	0.8±o.1	91.4
2	120.0±5	10.0±0.1	0.5±o.1	89.2
3	120.0±2	6.00±0.1	0.5±0.1	87.0
4	180.0±2	10.0±0.1	0.8±o.1	93.6
5	60.00±3	10.0±0.1	0.8±0.1	91.0
6	180.0±3	8.00±0.1	1.1±0.1	90.9
7	60.00±2	8.00±0.1	1.1±0.1	88.5
8	180.0±3	6.00±0.1	0.8±0.1	90.2
9	120.0±5	10.0±0.1	1.1±0.1	90.8
10	120.0±3	8.00±0.1	0.8±0.1	91.8
11	180.0±3	8.00±0.1	0.5±0.1	90.0
12	60.00±2	6.00±0.1	0.8±0.1	88.0
13	120.0±3	8.00±0.1	0.8±0.1	91.5
14	120.0±5	6.00±0.1	1.1±0.1	87.4
15	60.00±5	8.00±0.1	0.5±0.1	88.0

Table 5. Analysis of variance for Reactive Blue 19 removal percentage.

Model terms	Mean square error	Sum of the error squares	Degree of freedom	F-value	P-value	Status
Model	5.840	52.60	9	28.060	<0.0001	Significant
A:dye concentration	9.220	9.220	1	44.280	<0.0001	Significant
B: pH	23.94	23.94	1	114.95	<0.0001	Significant
C: P	2.050	2.050	1	9.8400	0.0257	Significant
A×B	0.065	0.065	1	0.3100	0.6004	Not significant
A×C	0.001	0.001	1	0.0480	0.8352	Not significant
B×C	0.440	0.440	1	2.1200	0.2049	Not significant
A×A	0.013	0.013	1	0.0620	0.8132	Not significant
B×B	1.870	1.870	1	8.9800	0.0302	Significant
C× C	15.66	15.66	1	75.170	0.0003	Significant
Lack of fit	0.270	0.810	3	2.2900	0.3184	Not significant

3.2. Effect of dye concentration

Increasing the dyes concentration resulted in higher removal efficiency due to the increase in size exclusion mechanism and space prevention [25]. As can be seen in Figs. 3a and 4a, Reactive Blue 19 removal efficiency is 6% more than that of Acid Black dye. As can be seen in Table 5 and 6, F-values for the concentration parameter are 293 and 44 for Acid Black 172 and Reactive Blue 19 respectively. Therefore, the concentration is the second important factor after pH for removal of both dyes. Dyes and the membrane surface charge at the NF operational pH are negative; as a result, by increasing the dyes concentrations the repulsive forces between the membrane surface and the dyes molecules increase.

3.3. Effect of pressure

In order to increase the pressure, the feed and permeate flow valves were adjusted to increase the feed side pressure while keeping recovery percentage constant. As shown in Figs. 3c and 4c, increasing the pressure from 0.5 to about 0.8 MPa led to an increase in removal efficiency for both dyes. However, further increase in pressure revealed a marginal reduction in the removal efficiency [26]. Operating NF processes at higher pressure increases the chance of concentration polarization which promotes fouling on the membrane surface and consequently increases the intrinsic membrane rejection and removal efficiency [26]. The maximum removal efficiency for dye removal from contaminated water was estimated 97.77%

by Box-Behnken method, which was attained at the optimum conditions of 180 mg l^{-1} of dye, pressure of 0.825 MPa and pH of 9.9 for Reactive Blue 19. However, the maximum efficiency based on the experimental data for Reactive Blue 19 was 95.7% at the concentration of 180 mg l^{-1}, pH of 10 and pressure of 0.8 MPa. The maximum estimated removal efficiency at the concentration of 178 mg/l of dye, pressure about 0.87 MPa and pH of 10 for Acid Black 172 from Box-Behnken method was 93.63% which is close to the experimentally obtained removal efficiency (93.6%, see Table 4)

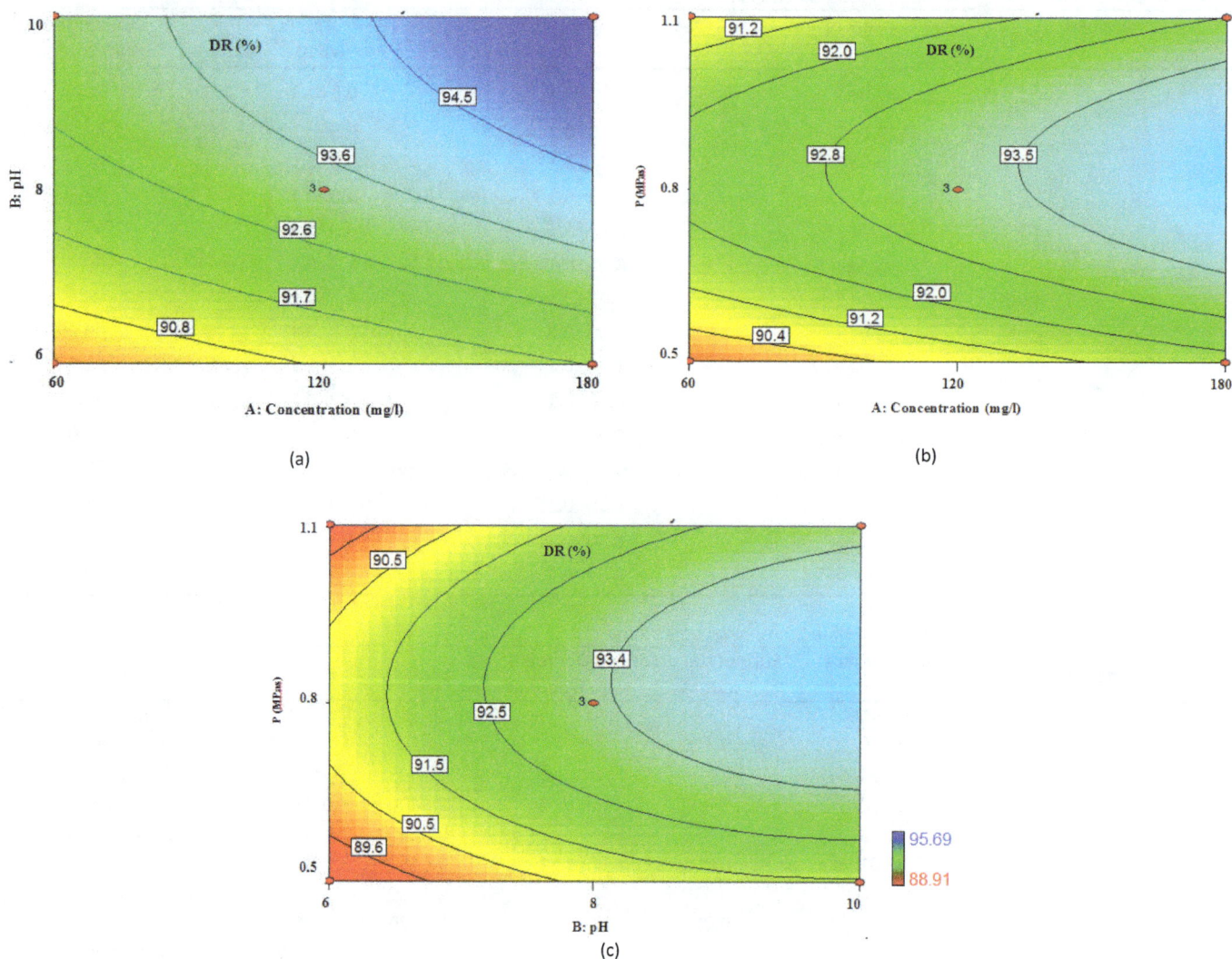

(a)

(b)

(c)

Fig. 3. Contour plots of the Reactive Blue 19 removal efficiency; (a): the effect of dye concentrations and pH on dye removal percentage (DR %) of Reactive Blue 19 at constant pressure. (b): the effect of pressure and dye concentration on the removal efficiency at constant pH (c): the effect of pH and pressure on the removal efficiency at constant dye concentration.

Table 6. Analysis of variance for Acid Blue 172 removal percentage.

Model terms	square Mean error	Sum of the error squares	Degree of freedom	F-value	P-value	Status
Model	5.360	48.22	9	156.95	<0.0001	Significant
A:dye concentration	10.10	10.10	1	293.30	<0.0001	Significant
B: pH	17.17	17.17	1	502.95	<0.0001	Significant
C: P	1.400	1.400	1	41.090	0.0014	Significant
A×B	0.130	0.130	1	3.8000	0.1089	Not significant
A×C	0.034	0.034	1	1.0000	0.3627	Not significant
B×C	0.420	0.420	1	12.380	0.0170	Not significant
A×A	0.013	0.013	1	0.1100	0.7522	Not significant
B×B	2.250	2.250	1	65.870	0.0005	Significant
C× C	17.43	17.43	1	510.67	<0.0001	Significant
Lack of fit	0.092	0.031	3	0.7800	0.6054	Not significant
Pure Error	0.079	0.039	2	-	-	-

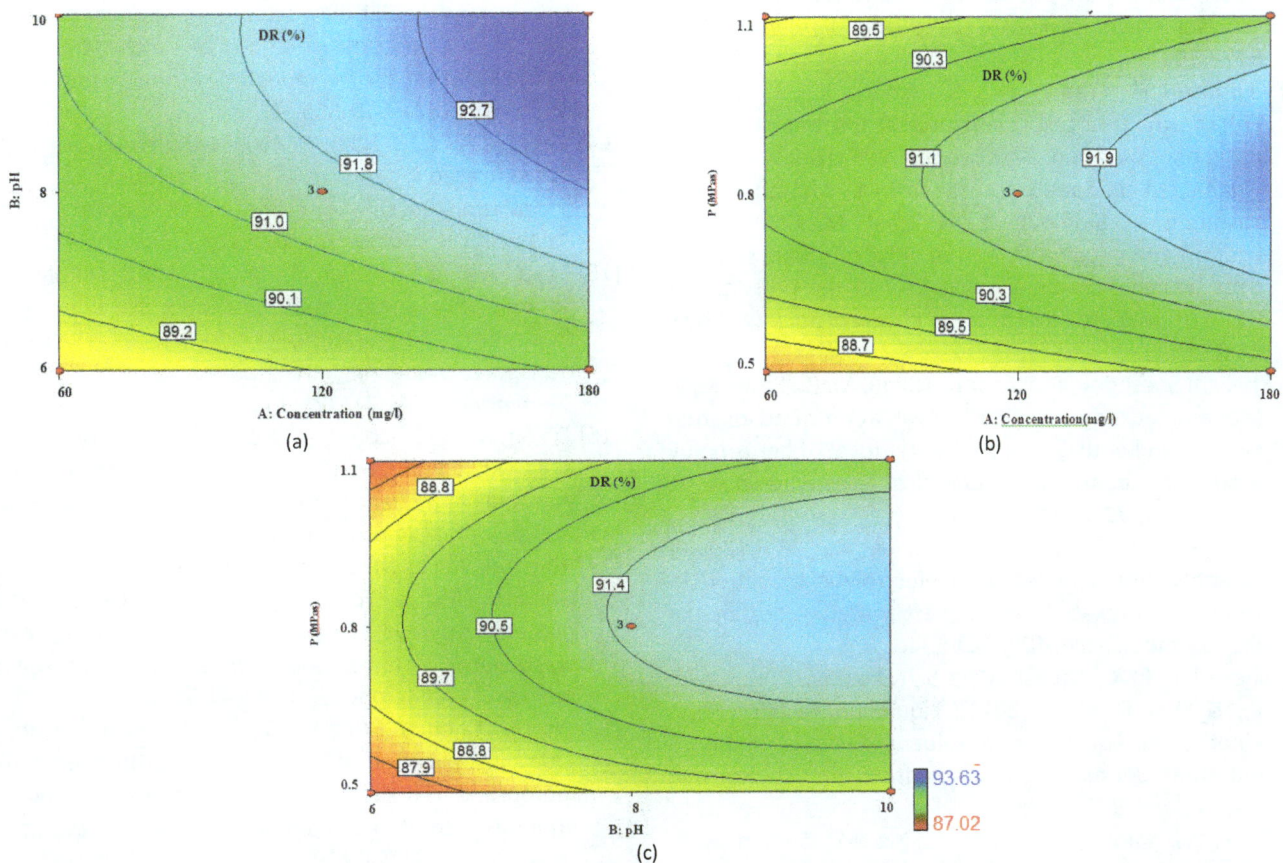

(a)

(b)

(c)

Fig. 4. Contour plots of the Acid Black 172 removal efficiency; (a): the effect of dye concentrations and pH on dye removal (DR %) efficiency of.Acid Black 172 at constant pressure. (b): the effect of P and dye concentration on the removal efficiency at constant pH (c): the effect of pH and pressure on the removal efficiency at constant dye concentration.

4. Conclusions

According to the obtained results, the commercial spiral wound polyamide nanofilter (TFC) was remarkably efficient for removing dyes from textile wastewater. pH had the most significant effect among the other factors (having the highest F-value) on the removal of both applied dyes (Reactive Blue 19 and Acid Black 172). The results indicated that with an increase in pH and dye concentration, the removal efficiency of both colors

increased. Also, our experimental data revealed that under optimum condition, operating pressure had a significant influence on the dye removal efficiency. The agreement between the estimated and obtained removal efficiency showed that the design of experiments using response surface method not only can be considered as a good choice for the experimental design and statistical analysis but also for the optimization of process parameters.

Acknowledgements

The authors of this work would like to gratefully acknowledge the Golnesar Woolen Co. (Isfahan, Iran) for the financial support and Environmental Research Institute (ERJ, University of Isfahan) which helped in the work.

References

[1] Aouni, A., Fersi, C., Cuartas-Uribe, B., Bes-Pía, A., Alcaina-Miranda, M. I., & Dhahbi, M. (2012). Reactive dyes rejection and textile effluent treatment study using ultrafiltration and nanofiltration processes. *Desalination, 297*, 87-96.

[2] Lucas, M. S., & Peres, J. A. (2007). Degradation of Reactive Black 5 by Fenton/UV-C and ferrioxalate/H 2 O 2/solar light processes. *Dyes and pigments, 74*(3), 622-629.

[3] Ellouze, E., Tahri, N., & Amar, R. B. (2012). Enhancement of textile wastewater treatment process using nanofiltration. *Desalination, 286*, 16-23.

[4] Mughal, M. J., Saeed, R., Naeem, M., Ahmed, M. A., Yasmien, A., Siddiqui, Q., & Iqbal, M. (2013). Dye fixation and decolourization of vinyl sulphone reactive dyes by using dicyanidiamide fixer in the presence of ferric chloride. *Journal of saudi chemical Society, 17*(1), 23-28.

[5] Khalighi Sheshdeh, R., Khosravi Nikou, M. R., Badii, K., & Mohammadzadeh, S. (2013). Evaluation of adsorption kinetics and equilibrium for the removal of benzene by modified diatomite. *Chemical engineering & technology, 36*(10), 1713-1720.

[6] Xu, L., Du, L. S., Wang, C., & Xu, W. (2012). Nanofiltration coupled with electrolytic oxidation in treating simulated dye wastewater. *Journal of membrane science, 409*, 329-334.

[7] Sheshdeh, R. K., Abbasizadeh, S., Nikou, M. R. K., Badii, K., & Sharafi, M. S. (2014). Liquid phase adsorption kinetics and equilibrium of toluene by novel modified-diatomite. *Journal of environmental health science and engineering, 12*(1), 148.

[8] Pi, K. W., Xiao, Q., Zhang, H. Q., Xia, M., Gerson, A. R. (2014). Decolorization of synthetic methyl orange wastewater by electro coagulation with periodic reversal of electrodes and optimization by RSM. *Process safety and environmental protection, 92*(6), 796-806.

[9] Sheshdeh, R. K., Nikou, M. R. K., Badii, K., Limaee, N. Y., & Golkarnarenji, G. (2014). Equilibrium and kinetics studies for the adsorption of Basic Red 46 on nickel oxide nanoparticles-modified diatomite in aqueous solutions. *Journal of the taiwan institute of chemical engineers, 45*(4), 1792-1802.

[10] Sinha, K., Saha, P. D., Datta, S. (2012). Response surface optimization and artificial neural network modeling of microwave assisted natural dye extraction from pomegranate rind. *Industrial crops and products, 37*(1), 408-414.

[11] Nabil, G. M., El-Mallah, N. M., Mahmoud, M. E. (2014). Enhanced decolorization of reactive black 5 dye by active carbon sorbent-immobilized-cationic surfactant (AC-CS). *Journal of industrial and engineering chemistry, 20*(3), 994-1002.

[12] Kadam, A. A., Kulkarni, A. N., Lade, H. S., Govindwar, S. P. (2014). Exploiting the potential of plant growth promoting bacteria in decolorization of dye Disperse Red 73 adsorbed on milled sugarcane bagasse under solid state fermentation. *International biodeterioration & biodegradation, 86*, 364-371.

[13] Shirzad-Siboni, M., Khataee, A., & Joo, S. W. (2014). Kinetics and equilibrium studies of removal of an azo dye from aqueous solution by adsorption onto scallop. *Journal of industrial and engineering chemistry,20*(2), 610-615.

[14] Zahrim, A. Y., & Hilal, N. (2013). Treatment of highly concentrated dye solution by coagulation/flocculation–sand filtration and nanofiltration. *Water resources and industry, 3*, 23-34.

[15] Lau, W. J., & Ismail, A. F. (2009). Polymeric nanofiltration membranes for textile dye wastewater treatment: preparation, performance evaluation, transport modeling, and fouling control-a review. *Desalination, 245*(1), 321-348.

[16] Dixon, M. B., Falconet, C., Ho, L., Chow, C. W., O'Neill, B. K., Newcombe, G. (2011). Removal of cyanobacterial metabolites by nanofiltration from two treated waters. *Journal of hazardous materials, 188*(1), 288-295.

[17] Hassani, A. H., Mirzayee, R., Nasseri, S., Borghei, M., Gholami, M., & Torabifar, B. (2008). Nanofiltration process on dye removal from simulated textile wastewater. *International journal of environmental science & technology, 5*(3), 401-408.

[18] Yu, S., Chen, Z., Cheng, Q., Lü, Z., Liu, M., Gao, C. (2012). Application of thin-film composite hollow fiber membrane to submerged nanofiltration of anionic dye aqueous solutions. *Separation and purification technology, 88*, 121-129.

[19] Sahinkaya, E., Uzal, N., Yetis, U., Dilek, F. B. (2008). Biological treatment and nanofiltration of denim textile wastewater for reuse. *Journal of hazardous materials, 153*(3), 1142-1148.

[20] Liu, M., Lü, Z., Chen, Z., Yu, S., Gao, C. (2011). Comparison of reverse osmosis and nanofiltration membranes in the treatment of biologically treated textile effluent for water reuse. *Desalination, 281*, 372-378.

[21] Zahrim, A. Y., Tizaoui, C., Hilal, N. (2011). Coagulation with polymers for nanofiltration pre-treatment of highly concentrated dyes: a review. *Desalination, 266*(1), 1-16.

[22] American public health association, American water, Works association, Water pollution control federation, & water environment Federation. (1915). *Standard methods for the examination of water and wastewater* (Vol. 2). American Public Health Association.

[23] Myers, R. H., Montgomery, D. C., & Anderson-Cook, C. M. (2009). *Response surface methodology: process and product optimization using designed experiments* (Vol. 705). John Wiley & Sons.

[24] Mahmoodi, P., Hosseinzadeh Borazjani, H., Farhadian, M., & Solaimany Nazar, A. R. (2015). Remediation of contaminated water from nitrate and diazinon by nanofiltration process. *Desalination and Water Treatment, 53*(11), 2948-2953.

[25] Ong, Y. K., Li, F. Y., Sun, S. P., Zhao, B. W., Liang, C. Z., & Chung, T. S. (2014). Nanofiltration hollow fiber membranes for textile wastewater treatment: Lab-scale and pilot-scale studies. *Chemical engineering science, 114*, 51-57.

[26] Razmjou, A., Mansouri, J., & Chen, V. (2011). The effects of mechanical and chemical modification of TiO_2 nanoparticles on the surface chemistry, structure and fouling performance of PES ultrafiltration membranes. *Journal of membrane science, 378*(1), 73-84.

20

Evaluation of sequencing batch reactor performance for petrochemical wastewater treatment

Mina Salari, Seyed Ahmad Ataei*, Fereshteh Bakhtiyari
Department of Chemical Engineering, Faculty of Engineering, Shahid Bahonar University of Kerman, Kerman, Iran

Keywords:
Sequencing Batch Reactor (SBR)
Petrochemical wastewater
Optimum operation time

A B S T R A C T

Sequencing batch reactor (SBR) technology has found many applications in industrial wastewater treatment in recent years. The aim of this study was to determine the optimal time for a cycle of the sequencing batch reactor (SBR) and evaluate the performance of a SBR for petrochemical wastewater treatment in that cycle time. The reactor was operated with a suspended biomass configuration under aerobic conditions. Carbon removal and operating parameters such as pH, temperature and dissolved oxygen (DO) were monitored during the wastewater treatment. The SBR was run at different cycle times and amongst the cycle times tested, the best performance was obtained with a 7 h cycle time composed of a fill time of 15min, reaction of 6 h, settling of 30 min, and withdrawal of 15 min. The SBR with the determined cycle time was used to study the treatment of wastewater with various organic loading rates (12.88 gr COD/L.d, 18.02 gr COD/L.d and 31.39 gr COD/L.d). The SBR performance was evaluated by chemical oxygen demand (COD), total solids (TS) total suspended solids (TSS) removal efficiencies. During the shock loading tests, the maximum COD, TS and TSS removal efficiencies were 84%, 67% and 92%, respectively.

1. Introduction

Oil industries require large amounts of water for different purposes such as cooling [1]. Industrial wastewater such as petrochemical wastewater usually has a high COD, low biochemical oxygen demand (BOD), and high total dissolved solids (TDS) as well as containing color, heavy metal and toxic materials. One of the economical and efficient methods for controlling and protecting the environment is the biological treatment of industrial wastewater [2,3]. However, toxic materials in industrial wastewater have an inhibiting effect on the growth of microorganisms. Batch mode operations like sequencing batch reactors can be a viable solution for this problem [4]. SBR is a fill and draw type batch activated sludge process. Wastewater is added to the reactor in a time known as fill time. In the reaction time, the microorganisms under aerobic/anoxic conditions use the pollutant as substrate. In the settling time, the activated sludge is allowed to settle and the effluent is withdrawn from the reactor. In this

process, equalization, aeration and sedimentation are done in the same tank, whereas in continuous flow systems, the operations are conventionally done in separate tanks [5-7]. The advantages of SBR include flexibility in sequence time, minimum space requirement, elimination of a clarifier, and no need for a sludge return pump; nonetheless, it has some disadvantages such as the need for frequent start/stop equipment and a higher pressure drop due to changing liquid levels [8,9]. Previous research has shown that SBR is a suitable activated sludge process for domestic wastewater treatment [7,8]. The SBR has shown relatively efficient performance compared to a conventional activated sludge system in treating complex chemical effluent [4]. This system is useful for treating pharmaceutical [10], dairy [11,12], brewery [13], petroleum refinery [3], wood fiber [14] and chemical wastes [4]. In this study, the treatment of actual petrochemical wastewater at varying organic loading rates in a sequencing batch reactor was studied.

*Corresponding author
E-mail address: ataei@mail.uk.ac.ir

2. Material and methods

2.1. Wastewater characteristics

The wastewater used in this study was collected from an olefin plant in the Pars special economic zone in the south of Iran. This plant consumes ethane and naphtha as feed. The pollutants expected in the feed were components that were derived from steam cracking ethane and naphtha as saturated hydrocarbons, aromatics, phenol and benzene. The characteristics of the feed are shown in Table 1.

Table 1. The characteristics of wastewater in the feed

Sample No.	pH	COD(ppm)	TS(ppm)	TSS(ppm)
1	13	6280	530	210
2	13	5260	2568	303
3	13	7360	2895	790
4	13	12820	4146	1026

2.2. SBR configuration and operation

A sequencing batch reactor was fabricated from Plexiglas material. The reactor had an internal diameter of 13.5 cm and a height of 38.8 cm. The operating volume and working volume was 4.2 L and 3 L, respectively. A schematic diagram of the SBR system used in this study is shown in Figure1. The reactor was operated in suspended growth configuration in sequencing batch mode under aerobic conditions. Feeding, withdrawing and sludge waste were accomplished by gravitational force. The reactor was fed with 3 L of the olefin plant wastewater and the influent flow rate of feed was 3 L/d. The suspended biomass concentration of feed was 3000 mg/L. The COD influent and SRT were 6280 mg/L and 10 days, respectively. Carbon removal and operating parameters such as temperature, pH and dissolved oxygen (DO) were monitored during the wastewater treatment. The air supply was provided by an air compressor (Resun ACO-010) with a constant flow of 4.5 L/min. Influent neutralization and pH was adjusted by sulfuric acid in the feed tank before feeding. The reactor was seeded by activated sludge prepared from an aerobic chamber of an industrial wastewater treatment plant located in the Pars special economic zone in the south of Iran. During the initial 20 days, the SBR was operated with no sludge withdrawal to acclimatize the microbial population to the wastewater. The cycle time and SRT was set at 10 days. The mixed liquor suspended solids (MLSS) at the start of the SBR operation was 3000 mg/L. Then for the first run, the reactor was operated under the following conditions: 15min: fill, 23 h: reaction, 90 min: settling and 15 min: withdrawing cycle time. At the end of the aerobic period, the sludge was removed from the reactor to maintain sludge retention time (SRT) at 10 days. The temperature was almost constant at 22 ºc. The variation of COD, pH, oxidation reduction potential (ORP) and DO during the sequence were monitored. The ability of this system was tested by employing different organic loading rates (OLR) using the 3 synthetic samples presented in Table1. The COD, TS and TSS removal efficiency was monitored at the end of each cycle.

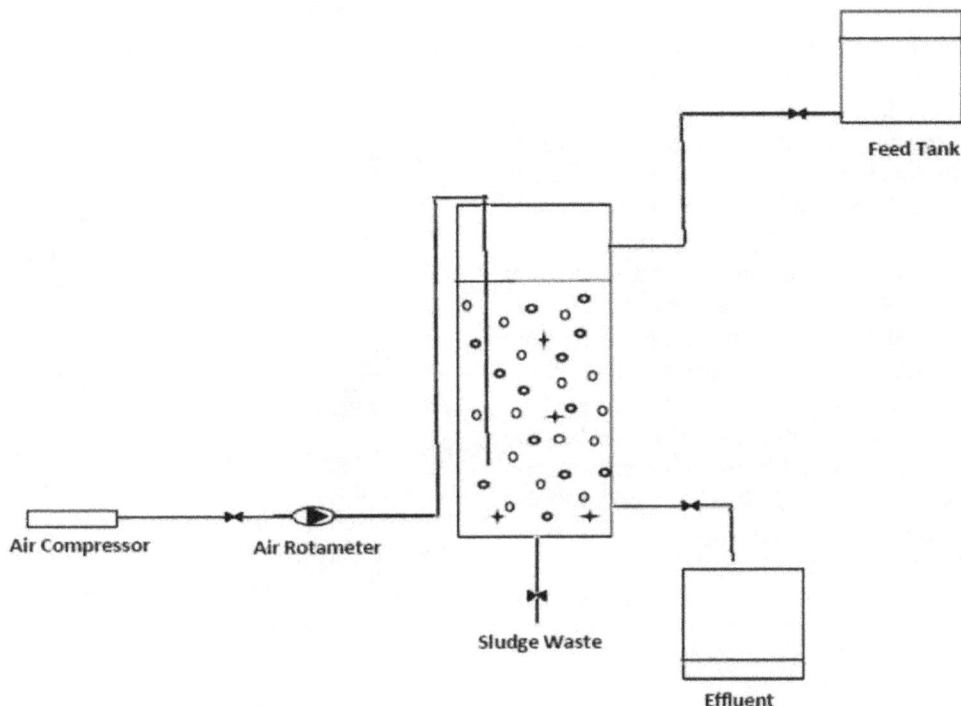

Fig. 1. Schematic of experimental set up

2.3. Analysis

All the analysis including COD (5220 D), TS (2540 B), TSS (2540D), MLSS (2540 G), TSS (2540 D), sludge value index (SVI) (2710D), Turbidity (2130 B), Dissolved Oxygen (DO) (4500-O G. Electrode method), pH (4500-H+ B, Electrometric method) and ORP (2580 B, Electrometric method) were carried out according to the Standard Methods for Water and Wastewater Examination [15].

3. Results and discussion

3.1. Optimum time of aeration and settling phases

Figure 2 shows the efficiency of COD removal. As shown in Figure 2, COD removal efficiency increased with rising aeration time up to 23 h, but the COD removal rate significantly decreased after 6 h of aeration. With regards to the reduction of COD concentration in the early hours, the chemical compounds in the wastewater were not resistant to biodegradation. Aeration is the main parameter in the cost operation, so a decrease in aeration time is very important. Increasing the aeration more than 6 h till 23 h increased the energy consumption almost 4 times, but the COD removal efficiency improved only 11% Based on this; 6 h was selected as the optimum aeration time. The COD removal efficiency increased to 61% in 6 hours and thereafter gradually increased to 72% at the end of the SBR cycle. The variation of turbidity with settling time was measured. Figure 3 shows the effect of settling time on the turbidity of the effluent. Based on the standard of treated wastewater in Iran (Nephlometric turbidity unit (NTU) <20 NTU), the optimum settling time selected was 30 minutes. The minimum time for fill and withdrawal was 15 min, so the optimum cycle time achieved was 15min: fill, 6 h: reaction, 30 min: settling, and 15 min: withdrawal. Thus, the total cycle time result was 7 h and the hydraulic retention time (HRT) was 9.8 h. At this time, the COD efficiency reached 61%. Effluent turbidity, TS and TSS were 17.9NTU, 70 mg/L and 50 mg/L, respectively. The DO concentration in the reactor was kept above 3 mg/L throughout the SBR cycle. For comparison of the results we were drawing out the related data from literature and presented in Table 2. As shown in Table 2, the initial effluent of COD concentration in this research with 6280 mg/L was greater than the other works. The high removal efficiency achieved in the Andereottola and Hudson study was due to its low COD in the feed 1400 and 1500 mg/L respectively and the high cycle time of operation 12 and 24 respectively.

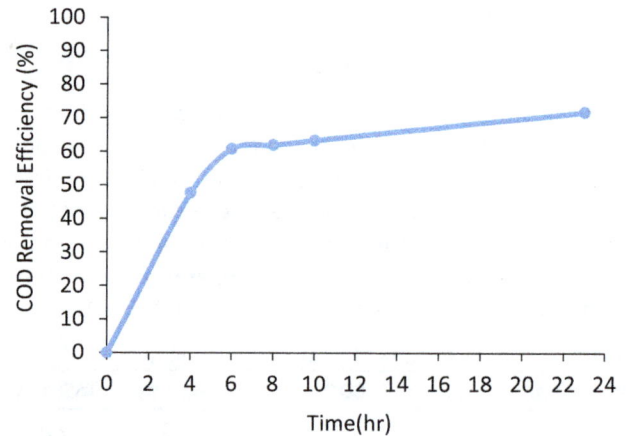

Fig. 2. COD removal efficiency during cycle operation

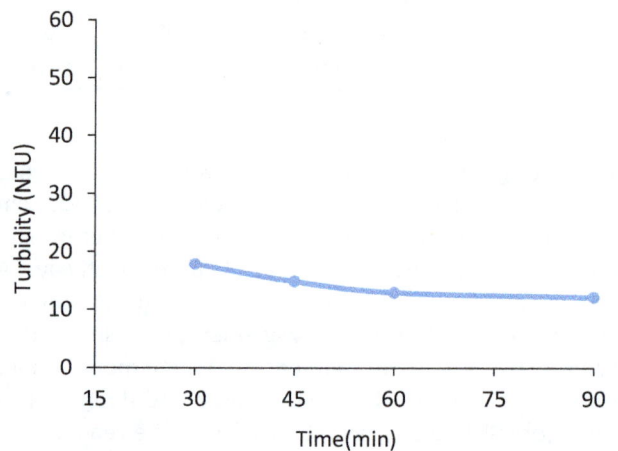

Fig .3. Turbidity variation during cycle operation

Table 2. Comparison with the results of other works on industrial wastewater

Removal efficiency (%)	Total cycle time(h)	COD (mg/L)	References
91	12	1400	Andereottola et al,2001 [16]
93	24	1500	Hudson et al, 2001 [17]
80	24	3500	Venkata et al, 2007 [4]
70	8	4000	Farina et al,2004 [18]
84	7	6280	This research

3.2. SRB Performance

The COD removal efficiency for different organic loading rates is shown in Figure 4. The OLR at each step increased. The efficiency of the reactor to treat the COD of the OLR in the SBR run was 84%. The obtained removal efficiencies are not steady state values and can only be used for comparative purposes.

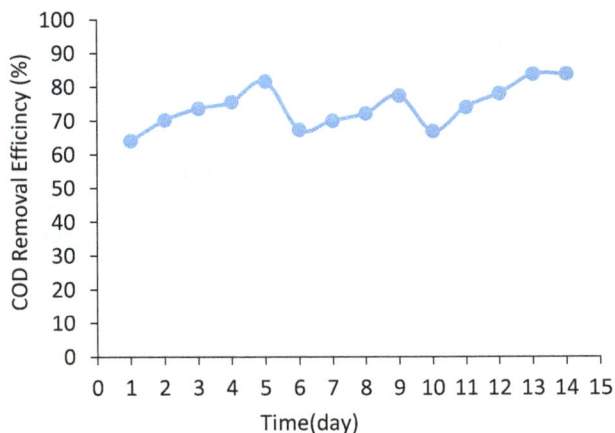

Fig. 4. COD removal efficiency in effluent

4. Conclusions

In this research the performance of sequencing batch reactor on COD removal of petrochemical wastewater was investigated and the operation parameters were optimized. The optimum cycle time for the sequences of SBR was 15min: fill, 6 h: reaction (aeration), 30 min: settling and 15 min: withdrawal. Furthermore, SBR efficiency in the achieved time and at different organic loading rates resulted in the removal of COD: 84%, TS: 67% and TSS: 92%. It can be concluded that the response of SBR to the variation of the organic load rate of wastewater was acceptable. Whereas the reduction of total time of operation is an effective economical factor (increasing the aeration time up to four times), this strategy for treatment of petrochemical wastewater is comparable with another works.

Nomenclature:

BOD	Biochemical Oxygen Demand
COD	Chemical Oxygen Demand
DO	Dissolved oxygen
HRT	Hydraulic Retention Time
MLSS	Mixed Liquor Suspended solid
NTU	Nephlometric Turbidity Unit
OLR	Oxygen Loading Rate
ORP	Oxidation Reduction Potential
SBR	Sequencing Batch Reactor
SRT	Sludge Retention Time
TDS	Total Dissolved Solid
TS	Total Solids
TSS	Total Suspended Solids

References

[1] Llop, A., Pocurull, E., Borrull, F. (2009). Evaluation of the removal of pollutants from petrochemical wastewater using a membrane bioreactor treatment plant. *Water, air, and soil pollution*, *197*(1-4), 349-359.

[2] Mohan, S. V., Rao, N. C., Sarma, P. N. (2007). Low-biodegradable composite chemical wastewater treatment by biofilm configured sequencing batch reactor (SBBR). *Journal of hazardous materials*, *144*(1), 108-117.

[3] Gasim, H. A., Kutty, S. R. M., Isa, M. H. (2010). Biodegradability of petroleum refinery wastewater in batch reactor. *International Conference on Sustainable Building and Infrastructure*, Kuala Lumpur.

[4] Mohan, S. V., Rao, N. C., Prasad, K. K., Madhavi, B. T. V., Sharma, P. N. (2005). Treatment of complex chemical wastewater in a sequencing batch reactor (SBR) with an aerobic suspended growth configuration. *Process biochemistry*, *40*(5), 1501-1508.

[5] Singh, M., Srivastava, R. K. (2011). Sequencing batch reactor technology for biological wastewater treatment: a review. *Asia-pacific journal of chemical engineering*, *6*(1), 3-13.

[6] Al-Rekabi, W. S., Qiang, H., Qiang, W. W. (2007). Review on sequencing batch reactors. *Pakistan Journal of nutrition*, *6*(1), 11-19.

[7] Debsarkar, A., Mukherjee, S., Datta, S. (2006). Sequencing batch reactor (SBR) treatment for simultaneous organic carbon and nitrogen removal-A laboratory study. *Journal of environmental science and engineering*, *48*(3), 169-174.

[8] Mahvi, A. H. (2008). Sequencing batch reactor: a promising technology in wastewater treatment. *Journal of environmental health science and engineering*, *5*(2), 79-90.

[9] Gerardi, M. H. 2010. Troubleshooting the Sequencing Batch Reactor, 1st ed. Wiley publication, New York.

[10] Altaf, M. S., Ali, T. A. (2010). Waste water treatment using sequential batch reactor and development of microbiological method for the analysis of relative toxicity. *Pakistan journal of nutrition*, *9*(6), 574-576.

[11] Bae, T. H., Han, S. S., Tak, T. M. (2003). Membrane sequencing batch reactor system for the treatment of dairy industry wastewater. *Process biochemistry*, *39*(2), 221-231.

[12] Mohseni-Bandpi, A., Bazari, H. (2004). Biological treatment of dairy wastewater by sequencing batch reactor. *Journal of environmental health science and engineering*, *1*(2), 65-69.

[13] Shao, X., Peng, D., Teng, Z., Ju, X. (2008). Treatment of brewery wastewater using anaerobic sequencing batch reactor (ASBR). *Bioresource technology*, *99*(8), 3182-3186.

[14] Ganjidoust, H., Ayati, B. (2004). Use of sequencing batch reactors (SBRs) in treatment of wood fiber wastewater. *Journal of environmental health science and engineering*, *1*(2), 91-96.

[15] A.P.H.A, A.W.W.A, W.P.C.F. (1995). Standard method for the examination of water and wastewater, 19th ed, Washington D.C.

[16] Andreottola, G., Foladori, P., Ragazzi, M. (2001). On-line control of a SBR system for nitrogen removal from industrial wastewater. *Water science and technology*, *43*(3), 93-100.

[17] Hudson, N., Doyle, J., Lant, P., Roach, N., de Bruyn, B., Staib, C. (2001). Sequencing batch reactor technology: the key to a BP refinery (Bulwer Island) upgraded environmental protection system-a low cost lagoon based retro-fit. *Water science and technology*, *43*(3), 339-346.

[18] Farina, R., Cellamare, C. M., Stante, L., Giordano, A. (2004, August). Pilot scale anaerobic sequencing batch reactor for distillery wastewater treatment. *X world congress on anaerobic digestion, Montreal, Canada* (Vol. 20, pp. 8-2).

21

A comparative study of Cu(II) and Pb(II) adsorption by Iranian bentonite (Birjand area) in aqueous solutions

Bahareh Sadeghalvad*, Mojtaba Torabzadehkashi, Amir Reza Azadmehr

Department of Mining & Metallurgical Engineering, Amirkabir University of Technology, Tehran, Iran

Keywords:
Cu(II) and Pb(II) removal
Comparative adsorption
Isotherm models

ABSTRACT

Heavy metals such as Cu(II) and Pb(II) are among the hazardous pollutants that lead to severe ecological problems and have a toxic effect on living organisms. The removal of Cu(II) and Pb(II) in the Iranian bentonite experiment were conducted in single component and multi component systems. The bentonite from the Birjand area was characterized by X-ray diffraction pattern and FTIR spectroscopy. The effects of initial Cu(II) and Pb(II) concentration were investigated on the adsorption process. An equilibrium study was performed and followed by five different isotherm models which included two-parameter (Langmuir, Frendlich, Temkin and D-R) and three-parameter (Khan) models. From the Langmuir isotherm, the equilibrium adsorption capacity for Cu(II) is 21.10 to 22.17 mg/g in single component and multi component systems, respectively; and that of Pb(II) is 57.803 to 40.49 mg/g in single component and multi component systems, respectively. The comparative adsorption of Cu(II) and Pb(II) onto bentonite showed that the affinity for Pb(II) to interact with bentonite is higher than Cu(II). Based on the value of the free energy of adsorption for Cu(II) and Pb(II), the interaction between these ions and Iranian bentonite is a chemical adsorption, that is to say, an ion exchange.

1. Introduction

The undesirable trend of hazardous pollutants such as heavy metals in the environment, which leads to severe toxic effects in living organisms and ecological problems, has gained significant global attention [1, 2]. Copper and lead are spread to the environment by natural and industrial activities. Although high concentrations of Cu(II) produces harmful effects, it is an essential nutrient needed by the body in trace amounts. In contrast, Pb(II) is void of any nutritional value and provides no essential function for the human body. High levels of copper also results in health problems such as kidney and liver damage, gastrointestinal disturbance, and disruption of the nervous system [3, 4]. Accordingly, to alleviate the negative health effects of Cu(II) and Pb(II), it is necessary to remove these metals from the wastewater.

In order to prevent the contamination of subsoil, groundwater, and surface water by these elements, a growing interest in the development of materials capable of adsorbing to heavy metals [1] has evolved. Different adsorbents such as red muds and fly ashes [5], waste and natural indigenous materials [6], calcite [7], natural carbonaceous materials [8], and chitosan immobilized on bentonite [3] have already been used for the removal of copper, lead, or both from aqueous solutions. Based on current research, bentonite is one of the significant adsorbents that remove and separate lead and copper from solutions because of its high cation exchange capacity (90–120 meq g^{-1}), high elasticity and plasticity, low cost, easy handling, high abundance as an adsorbent, and easy refining [9,10]. Bentonite is a volcanic clay with montmorillonite as its main clay mineral constituent and belongs to the Smectitic clay minerals with a (Na, Ca)$_{0.33}$(Al,

*Corresponding author
E-mail address: b_sadegholvad@yahoo.com

$Mg)_2Si_4O_{10}(OH)_2 \cdot nH_2O$ formula [11, 12]. Montmorillonite has a layer structure with one octahedral coordinated layer of aluminum as the basic structural unit, which is sandwiched between two tetrahedral coordinated layers of silicon. The electrostatic interaction between these layers is a weak Van Der Waals force, which facilitates the intercalation process. The net negative surface charge on the bentonite originates from the substitution of an aluminum ion for a silicon ion and magnesium or iron ions for an aluminum ion in the tetrahedral and octahedral sheets, respectively. The exchangeable cations such as H^+, Na^+, and Ca^{2+} on the surface layer are counterbalance charges of bentonite [4, 12, 13]. Since the suitability of Iranian bentonite (Birjand area) for Cu(II) and Pb(II) simultaneous adsorption has not been characterized as of yet, this study was conducted to investigate its effectiveness and evaluate the key factors involved in the adsorption process in the individual and simultaneous adsorption of these ions.

2. Material and Methods

2.1. Materials and physical instrument

A representative sample of bentonite from the Birjand area in southeastern Iran was used without any chemical pretreatment. The sample was ground and sieved by ASTM standard sieves to obtain a particle size of −150 μm in diameter. All chemical compounds were purchased from the Merck Company and used without further purification. The copper and lead solution was prepared by dissolving $Cu(NO_3)_2$ and $Pb(NO_3)_2$ in deionized water. The Pb(II) and Cu(II) concentration after adsorption was determined using a Unicom 939 atomic absorption spectrometer (AAS). The XRD pattern and XRF were obtained by a Philips X-ray diffract meter 1,140 (λ = 1.54 A, 40 kV, 30 mA, calibrated with Si-standard) and a Philips X-ray diffract meter Xunique II (80 kV, 40 mA, calibrated with Si-standard), respectively. The Fourier transform infrared (FTIR) spectra was recorded on a Shimadzu IR instrument from 4,000 to 400 cm^{-1} by using KBr pellets.

The adsorption percentage and results based on mg/g were calculated by the following equations:

$$\alpha = \frac{C_0 - C_e}{C_0} \times 100 \tag{1}$$

$$q_e = C_0 - C_e \times \frac{V}{m} \tag{2}$$

where α is adsorption percentage, C_0(mg/L) is the initial concentration, C_e(mg/L) is the concentration in the solution after the adsorption process, q_e(mg g^{-1}) is the amount of adsorption, V(L) is the volume of the solution, and m (g) is the mass of bentonite.

2.2. Experimental method

The lead and copper adsorption experiments were carried out by using batch equilibrium. All the adsorption experiments were conducted in a 250mL glass reactor using a magnetic stirrer bar at ambient temperature. In this study, the influence of lead and copper initial concentration (from 100 to 1250 mg/L) on total adsorption was investigated in the condition of 1.5 g bentonite with a particle size of -150 μm at 298 K with the stirring rate of 600 rpm during a contact time of 30 min. To investigate adsorption isotherms models (Langmuir, Freundlich, Temkin, Dubinin–Radushkevich and Khan), a weighted amount of bentonite (1 g) was added to 100mL of lead solution with different concentrations ranging from 100 to 1250 mg L^{-1}. These experiments were conducted in two part, in the individual solution of Cu(II) or Pb(II) and in the solution which contained both of Cu(II) and Pb(II) ions with the same initial concentration.

After each test, the solution was filtered off and the concentration of the remaining metal ions in the solution was determined by atomic absorption spectrometry.

2.3. Equilibrium studies

The capacity of a sorbent to adsorb a specific metal ion and relations between adsorbent and adsorbate in equilibrium are usually characterized by isotherm models. Isotherm models have an important role in the identification of an adsorption process and the design of an adsorption system. To achieve this goal, five different isotherm models which include two (Langmuir, Frendlich, Temkin and D-R) and three-parameter (Khan) models were used for analysis of the equilibrium adsorption of copper and lead ions onto the bentonite. All the models were analyzed in individual and simultaneous adsorption of Cu(II) and Pb(II). The equations of these models are represented in Table 1. The two-parameter models were fitted by the linear form of the equations and their parameters were calculated by linear regression; further, non-linear regression MATLAB software was employed for determination of the constant value in the three-parameter models.

In addition to the correlation coefficient (r^2), the data of the adsorption evaluation set to confirm the best fit of isotherm models, the Standard deviation (S.D.) was also calculated according to following equation:

$$S.D = \sqrt{\frac{\Sigma[(q_{e,exp} - q_{e,cal})/q_{e,exp}]2}{(n-1)}} \tag{3}$$

Where $q_{e,exp}$ is the experimental data of adsorption and $q_{e,cal}$ is the calculated data from the adsorption models and number of experimental data. If the obtained data from the model were similar to the experimental data, the S.D will be a small number near to zero; if the S.D tends towards 1, it shows that the obtained data were not fit to the selected model.

3. Result and Discussion

3.1. Characterization of the Bentonite

X-ray diffraction (XRD) is an essential tool in the identification and characterization of bentonite. Fig.1 shows the XRD pattern of bentonite. It reveals that the constituents of this sample are quartz, oligoclase, gypsum, montmorillonite, and illite. The diffractions at 8.93°, 19.89°, and 26.69° (2θ) correspond to the (001), (002), and (003) planes overlapped with the illite diffraction peak. The bentonite samples contained a modest amount of illite impurity. In addition, the main impurity in this sample was quartz, which is observed in the XRD pattern with the main diffraction peak at 27.73º (2θ) [11, 14]. Infrared spectroscopy is usually used to identify solid-state structures and functional groups of clay. The presence of two bands at 3,627, and 3,432 cm^{-1} indicate the stretching vibration of the O–H bond (Fig. 2). The band at 3,627 cm^{-1} is due to the hydroxyl linkage of the bentonite structure and the broad band at 3,432 cm^{-1} is due to the presence of the water molecule in the structure [15]. The bending vibration band of H–O–H for the water molecules is observed at 1,635 cm^{-1}. The strong and broad band at 1,040 cm^{-1} can be attributed to the Si–O stretching vibration of the Si–O–Si linkage of the tetrahedral sheets in the montmorillonite structure [16]. The two bands at 467 and 519 cm^{-1} represent the bending vibration of the Si–O–Al and Si–O–Si bonds, respectively. The vibration band at 693 cm^{-1} and its small shoulder (with a lower intensity, which is not assigned in Fig. 2) can be related to the deformation and bending modes of the Si–O–Si bond. The presence of quartz in this sample is indicated by a shoulder at 796–777 cm^{-1} [11, 17, 18].

3.2. Effect of Cu and Pb initial concentration

For set conditions of bentonite to a liquid ratio (m/V= 15 g/L) and the particle size of −150 μm, the adsorption of lead and copper ions onto the bentonite has been studied for different initial concentrations ranging from 100 to 1250 mg L^{-1}. The results are shown in Fig. 3. The removal percentage of metal ions was decreased by increasing the initial concentration from 100 to 1250 mg L^{-1}. This shows that the Cu (II) and Pb(II) uptake was limited to the active adsorption sites of bentonite

3.3. Effect of Cu and Pb initial concentration

In order to investigate equilibrium isotherm models in this study, five different models which include two (Langmuir, Frendlich, Temkin and D-R) or three-parameter models (Khan) were examined. The equations of the models and their parameters value were represented in Table.1 and 2, respectively.

3.4. Equilibrium studies of individual Cu and Pb adsorption onto bentonite

In order to investigate equilibrium isotherm models in this study, five different models which include two (Langmuir, Frendlich, Temkin and D-R) or three-parameter models (Khan) were examined. The equations of the models and their parameters value were represented in Table 1 and 2, respectively.

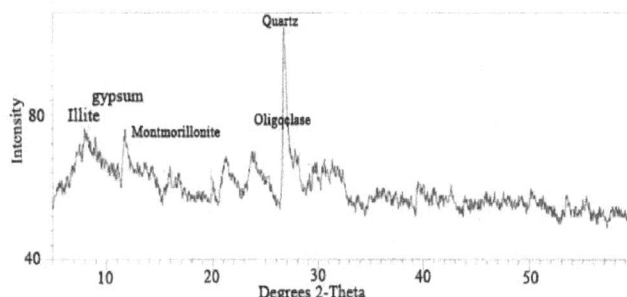

Fig. 1. XRD pattern of bentonite.

• *Langmuir isotherm model*

The linear form of the Langmuir isotherm model, presented in Table 1, is assuming that the adsorption of metal ions is homogeneous (uniform) onto the surface with a finite number of adsorption sites and identical energy. In Langmuir's equation, KL is a constant that is related to the binding energy of the active site of the adsorbent. The value of the adsorption coefficients K$_L$, that are related to the apparent energy of sorption for Pb(II), was greater than Cu(II) which indicates that the Pb(II) ions and bentonite are more significantly interacted (Table 2). The maximum adsorption capacity is calculated from the slope of C$_e$/q$_e$ vs. C$_e$ (Table 1). The maximum adsorption capacity of Pb(II) and Cu(II) are 57.8 and 21.097 mg/g, respectively, with a correlation coefficient (r^2) greater than 0.94. These values of adsorption capacity showed that the energy of adsorption was not very favorable to Cu (II), hence not all binding sites may be available to Cu(II).

In addition to the correlation coefficient (r^2) and Standard Deviation (S.D.) for assessing the accuracy of estimation, the model can be displayed in terms of a dimensionless constant called separation factor R$_L$ (also called equilibrium parameter) which is presented by this equation:

$$R_L = \frac{1}{1 + K_L C_0} \tag{3}$$

Where K$_L$ is the Langmuir constant which represents adsorption intensity (L/mg). In the range of initial concentrations of Cu (II) and Pb(II) from 100 to 1250 mg L^{-1}, the separation factor is from 0.13 to 0.65 and 0.0064 to 0.074, respectively. This data indicates that the copper and lead adsorption onto bentonite for various initial concentrations is favorable and reversible because

0<R$_L$<1 [10]. The high value of K$_L$ in Pb(II) adsorption was a reason for the decrease of the separation factor and with an increase in the Pb(II) initial concentration, the separation factor approached to zero.

- *Freundlich isotherm model*

The Freundlich isotherm model describes the processes of non-ideal, multilayer adsorption on heterogeneous surfaces and assumes that the adsorption process is reversible. According to the presented linear form of equation in Table.1, the Freundlich isotherm has two constants, K$_F$ and 1/n, which express the relative adsorption capacity and adsorption intensity (or surface heterogeneity), respectively. This parameter determined from an intercept and slope of a plot of lnq_e against lnC_e. The value of these parameters (Table. 2) shows that 1/n is 0.35 and 0.14 for copper and lead adsorption onto bentonite, respectively. A 1/n between 0 and 1 indicate a favorable process and 1/n below unity implies a chemisorptions process. Accordingly, the Cu(II) and Pb(II) adsorption onto bentonite is favorable and chemisorption [19].

Table 1. Equations of different isotherm models.

models	Equation	Descriptions	ref
Two parameters:			
Langmuir	$$\frac{C_e}{q_e} = \frac{1}{q_m . K_L} + \frac{C_e}{q_m}$$	q_m maximum adsorption capacity (mg/g), K_L Adsorption intensity or Langmuir (L/mg).	[19]
Freundlich	$$\log q_e = \log K_F + \frac{1}{n} \log Ce$$	K_F relative adsorption capacity (mg$^{1-1/n}$ L$^{1/n}$g^{-1}), n adsorption intensity	[19]
Temkin	$$q_e = \frac{RT}{b_T} LnA_T + \frac{RT}{b_T} LnC_e$$	A$_T$ and b$_T$ temkin constant that b$_T$ is related to the heat of adsorption (J mol^{-1}), R gas constant (8.314 J mol K^{-1}),T absolute temperature (K)	[19]
Dubinin-Radushkevich	$$Lnq_e = Lnq_{max} - \beta\varepsilon^2$$ $$\varepsilon = RT\, Ln(1 + 1/C_e)$$ $$E = \frac{1}{\sqrt{2\beta}}$$	β activity coefficient related to sorption energy (mol2/KJ2), E free energy per molecule of adsorbate (KJ) which represent: — physical adsorption If E<8 kJ mol^{-1} — chemical absorption or ion exchange If 8<E<16 kJ mol^{-1} — particle diffusion if E>16 kJ mol$^-$	[19]
Three parameter:			
Khan	$$q_e = \frac{q_m b_K C_e}{(1 + b_K C_e)^{a_K}}$$	b_K Khan constant, a_K Khan exponent	[21,22]

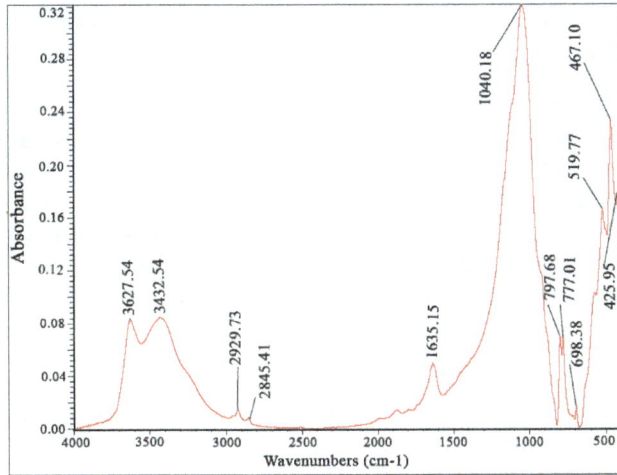

Fig. 2. FTIR spectrum of bentonite.

- *Temkin isotherm model*

The linear equation of the Temkin isotherm model is shown in Table 1. The binding energies of adsorbent–adsorbate interactions are assumed to be uniformly distributed, which causes a linear decrease in the heat of adsorption. On the other hand, the Temkin model has modified the Langmuir model by considering the effect of temperature in the adsorption processes. This isotherm equation has an A_T and a b_T constant which are determined from the intercept and slope of the linear plots of q_e versus $\ln C_e$ and are given in Table 2. The b_T constant is related to the heat of adsorption (J/mol) and A_T indicates the potential of adsorption. The amount of A_T, which are 0.1223 and 45.4028 for Cu(II) and Pb(II) respectively, confirms that the potential of Pb(II) adsorption is greater than the Cu(II) adsorption. However, the value of b_T represented that the heat of the Cu(II) adsorption was more than the Pb(II) adsorption onto bentonite.

Table 2. The isotherm model constants for Pb and Cu individual adsorption onto bentonite.

	model	parameter	Amount for Cu adsorption	Amount for Pb adsorption
Two parameter models	Langmuir	$q_m(mg\,g^{-1})$	21.097	57.803
		R^2	0.947	0.994
		R_L	0.13-0.65	0.006-0.074
		$K_L(L\,mg^{-1})$	0.005	0.125
		S.D	0.470	0.220
	Freundlich	$K_F(mg^{1-1/n}\,L^{1/n}\,g^{-1})$	1.678	23.850
		$1/n$	0.346	0.142
		R^2	0.982	0.935
		S.D	0.022	0.020
	Temkin	$b_T(J/mol)$	694.312	441.504
		$A_T(L/mg)$	0.122	45.403
		R^2	0.902	0.897
		S.D	0.140	0.130
	Dubinin-Radushkevich	$q_m(mg\,g^{-1})$	27.990	72.167
		$\beta(mol2\,KJ2^{-1})$	5.00E-09	1.00E-09
		$E(KJ\,mol^{-1})$	10	22.361
		R^2	0.960	0.890
		S.D	0.014	0.020
Three parameter model	Khan	q_m	0.609	9.856
		a_K	0.628	0.841
		b_K	10.117	163.040
		R^2	0.970	0.835

- *Dubinin–Radushkevich isotherm model*

The empirical Dubinin–Radushkevich model isotherm assumes homogenous surface adsorption [20] and its linear equation is given in Table.1. The q_{max} and β are calculated from the intercepts and slopes of the linear plots of $\ln q_e$ versus ε^2, respectively, and are given in Table 2. The calculated maximum adsorption capacity (q_{max}) is 27.99 and 72.17 mg/g for Cu (II) and Pb(II) adsorption onto bentonite, respectively, that is confirmed this parameter in the Langmuir isotherm. This model is usually applied to distinguish between physical and chemical adsorption of metal ions with the calculation of adsorption free energy. The adsorption free energy (E) is 10 and 22.36 kJ, respectively (Table 2), which means that the adsorption of Cu(II) onto bentonite is a chemical absorption or ion exchange process while the adsorption of Pb(II) onto bentonite is a particle diffusion process.

- *Khan isotherm model*

The Khan isotherm model is a three-parameter isotherm that was offered in 1997 for aromatic adsorption by activated carbon [21]. The equation of this model is presented in Table 1 which a_K and b_K are two constants and are provided in Table 2. The Khan isotherm equation also reflects the combined features of the Langmuir and Freundlich isotherm equations. When a_K verges to unity, it converges to the Langmuir isotherm. The equation can be converted to a Freundlich-type isotherm when the value of the term b_K becomes much greater than unity [22]. As shown in Table 2, a_K is 0.63 and 0.84 for Cu(II) and Pb(II) adsorption onto bentonite, respectively. It was seen that the Pb(II) adsorption verged to unity and the khan isotherm model was approached to Langmuir for its adsorption onto bentonite. Thus, the parameters of these two models can be adapted to each other. The b_K parameter is related to binding energy, similar to K_L of Langmuir, so that the value of b_K confirmed that the Pb(II) adsorption performed better than Cu(II) onto the bentonite. Furthermore, the maximum adsorption capacity of the khan isotherm is lower than the Langmuir model. According to these two models shown in Table 2, the Langmuir model with a higher correlation coefficient is better fitted to the experimental data than the khan model and it is more acceptable.

Table 3. The isotherm model constants for Pb and Cu simultaneous adsorption onto bentonite.

model	parameter	Cu in presence of Pb	Pb in presence of Cu
Langmuir	q_m(mg/g)	22.173	40.486
	R^2	0.968	0.995
	R_L	0.270-0.830	0.040-0.360
	K_L(L mg^{-1})	0.002	0.0175
	S.D	0.330	0.164
Freundlich	K_F(mg$^{1-1/n}$ L$^{1/n}$ g^{-1})	0.472	6.089
	1/n	0.505	0.255
	R^2	0.983	0.966
	S.D	0.030	0.400
Temkin	b_T(J mol^{-1})	482.253	382.913
	A_T(L mg^{-1})	0.0183	0.531
	R^2	0.967	0.980
	S.D	0.130	0.037
Dubinin-Radushkevich	q_m(mg g^{-1})	31.649	59.531
	β(mol2/KJ2)	7.00E-09	3.00E-09
	E(KJ mol^{-1})	8.451	12.909
	R^2	0.977	0.976
	S.D	0.010	0.012

3.5. Evaluation of individual Adsorption Isotherm Models of cu and Pb onto bentonite

According to the results that are summarized in Table 2, it was found that the affinity for Pb(II) to interact with clay is greater than Cu(II). The regression coefficients r^2 values represent that the five isotherm models provide a good correlation for the sorption of the two metal ions. The good correlation coefficient confirms the monolayer adsorption of Pb(II) and multilayer adsorption of Cu(II) onto the surface of bentonite.

3.6. Equilibrium studies of Cu and Pb simultaneous adsorption onto bentonite

As mentioned in the experimental method section, Cu(II) and Pb(II) were mixed and the equilibrium adsorption was investigated similar to the individual adsorption. The results of these analysis were represented in Table 3. It demonstrated that the amount of Pb(II) adsorption in the presence of Cu(II) in comparison to its individual adsorption, decreased from 57.803 to 40.486. This indicates that in a mixture of these two ions, Cu(II) could inhibit the Pb(II) adsorption. However, the Cu(II) adsorption in the presence of the Pb(II) ion is a little more than the individual adsorption of Cu(II) onto bentonite. Generally, competition between Cu(II) and Pb(II) for adsorption onto Iranian bentonite has showed the following affinity order : Pb(II) > Cu(II), which has the same affinity as their adsorption onto Chinese natural bentonite [12]. When comparing the performance of Iranian bentonite with Turkish bentonite, the Iranian bentonite efficiency in Cu(II) and Pb(II) adsorption was much better than the Turkish bentonite [4]. Furthermore, according to the correlation coefficients which are presented in Table 3, similar to individual adsorption, Pb(II) adsorption in the presence of Cu(II) was fitted to the Langmuir model better than the Freundlich model. So, in the presence of Cu(II), only a monolayer of Pb(II) was adsorbed onto the bentonite. By the way, the adsorption of Cu(II) onto the bentonite in the presence of Pb(II) was a multilayer process. According to the adsorption free energy (E) of the Dubinin–Radushkevich model, the adsorption of Cu(II) in the presence of Pb(II) and also Pb(II) in the presence of Cu(II) onto the bentonite are a chemical adsorption or ion exchange process.

4. Conclusion

The removal of Cu(II) and Pb(II) by Iranian bentonite was investigated in an individual and simultaneous system by the batch adsorption technique. In comparing the individual adsorption of Cu(II) and Pb(II) onto bentonite, it was found that the affinity for Pb(II) to interact with bentonite was higher than Cu(II). The Pb(II) ion was adsorbed homogeneously in the monolayer process onto the bentonite and the Cu(II) had multilayer adsorption on the heterogeneous surfaces of bentonite. The simultaneous system competition between Cu(II) and Pb(II) in adsorption onto the bentonite caused the adsorption capacity of Pb(II) to decrease from 57.803 mg/g to 40.49 mg/g; however, the adsorption of Cu(II) increased from 21.10 to 22.17 mg/g which showed that in the simultaneous system, Cu(II) can prevail over Pb(II) in adsorption onto the bentonite. In the simultaneous system, like the individual ones, the adsorption of Pb(II) was done homogeneously onto the bentonite in a monolayer process and Cu(II) had multilayer adsorption on the heterogeneous surfaces of the bentonite. Also, both of the adsorptions were done on a chemical absorption or an ion exchange process.

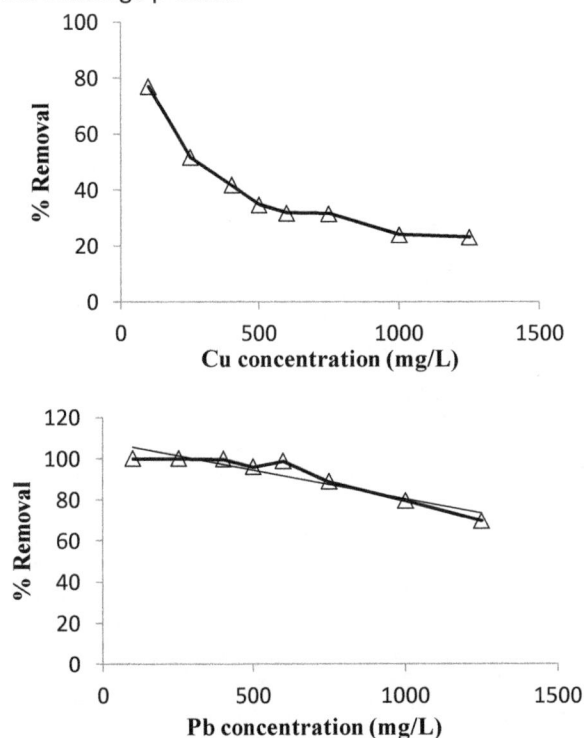

Fig. 3. Effect of Pb(II) and Cu(II) ion concentration on their adsorption onto bentonite, m/V: 15 g/L, T: 298 K, particle size: 150 μm, stirring speed: 600 rpm, contact time: 30 min.

Reference

[1] Demirbas, A. (2004). Adsorption of lead and cadmium ions in aqueous solutions onto modified lignin from alkali glycerol delignication. *Journal of hazardous materials*, *109*(1), 221-226.

[2] Lin, S. H., Lai, S. L., Leu, H. G. (2000). Removal of heavy metals from aqueous solution by chelating resin in a multistage adsorption process. *Journal of hazardous materials*, *76*(1), 139-153.

[3] Futalan, C. M., Kan, C. C., Dalida, M. L., Hsien, K. J., Pascua, C., Wan, M. W. (2011). Comparative and competitive adsorption of copper, lead, and nickel using chitosan immobilized on bentonite. *Carbohydrate polymers*,*83*(2), 528-536.

[4] Korkut, O., Sayan, E., Lacin, O., Bayrak, B. (2010). Investigation of adsorption and ultrasound assisted desorption of lead (II) and copper (II) on local bentonite: A modelling study. *Desalination*, *259*(1), 243-248.

[5] Apak, R., Tütem, E., Hügül, M., Hizal, J. (1998). Heavy metal cation retention by unconventional sorbents (red muds and fly ashes). *Water research*, *32*(2), 430-440.

[6] Rahman, M.A., S. Ahsan, S. Kaneco, H. Katsumata, T. Suzuki, and K. Ohta. (2005). Wastewater treatment with multilayer media of waste and natural indigenous materials. *Journal of environmental management*, 74, 107-110.

[7] Yavuz, O., Guzel, R., Aydin, F., Tegin, I., Ziyadanogullari, R. (2007). Removal of cadmium and lead from aqueous solution by calcite. *Polish journal of environmental studies*, *16*(3), 467.

[8] Hanzlik, P., Jehlicka, J., Weishauptova, Z., Sebek, O. (2004). Adsorption of copper, cadmium and silver from aqueous solutions onto natural carbonaceous materials. *Plant soil and environment.*, *50*(6), 257-264.

[9] Bergaya, F., Lagaly, G. (2013). *Handbook of clay science* (Vol. 5). Newnes.

[10] Sadeghalvad, B., Karimi, H. S., Hosseinzadegan, H., Azadmehr, A. R. (2014). A comparative study on the removal of lead from industrial wastewater by adsorption onto raw and modified Iranian Bentonite (from Isfahan area). *Desalination and water treatment*, *52*(34-36), 6440-6452.

[11] Sadeghalvad, B., Armaghan, M., Azadmehr, A. (2014). Using iranian bentonite (Birjand area) to remove cadmium from aqueous solutions. *Mine water and the environment*, *33*(1), 79-88.

[12] Liu, Y., Shen, X., Xian, Q., Chen, H., Zou, H., Gao, S. (2006). Adsorption of copper and lead in aqueous solution onto bentonite modified by 4′-methylbenzo-15-crown-5. *Journal of hazardous materials*, *137*(2), 1149-1155.

[13] Hu, Q. H., Qiao, S. Z., Haghseresht, F., Wilson, M. A., Lu, G. Q. (2006). Adsorption study for removal of basic red dye using bentonite. *Industrial & engineering chemistry research*, *45*(2), 733-738.

[14] Caglar, B., Afsin, B., Tabak, A., Eren, E. (2009). Characterization of the cation-exchanged bentonites by XRPD, ATR, DTA/TG analyses and BET measurement. *Chemical engineering journal*, *149*(1), 242-248.

[15] Wang, S., Dong, Y., He, M., Chen, L., Yu, X. (2009). Characterization of GMZ bentonite and its application in the adsorption of Pb (II) from aqueous solutions. *Applied clay science*, *43*(2), 164-171.

[16] Wang, Q., X. Chang, D. Li, Z. Hu, R. Li, Q. He. (2011). Adsorption of chromium (III), mercury (II) and lead (II) ions onto 4-aminoantipyrine immobilized bentonite. *Journal of hazardous materials*, 186, 1076-1081.

[17] Klinkenberg, M., R. Dohrmann, S. Kaufhold, and H. Stanjek. (2006). A new method for identifying Wyoming bentonite by ATR-FTIR. *Applied clay science*, 33, 195-206.

[18] Yang, S., Zhao, D., Zhang, H., Lu, S., Chen, L., Yu, X. (2010). Impact of environmental conditions on the sorption behavior of Pb (II) in Na-bentonite suspensions. *Journal of hazardous materials*, *183*(1), 632-640.

[19] Foo, K.Y., Hameed, B. H. (2010). Insights into the modeling of adsorption isotherm systems. *Chemical engineering journal*, *156*(1), 2-10., *73*(8), 2720-2727.

[20] Hobson, J. P. (1969). Physical adsorption isotherms extending from ultrahigh vacuum to vapor pressure. *The Journal of physical chemistry,* 73(8), 2720-2727.

[21] Khan, A. R., Ataullah, R., Al-Haddad, A. (1997). Adsorption studies of some aromatic pollutants from dilute aqueous solutions on activated carbon at different temperatures. *Journal of colloid and interface science*, *194*(1), 154-165.

[22] Liu, Y., Liu, Y. J. (2008). Biosorption isotherms, kinetics and thermodynamics. *Separation and purification technology*, *61*(3), 229-242

22

Removal of copper (II) from aqueous solutions by sodium alginate/hydroxy apatite hydrogel modified by Zeolite

Afsaneh Barekat[1]*, Masoomeh Mirzaei[2]

[1] Department of Chemistry, Mahshahr Branch, Islamic Azad University, Mahshahr, Iran
[2] Department of Chemical Engineering, Mahshahr Branch, Islamic Azad University, Mahshahr, Iran

Keywords:
Copper
Hydrogel
Zeolite
Sodium alginate
Adsorption

ABSTRACT

The study presented in this article investigated the removal of copper ions from aqueous solutions by a synthetic hydrogel-forming adsorbent polymer based on sodium alginate (SA) and hydroxy apatite (HA) nanoparticles. The effect of adding Zeolite on the adsorption performance of this hydrogel was also investigated, and the optimum amount of Zeolite was determined by changing its quantity. The FTIR spectrum determined the structure of the synthesized adsorbent; non-continuous adsorption tests were performed to study the kinetics and thermodynamics of adsorption and also the recovery of the adsorbent. The degree of adsorption of the synthesized nanocomposite was compared with that of Zeolite, and the results showed that the maximum adsorption capacities of Zeolite and the nanocomposite for Cu ions were 29.7 and 75.8 mg/g, respectively. The kinetic studies indicated that the process of adsorption of Cu ions on both absorbents followed a pseudo second order kinetic equation. It took the Zeolite and the hydrogel 90 and 120 minutes, respectively, to reach equilibrium. The thermodynamic studies showed that Cu absorption by both adsorbents matched the Langmuir isotherm very well ($R^2=0.99$). Since adsorbent recovery and its lifespan are of significant importance in absorption processes, recovery was carried out by hydrochloric acid (2% by weight). The repulsion coefficient of the recovered adsorbent and its efficiency in five recovery cycles were measured. The results of the tests indicated that the repulsion coefficient of Cu was 70-82.75 percent and the adsorption efficiency of Cu after 5 recovery cycles was 75 percent of the initial adsorbent.

1. Introduction

At present, various sources of industrial wastewater, including that of petrochemical, paper and pesticide industries, tanneries, etc., contain impermissible amounts of heavy metals such as chromium, mercury, cadmium, lead, etc. The accumulation of these heavy metals has created a massive problem for microorganisms [1]. The human body needs Cu, but excess levels results in problematic conditions such as nervous disorders, various types of cancer, anemia, fatigue, nausea and vomiting, weakened immune system, etc. [2]. Researchers have introduced various methods including ion exchange,

electrochemical reactions, membrane filtration, and adsorption for removing and reducing heavy metals present in aqueous solutions. The cost and environmental effects of adsorbents are factors that must be considered in their selection. Adsorption has been widely used for water purification because it is an inexpensive and non-toxic method that is easy to design and implement [3-6]. Hydrogels are a class of polymeric materials with a three-dimensional network structure (physical or chemical cross-linking). In recent studies, hydrogels have exhibited very high adsorption capacities for removing pollutants such as heavy metals [7]. SA is a hydrophilic polysaccharide with an anionic nature, mainly composed of the cell walls of marine

*Corresponding author
E-mail address: afsaneh.barekat@yahoo.com

brown algae. This is an environmentally friendly/biodegradable and non-toxic natural polymer. The physical and chemical properties of alginates are directly related to their molecular structures. In 2016, Nichelle et al. produced hydrogels that were based on sodium alginate and used them to remove methylene blue dye. They performed continuous and discontinuous experiments, obtained the total pollutant adsorption in the column of 255.5 mg/g of the adsorbent, and the discontinuous experiments indicated that the adsorption data matched very well with Langmuir's isotherm [8]. Xiao et al. (2016) produced the SA adsorbent by adding graphene oxide nanoparticles and used it to adsorb heavy metals in discontinuous systems. SEM images showed that graphene oxide increased the porosity of the adsorbent; the adsorption tests indicated that the maximum adsorption of Cu and lead ions were 98 and 257 mg/g of the adsorbent, respectively [9]. Anne et al. (2015) studied the competitive adsorption of bivalent metal ions of metals such as Pb, Ni, Ca, and Cu by using hydrogel granules in continuous and discontinuous experiments. The kinetic experiments showed that 80 percent of the adsorption took place within four hours, equilibrium was reached in 48 hours, the adsorbent could be reused by using 4 percent hydrochloric acid, and the adsorbent could be recovered for nine times with 90 percent of its adsorption capacity [10]. K. Vijayalakshmi studied the use of biopolymer grains consisting of three nanochitosan adsorbents NCS/SA/microcrystal cellulose (MC) in a 1:8:2 ratio as an inexpensive adsorbent for removing Cu from aqueous solutions using the discontinuous method [11]; their laboratory results matched the pseudo second order kinetic model very well (R^2=0.999). The present research conducted a laboratory study of Cu removal from aqueous solutions by means of a synthetic hydrogel-forming adsorbent polymer based on SA/HA nanoparticles with the Zeolite additive and compared the performance of Zeolite and the synthesized hydrogel.

2. Theoretical relations

The parameter of adsorption capacity (mg/g) and adsorption percentage (R) are used in adsorption calculations using the following

$$q = \frac{C_0 - C_e}{m} \times V \qquad (1)$$

where C0 (ppm) and C_e are pollutant concentrations before and after the adsorption process, V is the volume of the solution in liters, and m is the mass of the hydrogel in mass units.

$$R = \frac{C_0 - C_e}{C_0} \times 100 \qquad (2)$$

where C_0 is the incoming concentration, C_e is the final concentration in mg/l, and R is the percent adsorption.

2.1. Pseudo first-order kinetic model

The linear form of the pseudo first-order kinetic model is as follows:

$$Ln(q_e - q_t) = Ln\ q_e - k_1 t \qquad (3)$$

In the above relation, q_e and q_t are metal adsorption (mg/g) at equilibrium and at a given time t (min); k_1 is the kinetic constant of the pseudo first-order kinetic model.

2.2. Pseudo second-order kinetic model

The linear form of the pseudo second-order kinetic model is as follows:

$$\frac{t}{q_t} = \frac{1}{k_2\, q_e^2} + \frac{t}{q_e} \qquad (4)$$

In the above relation, q_t (mg/g) is metal adsorption at time t (min), q_e is adsorption at equilibrium, and k_2 is the kinetic constant of the pseudo second-order kinetic model.

2.3. Langmuir's isotherm

$$q_e = \frac{q_m\, K_L\, C_e}{1 + K_L C_e} \qquad (5)$$

In the relation above, the Langmuir constants q_m (mg/g) and K_L (L/mg) are related to the maximum single-layer adsorption and the adsorption energy, respectively. The larger the value of K_L, the greater the attraction between the adsorbed component and the adsorbent will be. Moreover, q_e (mg/g) and C_e (mg/L) represent adsorption at equilibrium and equilibrium concentration of the adsorbed component, respectively. The parameters of Langmuir's isotherm can be obtained by drawing its linear diagram:

$$\frac{C_e}{q_e} = \frac{C_e}{q_{max}} + \frac{1}{b q_{max}} \qquad (6)$$

As can be seen in relation 6, if the C_e/q_e line is drawn against C_e, the slope of the line and the y-intercept will be q_m and K_L, respectively.

2.4. Freundlich isotherm

$$q_e = K C_e^{\frac{1}{n}} \qquad (7)$$

In relation 7, K is the Freundlich constant related to adsorption capacity ; n (the Freundlich dimensionless constant related to the strength of adsorption) has a value between zero and one and the larger its value is, the greater the attractive forces between the adsorbent and the adsorbed component will be. The Freundlich constants (n and k) are related to adsorption intensity and adsorption capacity, respectively. The linear of the Freundlich isotherm is used to make the calculations easier. The linear form of the Freundlich isotherm is derived as follows by taking the logarithm of relation 7:

$$\ln q_e = \ln K + \frac{1}{n} \ln C_e \qquad (8)$$

The constants in the Freundlich isotherm will be derived from the slope and the y-intercept of the linear diagram ln (q_e) against ln (C_e), as shown in relation 8.

3. Materials and methods

3.1. Materials and equipment

The SA and HA supplied by the Sigma-Aldrich Company, commercial Zeolite, and Cu sulfate pentahydrate, ammonia, and sodium chloride provided by the Merck Company were used together with a UV absorption spectrometer, model 7115, made by the JINWAY Company. Other equipment included a model DK203H ultrasonic instrument produced by the German Company BADDELIN and a RH b2 model magnetic heater made by the German Company IKA.

The phase analysis of fabricated framework was done by an X-ray Diffraction (XRD) diffractometer, model: Equinox 3000, Intel Co. The micrographs and elemental analysis were obtained from Field Emission Scanning Electron Microscopy (FE-SEM) model: Mira II, Tescan Co. and Energy Dispersive Spectroscopy (EDS) model: Mira II, Tescan Co. instruments. Ultrasonic bath , model: S 4000, Misonix Co. with output power: 600W was used in the preparation process.

3.2. Description of the experiment

The adsorbent was first synthesized. The adsorption experiments were then conducted to study the effects of contact time, initial concentration, and adsorbent recovery. The performance of Zeolite in removing Cu was compared with that of the hydrogel that was synthesized under identical conditions.

3.2.1. Adsorbent synthesis

First, 0.2g of HA was mixed with 50 ml of water and placed in the ultrasonic instrument for 15 minutes for the HA to completely disperse in the water. Subsequently, 1.5g of SA were added to the mixture and placed on the mixer until a homogeneous gel mixture was formed. Various quantities of Zeolite in weight ratios of zero to 25 percent were then added to the mixture. Each of the obtained gels was injected into 0.1M calcium chloride, and the granules were placed in a saline solution for two hours in order for them to acquire their stable form; then, they were dried.

3.2.2. Contact time

The adsorbent Zeolite (0.1 g) and also the synthetic hydrogel SA (0.1g) were put in contact with 25 ml of 300 ppm Cu for periods from zero up to 150 minutes; the amount of the Cu in the solution after each contact time was measured.

3.2.3. The effect of initial concentration

Twenty five ml of each of 100 to 500 ppm Cu solutions were added to the adsorbent Zeolite (0.1g) and also the synthetic hydrogel SA (0.1g); the mixtures were placed on the magnetic mixer for 150 minutes. The solutions were then removed from the adsorbents and Cu contents of the solutions were measured.

3.2.4. Adsorbent recovery

The employed hydrogels were put in 2% hydrochloric acid in order for the repulsion process to take place and the adsorbent to be recovered. The recovered adsorbent was then washed several times with distilled water to remove the acid and for the adsorbent to become neutralized for reuse. This process was repeated five times. The following equation for the repulsion coefficient was used to study the efficiency of the recovered adsorbent and calculate the coefficient of adsorbent repulsion and adsorption percentage in the various recovery cycles of the adsorbent:

D% = (Weight of heavy metal (mg) repulsed by the $\qquad$ (9) hydrochloric acid/Metal ions (mg) adsorbed on the adsorbent) × 100

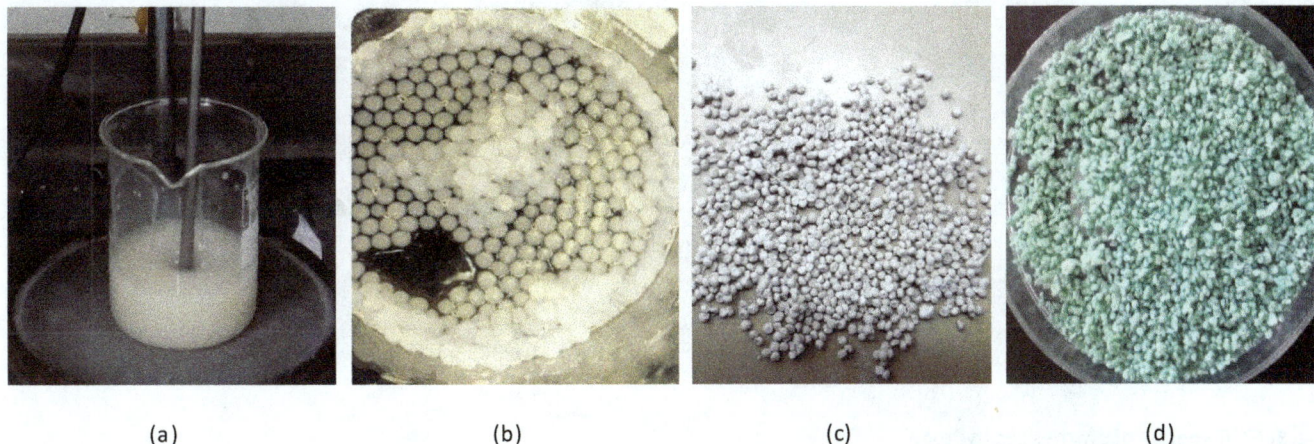

(a)   (b)   (c)   (d)

Fig. 1. Hydrogel synthesis process (a) prepared gel, (b) injected granules, (c) dried adsorbent beads, (d) hydrogels after Cu adsorption

4. Results and discussion

The results of the tests on adsorbent identification and on the effects of salinity, competitive ion, and adsorbent recovery are presented below:

4.1. Adsorbent identification

The FTIR tests were used to confirm the chemical structure of the hydrogels. Spectra 1 and 2 shown in Figures. 2 and 3 are those of the pure SA and the composite hydrogel, respectively.

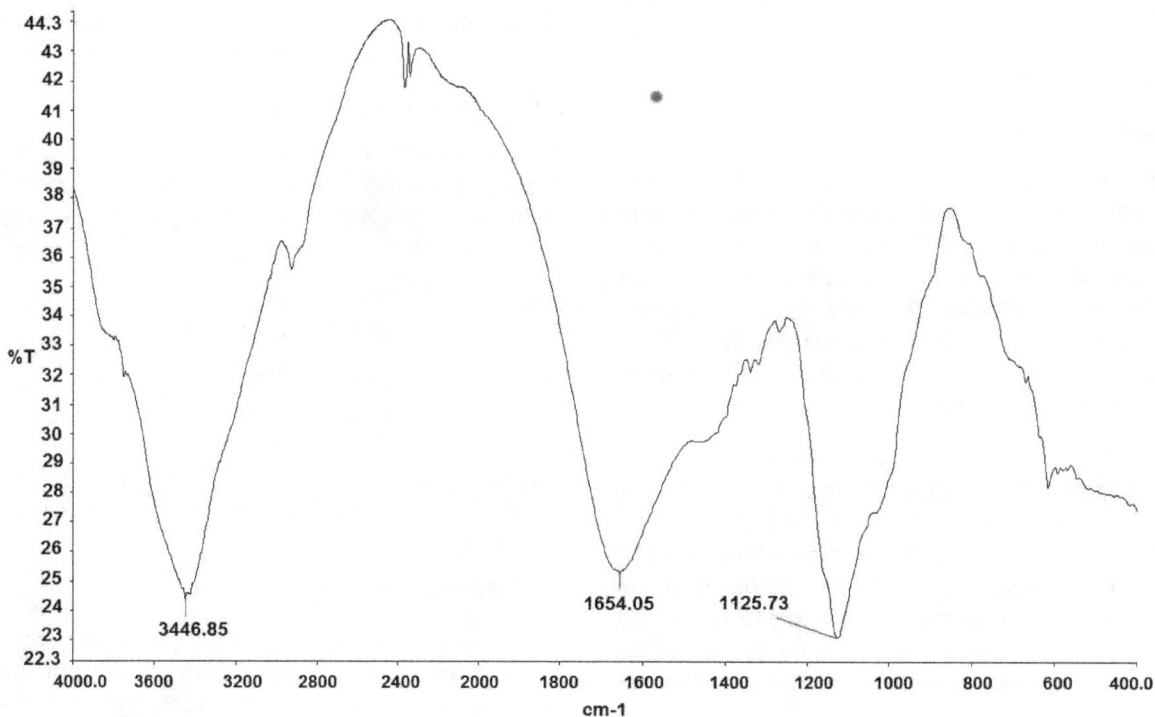

Fig. 2. FTIR spectra of pure sodium alginate

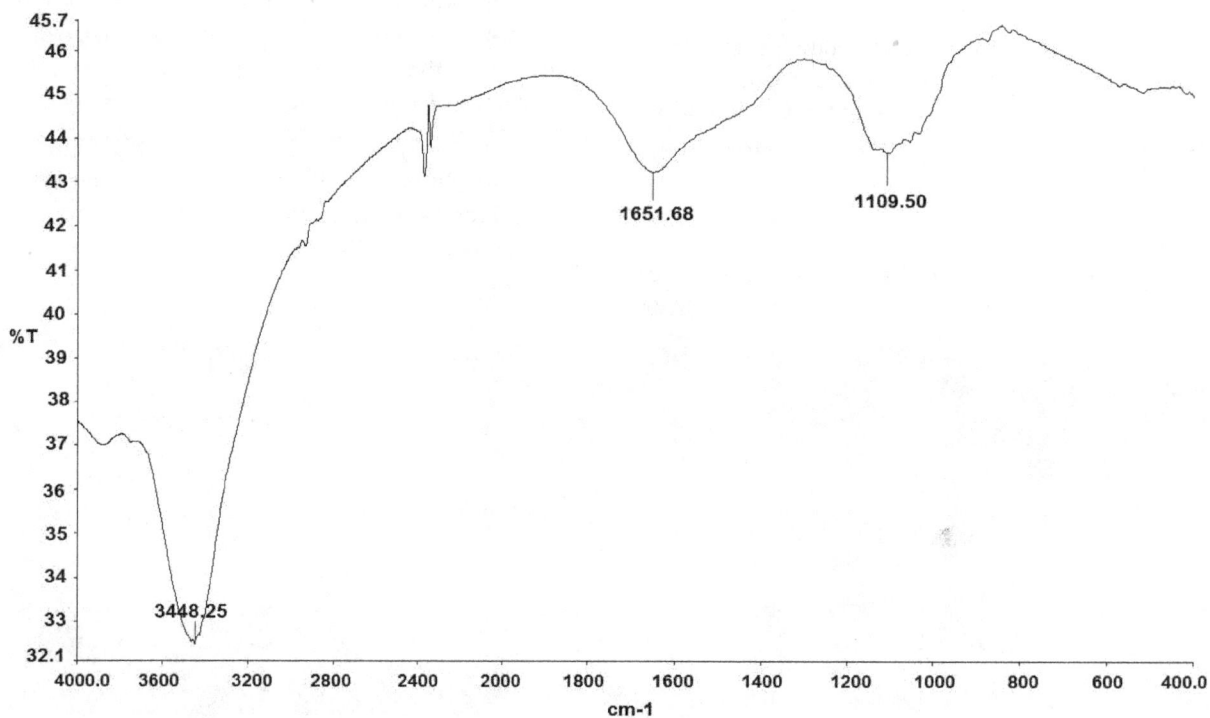

Fig. 3. FTIR spectra of synthesized hydrogel

Figure 2 displays the pure SA with peaks at 3448.25 cm^{-1} related to tensile vibrations of the hydroxyl group; at 1654.05 cm^{-1} to asymmetric tensile vibrations of the carboxylate group; and at 1125.73 cm^{-1} to tensile vibrations of the ether group in the polysaccharide. Figure 3 presents the composite hydrogel with a peak at 3448.25 cm^{-1} related to tensile vibrations of the hydroxyl group in the alginate chain, one at 1651.68 cm^{-1} to asymmetric vibrations of the functional carboxylate group, and another at 1109.50 cm^{-1} to vibrations of the C-O group. The adsorbent was produced in various weight ratios of Zeolite to SA (from zero to 25 percent). The adsorbent (0.1 g) was put in contact with 25 ml of a 300 ppm Cu solution for three hours. The results of this experiment are presented in Figure 4.

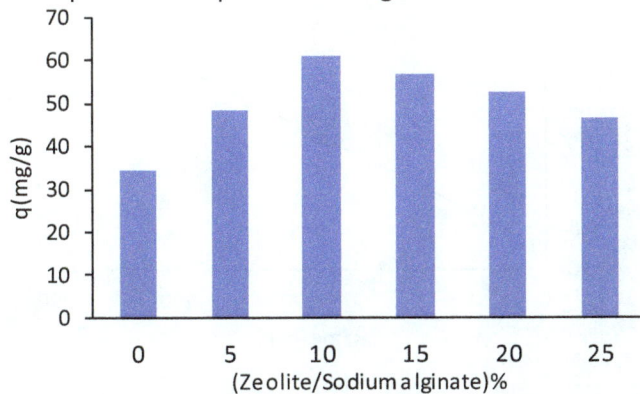

Fig. 4. Effect of Zeolite/Sodium alginate weight percent on adsorption capacity of hydrogel

As can be seen in Figure 4, the adsorption capacity of the hydrogel increased when the weight ratio of Zeolite was raised from zero to 10 percent. This could be due to the capacity of Zeolite in adsorbing Cu. In other words, the ability of Zeolite in adsorbing heavy metals in addition to the functional groups in the hydrogel increased the adsorption capacity. However, the adsorption capacity declined in samples with weight ratios of 15-25 percent because the porosity and swelling capability of the hydrogels in these samples decreased due to their high content of Zeolite. Therefore, we concluded that if clay and other mineral

additives were added to the hydrogel structure, their optimum ratios must be determined. The 10 percent weight ratio was selected for the synthesized hydrogel sample in this research.

4.2. Contact time

Adsorption experiments were carried out at various intervals for both Zeolite and the hydrogel to determine the related equilibrium times. The results are presented in Table 1.

The data related to the calculated capacities of Zeolite and the hydrogel for adsorbing Cu is plotted against time in Figure 5.

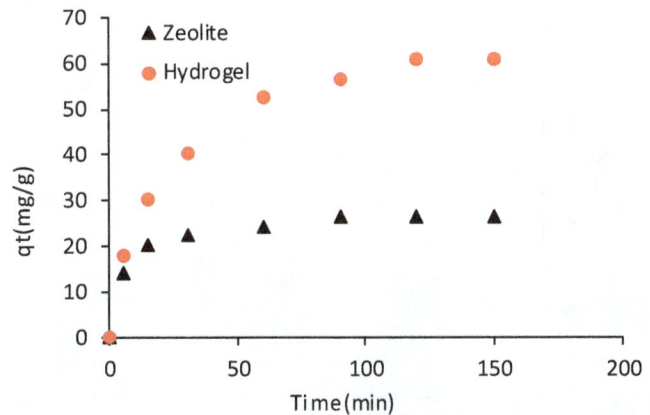

Fig. 5. Contact time effect for hydrogel ad Zeolite

As shown in Table 1 and in Figure 5, the Zeolite achieved the maximum adsorption of 26.35 mg/g in 90 minutes, but the produced hydrogel reached equilibrium in 120 minutes and its maximum adsorption was 60.81 mg/g. The results showed that the Zeolite particles had a greater adsorption speed than the hydrogel granules. This could be due to the smaller size of the Zeolite particles and their greater surface area because Zeolite was used as a powder but the employed hydrogel had larger particles. However, it was observed that the adsorption capacity of the hydrogels was 2.3 times greater than that of the Zeolite.

Table 1. Experimental results of contact time effect for Zeolite and synthesized hydrogel

Hydrogel					Zeolite				
Time (min)	Ce (ppm)	Q (mg/g)	ln(qe-qt)	t/qt	Time (min)	Ce (ppm)	Q (mg/g)	ln(qe-qt)	t/qt
0	300	0			0	300	0		
5	227/02	18/24	3/75	0/27	5	243/24	14/18	2/49	0/35
15	178/37	30/40	3/41	0/49	15	218/91	20/27	1/80	0/7ɛ
30	137/83	40/54	3/00	0/74	30	210/81	22/29	1/40	1/34
60	89/18	52/70	2/09	1/13	60	202/70	24/32	0/71	2/46
90	72/97	56/75	1/40	1/58	90	194/59	26/35	-4/75	3/41
120	56/75	60/8ʻ	-4/68	1/97	120	194/59	26/35	-4/7°	4/55
150	56/75	60/8ʻ	-4/68	2/46					

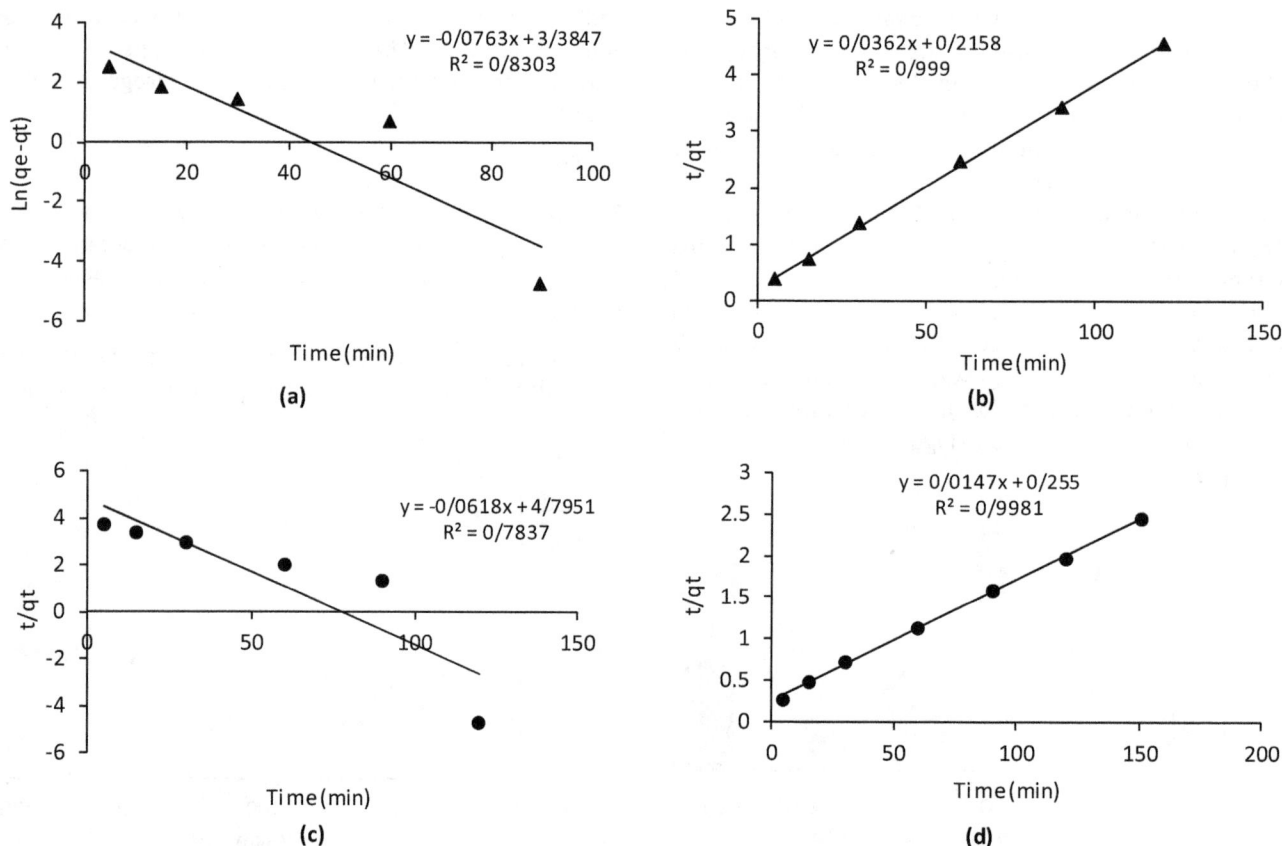

Fig. 6. Kinetics models (a) Pseudo first order for Zeolite, (b) Pseudo first order for hydrogel, (c) Pseudo second order for Zeolite, (d) Pseudo second order for hydrogel

In Table 1, the values of Ln (q_e-q_t) and of t/q_t were calculated for drawing and studying the pseudo first and second order kinetic curves. In Figure 6, these values were plotted against time for both of the adsorbents. After drawing the kinetic curves, the constants of these equations were also calculated. The calculated values are listed in Table 2. The results showed that the pseudo second order kinetic curves for both adsorbents matched the empirical data very well.

4.3. Initial adsorbent concentration

Adsorption experiments were performed at various initial concentrations in the 100-500 ppm range to obtain the isotherm. The results are presented in Table 3.

The results indicated that the maximum adsorption of Cu was 29.71 mg/g for Zeolite and 75.82 mg/g for the hydrogel. Therefore, the synthetic hydrogel had a greater adsorption capacity than the Zeolite under identical conditions, concentration and the amount of the adsorbent. Using the results in Table 3, the Freundlich and Langmuir isotherm curves were drawn for both the Zeolite and the hydrogel (Figure 7) and their constants were calculated; they are listed in Table 4. The results showed that both adsorbents followed the Langmuir model.

Table 2. Kinetic models constants for hydrogel and Zeolite

Kinetic model order	Hydrogel			Zeolite		
	q_e(mg/g)	K (1/min)	R2	Qe (mg/g)	K (g/mg)	R^2
Pseudo first	120.3	0.062	0.783	29.38	0.076	0.830
Pseudo second	68.03	000084	0.998	27.62	0.006	0.999

Table 3. Experimental effect of initial concentration effect for hydrogel and Zeolite

	Hydrogel					Zeolite				
C_0 (ppm)	C_e(ppm)	q (mg/g)	Log (C_e)	Log (q_e)	C_e/q_e	C_e (ppm)	q (mg/g)	Log (C_e)	Log (q_e)	C_e/q_e
100	3/8	24/0	0/59	1/38	0/16	15	21/15	1/19	1/33	0/73
200	8	48/0	0/90	1/68	0/17	104	24/00	2/02	1/38	4/33
300	56	60/8	1/75	1/78	0/93	186	28/37	2/27	1/45	6/57
400	130	67/3	2/12	1/83	1/94	281	29/59	2/45	1/47	9/52
500	196	75/8	2/29	1/88	2/59	381	29/71	2/58	1/47	12/83

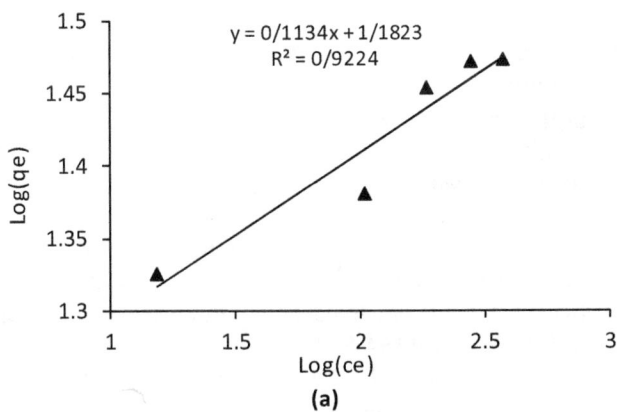

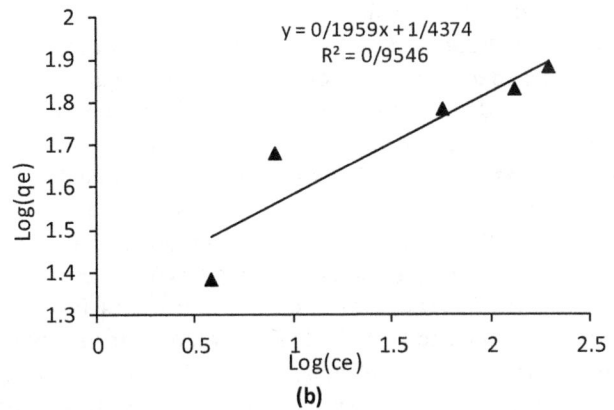

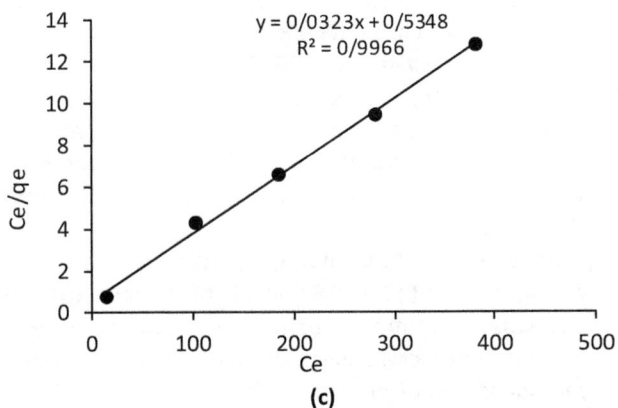

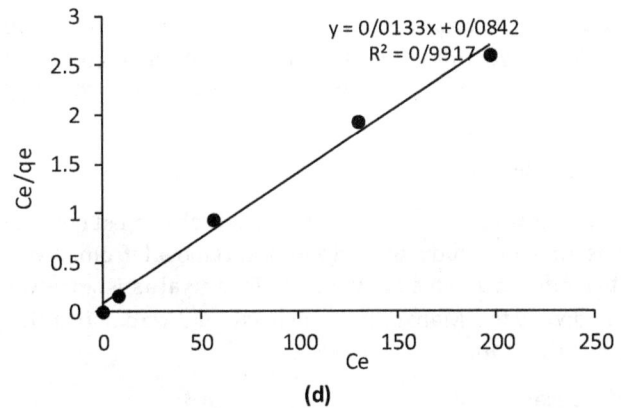

Fig. 7. Linear form of isotherms (a) Freundlich for Zeolite, (b) Freundlich for hydrogel, (c) Langmuir for Zeolite, (d) Langmuir for hydrogel.

Table 4. Thermodynamic constants of Freundlich and Langmuir isotherms for Zeolite and hydrogel

Langmuir			Freundlich			
q_m(mg/g)	K_L	R^2	n	K_f	R^2	
30.96	0.061	0.996	0.113	15.2	0.922	Zeolite
75.18	0.158	0.991	0.196	27.5	0.955	Hydrogel

4.4. Adsorbent recovery

The synthesized adsorbents were recovered after performing one stage of the process and were used in four successive stages. The table below shows the amounts adsorbed by the unused adsorbents and the quantities of the recovered adsorbents. As shown in Table 5, the efficiency of the adsorption process using recovered adsorbents for five consecutive adsorption stages remained at about 75 percent, which shows a good capability in recovering the adsorbent and in reusing it. Formula 9 shows that the adsorbent recovered by 2 percent of HCl had the repulsion coefficient of 70-82.75 percent.

Table 5. Comparison of fresh and reduced adsorbent

Recovery cycle no.	Removal %
Fresh adsorbent	81/2
1	77/4
2	72/1
3	66/5
4	61/2

5. Conclusions

The SA and Zeolite synthesized hydrogel had an adsorption capacity of about 2.3 times greater than that of Zeolite so that under identical conditions in regard to the amount of adsorbent and initial concentration, the adsorption capacity of Zeolite was 29.7 mg/g of the adsorbent and that of the hydrogel was 75.8 mg/g. The kinetic studies showed that Cu adsorption on the SA hydrogel and activated carbon followed the linear form of the pseudo second order model with $R^2 = 0.991$. The thermodynamic studies for determining the isotherm indicated that the laboratory data for both adsorbents matched with the Langmuir isotherm very well. The synthesized hydrogel had greater adsorption capacity but lower adsorption speed so that the equilibrium time was 90 minutes for Zeolite and 120 minutes for the hydrogel. It was possible to recover and reuse the adsorbent five times with the efficiency of 75 percent, and the repulsion coefficient of the adsorbent was 70-82.75 percent.

Acknowledgement

This paper was prepared from the research project entitled "Experimental study of cationic ion removal from water with modified polymeric hydrogel". Financial assistance was provided by the Mahshahr Branch, Islamic Azad University, Mahshahr, Iran.

References

[1] Liu, T., Yang, X., Wang, Z. L., Yan, X. (2013). Enhanced chitosan beads-supported Fe 0-nanoparticles for removal of heavy metals from electroplating wastewater in permeable reactive barriers. *Water research*, 47(17), 6691-6700.

[2] Hossain, M. A., Ngo, H. H., Guo, W. S., Setiadi, T. (2012). Adsorption and desorption of copper (II) ions onto garden grass. *Bioresource technology*, 121, 386-395.

[3] Qaiser, S., Saleemi, A. R., Umar, M. (2009). Biosorption of lead from aqueous solution by Ficus religiosa leaves: batch and column study. *Journal of hazardous materials*, 166(2), 998-1005.

[4] Wu, G., Kang, H., Zhang, X., Shao, H., Chu, L., Ruan, C. (2010). A critical review on the bio-removal of hazardous heavy metals from contaminated soils: issues, progress, eco-environmental concerns and opportunities. *Journal of hazardous materials*, 174(1), 1-8.

[5] Nayek, S., Gupta, S., Saha, R. N. (2010). Metal accumulation and its effects in relation to biochemical response of vegetables irrigated with metal contaminated water and wastewater. *Journal of hazardous materials*, 178(1), 588-595.

[6] Vijayaraghavan, K., Prabu, D. (2006). Potential of Sargassum wightii biomass for copper (II) removal from aqueous solutions: Application of different mathematical models to batch and continuous biosorption data. *Journal of hazardous materials*, 137(1), 558-564.

[7] Zhuang, Y., Yu, F., Chen, J., Ma, J. (2016). Batch and column adsorption of methylene blue by graphene/alginate nanocomposite: Comparison of single-network and double-network hydrogels. *Journal of environmental chemical engineering*, 4(1), 147-156.

[8] Mohammed, N., Grishkewich, N., Waeijen, H. A., Berry, R. M., Tam, K. C. (2016). Continuous flow adsorption of methylene blue by cellulose nanocrystal-alginate hydrogel beads in fixed bed columns. *Carbohydrate polymers*, 136, 1194-1202.

[9] Lawrie, G., Keen, I., Drew, B., Chandler-Temple, A., Rintoul, L., Fredericks, P., Grøndahl, L. (2007). Interactions between alginate and chitosan biopolymers characterized using FTIR and XPS. *Biomacromolecules*, 8(8), 2533-2541.

[10] An, B., Lee, H., Lee, S., Lee, S. H., Choi, J. W. (2015). Determining the selectivity of divalent metal cations for the carboxyl group of alginate hydrogel beads during competitive sorption. *Journal of hazardous materials*, 298, 11-18.

[11] Vijayalakshmi, K., Gomathi, T., Latha, S., Hajeeth, T., Sudha, P. N. (2016). Removal of copper (II) from aqueous solution using nanochitosan/sodium alginate/microcrystalline cellulose beads. *International journal of biological macromolecules*, 82, 440-452.

23

New Chitosan/Ag/Carbacylamidophosphate nanocomposites: Preparation and antibacterial study

Nasrin Oroujzadeh

Department of Chemical Technologies, Iranian Research Organization for Science and Technology (IROST), Tehran, Iran

Keywords:
Carbacylamidophosphate
Chitosan
Ag Nanoparticles
Nanocomposite
Antibacterial activity

ABSTRACT

Two new Chitosan-based nanocomposite films were prepared: Chitosan /7% Ag nanoparticles (NPs) (NC1) and Chitosan/7% Ag NPs/5%Carbacylamidophosphate(NC2), in which the carbacylamidophosphate derivitive is N-Nicotinyl-N',N''- bis(hexamethylenyl) phosphorictriamide (NHE) with the formula: $C_5H_4NC(O)NHP(O)(NC_6H_{12})_2$. X-ray Powder Diffraction (XRD), Field Emission Scanning Electron Microscopy (FE-SEM) and Energy Dispersive X-ray Spectroscopy (EDS) methods were used to characterize and confirm the prepared frameworkrs. XRD graph of the two nanocomposites showed all the characteristic peaks of NHE, Ag NPs, and chitosan, indicating the fact that the preparing process has not made any changes in the phases of the nanocomposites components. All the SEM micrographs and EDS analysis results also confirmed the desired structures. To study the effect of the additive NHE on the antibacterial activity of the films, *in vitro* antibacterial tests were done on the prepared nanocomposites against two Gram-positive (Staphylococcus aureus, Bacillus cereus) and two Gram-negative bacteria (Escherichia coli, Pseudomonas aeruginosa) in Brain-Heart Infusion(BHI) medium. Results showed that the antibacterial effects of the nanocomposite containing NHE on each of the four bacteria is stronger than those for the nanocomposite without NHE.

1. Introduction

Our environment is contaminated with various microorganisms, many of which are harmful to human health. Microbiological contamination of water has long been a concern to the public. Water-borne pathogen contamination in water resources and related diseases are a major water quality concern throughout the world. From the 1920's-1960's, the bacillus which causes typhoid fever was considered a major problem in the water supply [1]. There are also dangerous soil-sourced microorganisms that, beside water- born harmful microbes, can cause many diseases through the food. Staphylococcus aureus (S. aureus), Bacillus cereus (B. cereus), Pseudomonas aeruginosa (P. aeruginosa) and Escherichia coli (E. coli) are examples of common bacterial food pathogens. Staphylococcus aureus is a gram-positive, round-shaped bacterium that is one of the most common causes of bacteremia and infective endocarditis. Additionally, it can cause various skin and soft tissue infections [2,3]. Escherichia coli is a gram-negative, facultatively anaerobic, rod-shaped, coliform bacterium of the genus Escherichia that is commonly found in the lower intestine of warm-blooded organisms (endotherms) [4] Escherichia coli O157:H7 enterohemorrhagic (EHEC) can cause hemolytic-uremic syndrome [5]. Bacillus cereus is a Gram-positive bacterium commonly found in soil and food. It is the cause of "fried rice syndrome". Some Bacillus cereus strains are very harmful to humans and cause foodborne illness [6-8]. Pseudomonas aeruginosa is a common Gram-negative, rod-shaped bacterium that can cause disease in plants and animals, including humans. An opportunistic, nosocomial pathogen of immunocompromised individuals, P. aeruginosa typically infects the airway, urinary tract, burns, and wounds, and also causes other blood infections [9]. Increasing interest in controlling environmental pathogens

*Corresponding author
E-mail address: n_oroujzadeh@irost.ir

evidenced by a large number of recent publications [10] clearly corroborates to the need for studies that synthesize knowledge from multiple fields covering comparative aspects of pathogen contamination. Currently, there has been growing interest in antibacterial material research area. Among polymer-based potential antimicrobial agents Chitosan-based materials, especially those contain silver in their structure, have shown excellent antibacterial properties. The antibacterial behavior of various nanocomposites based on Chitosan/Ag NPs have been studied by many researchers [11–14]. Chitosan, because of its biodegradable, biocompatible and nontoxic properties has been applied in several biomedical areas such as wound healing [15], tissue engineering [16,17], drug delivery and gene delivery [18- 20]. In the last few years, carbacylamidophosphates, as an important biologically active class of organophosphorus compounds, has become very attractive [21-30]. Due to the unique physicochemical properties and existence of a peptide like group -C (O) NHP (O) - in their skeleton, these materials have valuable biological activities and critical roles in catalytic and metabolism processes [31-34]. Some carbacylamidophosphates have potential applications as antimicrobial agents [35], anti-tumor drugs [36] and also inhibitors for cholinesterase and butyrylcholinesterase enzymes [37,38]. Here, two new nanocomposite films with the composition: Chitosan/7%Ag NPs (NC1) and Chitosan/7%Ag/5%Carbacylamidophosphate (NC2), in which the carbacylamidophosphate derivative is N-Nicotinyl-N', N''-bis(hexamethylenyl) phosphoric triamide(NHE), were prepared and characterized by X-ray Powder Diffraction (XRD), Field Emission Scanning Electron Microscopy (FE-SEM) and Energy Dispersive X-ray Spectroscopy(EDS) analysis methods. In vitro antibacterial activities of these nanocomposites were also evaluated against two Gram-positive bacteria: S. aureus, B. cereus and two Gram-negative bacteria: E. coli, P. aeruginosa in Brain-Heart Infusion (BHI) medium.

2. Materials and methods

2.1. Materials

Silver nitrate, Sodium citrate, PCl$_5$, Hexamethyleneimine, CCl$_4$, Acetonitrile, Eethanol, Acetic acid, Nicotinamide and distilled water obtained from Merck Co. High molecular weight Chitosan was purchased from Loba Chemie Pvt. Ltd. and Glutar aldehyde (25 wt% in water) obtained from Daejong Co. All compounds were used without any more purification.

2.2. Synthesis

2.2.1. N-Nicotinyl-N', N''-bis (hexamethylenyl) phosphoric triamide(NHE)

This compound was synthesized according to our previously reported method [39]. Briefly, to a mixture of N-

Nicotinyl phosphoramidic dichloride (1 mmol), Hexamethyleneimine (4 mmol) and acetonitrile was added at 0 °C. After 8 h, the precipitate was filtered and washed with distilled water and dried at 25 °C to obtain the final product.

2.2.2. Synthesis of Ag nanoparticles

The Ag nanoparticles were synthesized by citrate reduction method [40]. Briefly, an aqueous solution 2 mmol of silver nitrate was added to 1 mmol aqueous sodium citrate solution, under atmospheric conditions. In order to avoid light-induced reduction of Ag, the reaction flask was heated in a dark environment, to about 140 C for 5 h. Then, it was cooled to the room temperature and the Ag nanoparticles were filtered and dried after washing with distilled water , for several times.

2.2.3. Preparation of Chitosan/ Ag NPs/ NHE nanocomposite films

For preparing the nanocomposite, at *first*, 0.5 g of chitosan powder was dispersed in 5 ml of distilled water and 3 ml acetic acid was added to the mixture. Next, the mixture was placed in an ultrasonic bath for 2 h to yield a homogenous brown viscose gel. Then, the carbacylamidophosphate (NHE) powder (0 and 5% w/w of chitosan, for the films NC1 and NC2 respectively) was dissolved in ethanol, added to the chitosan and placed into the ultrasonic bath for 30 min. After that, Ag NPs (7% w/w of chitosan) was added to the chitosan/ NHE flask and *placed in the ultrasonic bath for 30 min. The homogenous mixtures were then poured into proper plates and were allowed to be dried.* Finally, the prepared films were put in a desiccator containing glutar aldehyde steam with the pressure of 0.5 atm for 24 hours, for cross linking of chitosan and obtaining the desired nanocomposite films.

2.3. Instrumentations

The phase analysis of fabricated framework was done by an X-ray Diffraction (XRD) diffractometer, model: Equinox 3000, Intel Co. The micrographs and elemental analysis were obtained from Field Emission Scanning Electron Microscopy (FE-SEM) model: Mira II, Tescan Co. and Energy Dispersive Spectroscopy (EDS) model: Mira II, Tescan Co. instruments. Ultrasonic bath , model: S 4000, Misonix Co. with output power: 600W was used in the preparation process.

2.4. In vitro antibacterial test

The in vitro antibacterial activities of the films were evaluated by the filter paper disk method [41], with the disk diameter of 6.5 mm. The bacteria were cultured in BHI medium. About 0.005 g of each film was used in each test. *The* thickness of the BHI medium was kept equal in all Petri dishes. Four bacteria including two Gram-positive (S. aureus, B. cereus) and two Gram-negative bacteria

(E. coli, P. aeruginosa) were examined. The disks were incubated at 37 °C for 24 h. The inhibition zone of growth, which specifies the amount of inhibitory effect of the compounds on the growth of the bacteria, was measured and the average of three diameters was calculated for each sample (Table1).

3. Results and discussion

3.1. X-Ray Diffraction Analysis

The X-Ray Diffraction method was used to investigate the crystallinity and the phases of the material. XRD pattern of Ag NPs is given in Figure 1. Comparing XRD patterns of NHE, Chitosan and the two fabricated nanocomposite films, NC1 and NC2, are also shown in Figure 2. As it can be seen in the figures, owing to their high crystallinity, Ag NPs and carbacylamidophosphate (NHE) represent sharp peaks; but the chitosan film shows a broad pattern, which refers to its amorphous polymeric structure. XRD of Ag nanoparticles (Figure 1) shows a strong diffraction pattern,

relating to the hkl values of (1 1 1), (2 0 0), (2 2 0), (3 1 1) and (2 2 2) planes, that matches with the standard JCPDS no. [00–087-0719]. As NHE is a newly synthesized compound, no standard card can be found for its XRD pattern. This phosphoric triamide shows 12 peaks at $2\theta \approx 9$, 10, 13.5, 15.5, 16.5, 18, 18.5, 20.5, 22, 23, 25 and 35. The very intense and sharp peak at $2\theta \approx 10$ and the smaller one at $2\theta \approx 25$ could be considered as its characteristic peaks in the nanocomposite XRD pattern. The XRD graph of chitosan, include all the characteristics peak of chitosan, reported in literature [42,43]. In XRD graph of the nanocomposites, NC1 and NC2, (Figure 2), the characteristic peaks of NHE, Ag NPs, and chitosan are observed which confirms that the mixing process has not made any changes in the phases of the nanocomposites components. The intensity of some of components peaks have been changed, which is due to the using of ultrasonic waves in the mixing process.

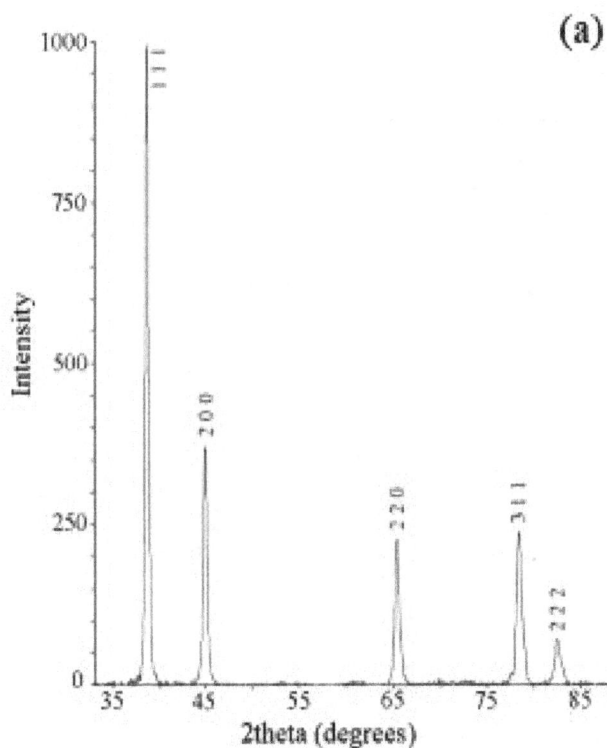

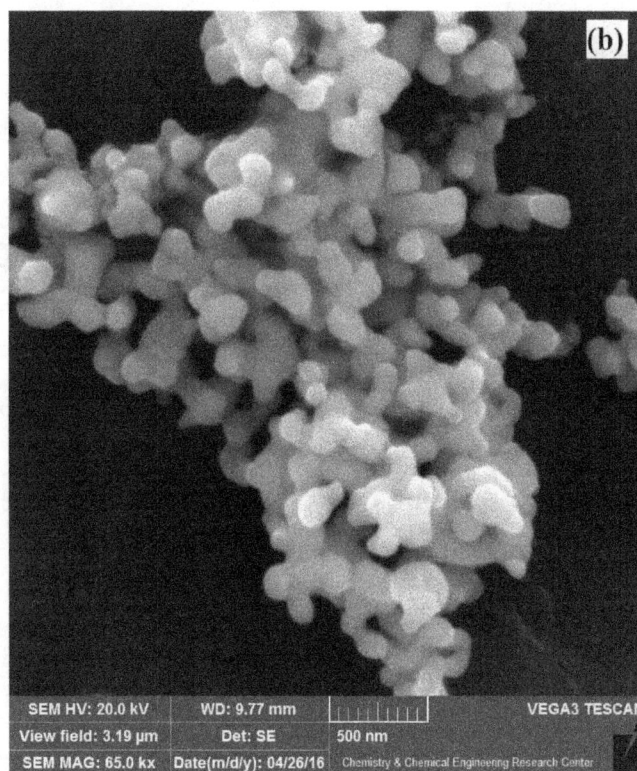

Fig. 1. The XRD pattern (a) and the FE-SEM (b) micrograph of Ag nanoparticles

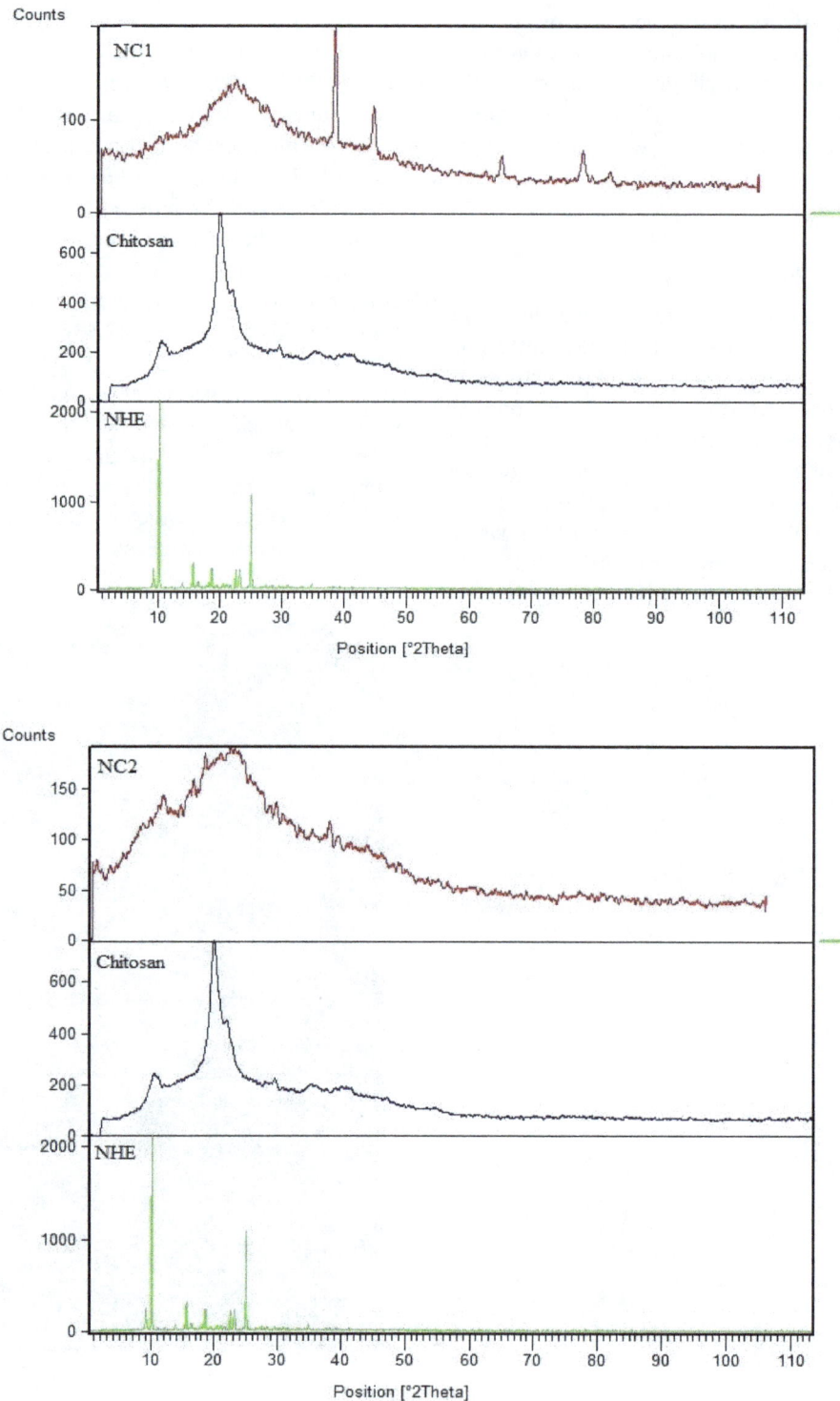

Fig. 2. The XRD pattern of nanocomposites NC1(0% NHE) and NC2 (5% NHE) comparing to their components

3.2. FE-SEM and EDS

To study the surface morphology of the nanocomposites and measuring the average particle size of the Ag NPs, the FE-SEM micrographs were obtained (Figures 1(b) and 3). FE-SEM micrograph of Ag NPs indicates that the average particle size of Ag NPs is about 80-90 nm and most of these fine particles have stuck together and formed agglomerates. FE-SEM micrographs of nanocomposites NC1 and NC2, containing 0% and 5% NHE, are shown in Figure 3. The presence of Ag NPs can be obviously observed in these images.

A comparison between the micrograph of NC1 and NC2 films indicates that NC2 has a better surface with a more homogenous dispersion of smaller Ag nanoparticles. This could be due to the more stirring time under ultrasonic waves for NC2, owing to the step of adding carbacylamidophosphate(NHE), which NC1 did not have.

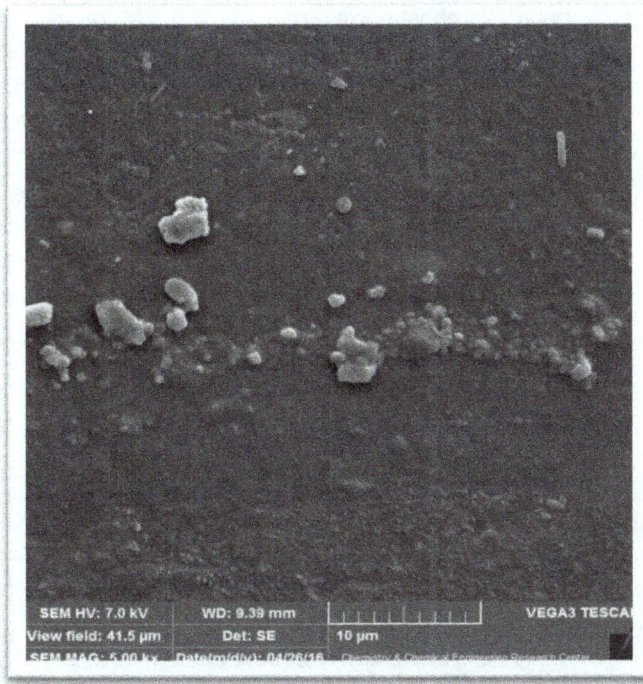

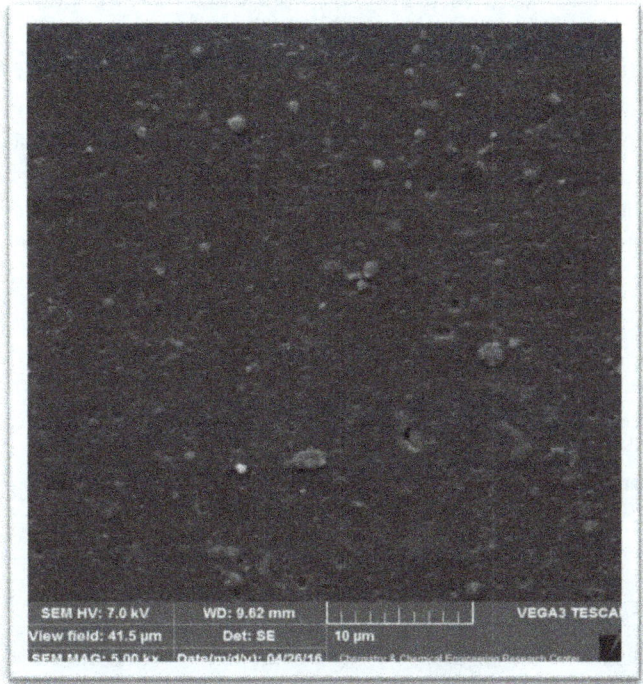

NC1   NC2

Fig. 3. The FE-SEM micrographs of nanocomposites NC1 and NC2

To confirm the presecnce of all initial components and all their elements in the nanocomposites framework Energy Dispersive Spectroscopy (EDS) was applied. EDS analysis of NC1 and NC2, containing 0% and 5% NHE, are shown in figure 4, which verifies the existence of all the designed parts within the nano films and confirms the desired percentages of the elements. The presence of carbacylamidophosphate (NHE) can also be obiously detected by observing the Phosphorus element (Figure4).

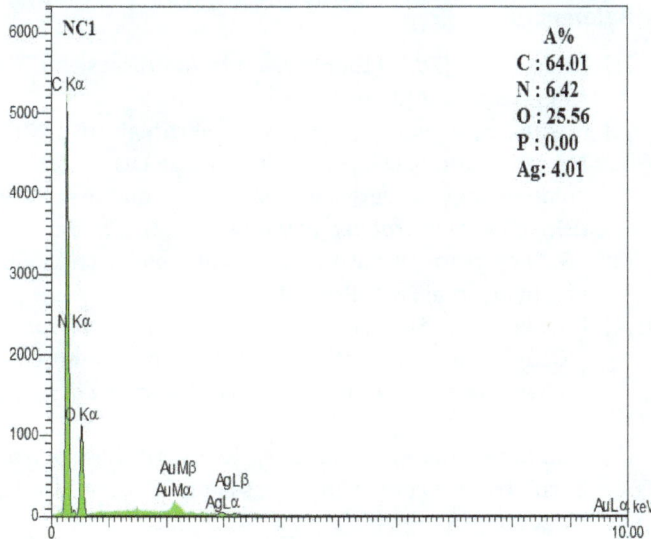

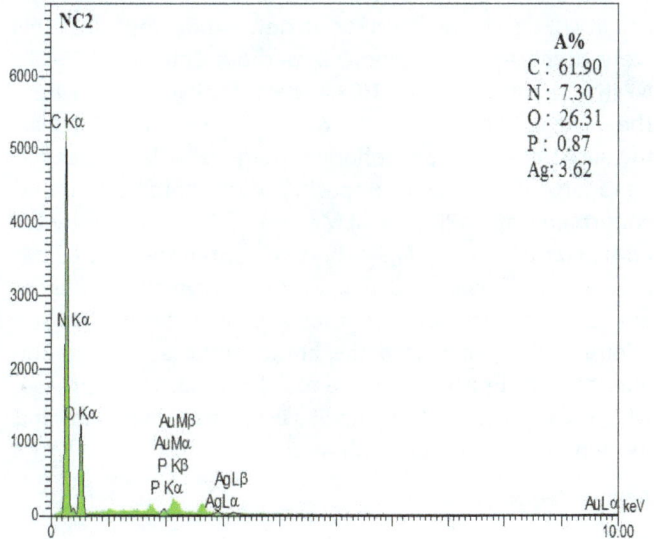

Fig. 4. The EDS of nanocomposites NC1 and NC2

3.3. Antibacterial Activity

The *in vitro* antibacterial activities of nanocomposite NC1 and NC2 were tested against four bacteria: two gram-positive (B. cereus, S. aureus), and two gram-negative (P. aeruginosa, E. coli). Each test was repeated for at least 3 times and the average of results were summarized in Table 1. As it can be seen from the table, the highest antibacterial activity of the both films is against B. cereus and the lowest is against S. aureus. Comparing the inhibition zones of the fabricated films indicates that, adding 5% of carbacylamidophosphate (NHE) to the structure of the nanocomposite results in the increasing of the antibacterial effect of the nanocomposite on all the four bacteria. This is in a good agreement with our previously reported data [41], which indicated the antibacterial improving of a chitosan based silver-contained nanocomposite by adding 5% of an alternative carbacylamidophosphate derivative, N-Nicontinyl-N',N"-bis(tert-butyl) phosphoric triamide(NPt). Moreover, as Table1 shows, the inhibitory zone (mm) for chitosan/5% NHE/7%Ag NPs nanocomposite (NC2) is even more than those for the previously reported chitosan/10%Ag NPs nanocomposite [41]. In the other word, the effect of 5% NHE on the antibacterial activity of the nanocomposite is more than that of the 3% Ag NPS. Since silver nanoparticles are well known strong antibacterial compounds [11-14], this could confirm remarkable antibacterial activity of the additive NHE.

Table 1. The inhibition zone (mm) measured for the antibacterial activities

Bacteria / Nanocomposite	Gram Negative		Gram Positive	
	P. aeruginosa 1074	E. coli 1330	S. aureus 1113	B. cereus 1254
Chitosan / 7% Ag NPs (**NC1**)	11.99± 0.54	8.61± 0.49	8.47± 0.58	12.02± 0.48
Chitosan /7% Ag NPs /5% NHE (**NC2**)	13.16± 0.42	10.45± 0.47	10.28± 0.51	13.53± 0.46
Chitosan/ 10% Ag NPs [41]	12.78± 0. 47	9.33± 0.57	9.11± 0.55	12.42± 0.51

4. Conclusions

To study the effect of an additive on the antibacterial activity of a chitosan/Ag NPs two new nanocomposite films were prepared, using crosslink method: Chitosan / 7% Ag NPs(NC1) and Chitosan /7% Ag NPs /5% NHE (NC2), where the additive NHE was N-Nicotinyl-N',N"-bis (hexamethylenyl) phosphoric triamide. The desired nanoparticles and nanocomposites were characterized and confirmed by XRD, FE-SEM and EDS. The in vitro antibacterial test against two Gram-negative (P. aeruginosa, E. coli) and two Gram-positive (B. cereus, S. aureus) bacteria indicated that the phosphoric triamide additive (NHE) increased the antibacterial activity of the nanocomposite against all the four bacteria. This increase, which was a result of adding 5%NHE, was remarkable and more than the addition of 3%AgNps.

Acknowledgment

The financial support of this work by Iran National Science Foundation (INSF) and Iranian Research Organization for Science and Technology (IROST) is gratefully acknowledged.

References

[1] Craun, G. F. (Ed.). (1986). Waterborne diseases in the United States. CRC Pressl Llc.

[2] Kluytmans, J., Van Belkum, A., & Verbrugh, H. (1997). Nasal carriage of Staphylococcus aureus: epidemiology, underlying mechanisms, and associated risks. *Clinical microbiology reviews*, 10(3), 505-520.

[3] U.S. Centers for Disease Control and Prevention (2016). Staphylococcal Food Poisoning.

[4] Tenaillon, O., Skurnik, D., Picard, B., & Denamur, E. (2010). The population genetics of commensal Escherichia coli. Nature reviews. *Microbiology, 8(3),* 207.

[5] Food Standards Agency (2016). Reducing the risk from E. coli 0157 – controlling cross-contamination. United Kingdom.

[6] Ryan, K. J., & Ray, C. G. (2004). Medical microbiology. McGraw Hill, 4, 370.

[7] Sanford, C. A., Jong, E. C., & Pottinger, P. S. (2016). The Travel and Tropical Medicine Manual E-Book. Elsevier Health Sciences.

[8] Asaeda, G., Caicedow, G., & Swanson, C. (2005). Fried rice syndrome. JEMS: a journal of emergency medical services, 30(12), 30-32.

[9] Todar, K. (2013). Online Textbook of Bacteriology. 2011. Bacterial Endotoxin.

[10] Pandey, P. K., Kass, P. H., Soupir, M. L., Biswas, S., & Singh, V. P. (2014). Contamination of water resources by pathogenic bacteria. AMB Express, 4(1), 51.

[11] Sanpui, P., Murugadoss, A., Prasad, P. D., Ghosh, S. S., & Chattopadhyay, A. (2008). The antibacterial properties of a novel chitosan–Ag-nanoparticle composite. International journal of food microbiology, 124(2), 142-146.

[12] Pinto, R. J., Fernandes, S. C., Freire, C. S., Sadocco, P., Causio, J., Neto, C. P., & Trindade, T. (2012). Antibacterial activity of optically transparent nanocomposite films based on chitosan or its derivatives and silver nanoparticles. Carbohydrate Research, 348, 77-83.

[13] Hu, L., Wu, X., Han, J., Chen, L., Vass, S. O., Browne, P., Hall, B.S., Bot, C., Gobalakrishnapillai, V., Searle, P.F., Knox, R. J. (2011). Synthesis and structure–activity relationships of nitrobenzyl phosphoramide mustards as nitroreductase-activated prodrugs. Bioorganic & medicinal chemistry letters, 21(13), 3986-3991.

[14] Wang, B. L., Liu, X. S., Ji, Y., Ren, K. F., & Ji, J. (2012). Fast and long-acting antibacterial properties of chitosan-Ag/polyvinylpyrrolidone nanocomposite films. Carbohydrate polymers, 90(1), 8-15.

[15] Ueno H., Mori, T., Fujinaga, T. (2001). Topical formulations and wound healing applications of chitosan. Advanced Drug Delivery Reviews, 52,105–15.

[16] Jabeen, S., Kausar, A., Saeed, S., Muhammad, B., & Gul, S. (2016). Poly (vinyl alcohol) and chitosan blend cross-linked with bis phenol-F-diglycidyl ether: mechanical, thermal and water absorption investigation. Journal of the Chinese Advanced Materials Society, 4(3), 211-227.

[17] Kast, C. E., Frick, W., Losert, U., & Bernkop-Schnürch, A. (2003). Chitosan-thioglycolic acid conjugate: a new scaffold material for tissue engineering. International journal of pharmaceutics, 256(1), 183-189.

[18] Yoo, H. S., Lee, J. E., Chung, H., Kwon, I. C., & Jeong, S. Y. (2005). Self-assembled nanoparticles containing hydrophobically modified glycol chitosan for gene delivery. Journal of Controlled Release, 103(1), 235-243.

[19] Park, J. H., Saravanakumar, G., Kim, K., & Kwon, I. C. (2010). Targeted delivery of low molecular drugs using chitosan and its derivatives. Advanced drug delivery reviews, 62(1), 28-41.

[20] Chan, P., Kurisawa, M., Chung, J. E., & Yang, Y. Y. (2007). Synthesis and characterization of chitosan-g-poly (ethylene glycol)-folate as a non-viral carrier for tumor-targeted gene delivery. Biomaterials, 28(3), 540-549.

[21] Grimes, K. D., Lu, Y. J., Zhang, Y. M., Luna, V. A., Hurdle, J. G., Carson, E. I., ... & Lee, R. E. (2008). Novel Acyl Phosphate Mimics that Target PlsY, an Essential Acyltransferase in Gram-Positive Bacteria. ChemMedChem, 3(12), 1936-1945.

[22] Oroujzadeh, N., Gholivand, K., & Shariatinia, Z. (2013). The Spectroscopy and Structure of New 1, 3, 2-Diazaphospholes and 1, 3, 2-Diazaphosphorinanes. Phosphorus, Sulfur, and Silicon and the Related Elements, 188(1-3), 183-191.

[23] Adams, L. A., Cox, R. J., Gibson, J. S., Mayo-Martín, M. B., Walter, M., & Whittingham, W. (2002). A new synthesis of phosphoramidates: inhibitors of the key bacterial enzyme aspartate semi-aldehyde dehydrogenase. Chemical Communications, (18), 2004-2005.

[24] Gholivand, K., Oroujzadeh, N., & Shariatinia, Z. (2010). New phosphoric triamides: Chlorine substituents effects and polymorphism. Heteroatom Chemistry, 21(3), 168-180.

[25] Znovjyak, K. O., Moroz, O. V., Ovchynnikov, V. A., Sliva, T. Y., Shishkina, S. V., & Amirkhanov, V. M. (2009). Synthesis and investigations of mixed-ligand lanthanide complexes with N, N'-dipyrrolidine-N''-trichloracetylphosphortriamide, dimethyl-N-trichloracetylamidophosphate, 1, 10-phenanthroline and 2, 2'-bipyrimidine. Polyhedron, 28(17), 3731-3738.

[26] Amirkhanov, V. M., Ovchynnikov, V. A., Trush, V. A., Gawryszewska, P., & Jerzykiewicz, L. B. (2014). Powerful new ligand systems: carbacylamidophosphates (CAPh) and sulfonylamidophosphates (SAPh). Chapter, 7, 199-248.

[27] Gholivand, K., Oroujzadeh, N., & Shariatinia, Z. (2010). N-2, 4-dichlorobenzoyl phosphoric triamides: Synthesis, spectroscopic and X-ray crystallography studies. Journal of chemical sciences, 122(4), 549-559.

[28] Litsis, O. O., Ovchynnikov, V. A., Shishkina, S. V., Sliva, T. Y., & Amirkhanov, V. M. (2013). Dinuclear 3D metal complexes based on a carbacylamidophosphate ligand: redetermination of the ligand crystal structure. Transition Metal Chemistry, 38(4), 473-479.

[29] Amirkhanov, O. V., Moroz, O. V., Znovjyak, K. O., Sliva, T. Y., Penkova, L. V., Yushchenko, T., ... & Amirkhanov, V. M. (2014). Heterobinuclear Zn–Ln and Ni–Ln Complexes with Schiff-Base and Carbacylamidophosphate Ligands: Synthesis, Crystal Structures, and Catalytic Activity. European Journal of Inorganic Chemistry, 23, 3720-3730.

[30] Gubina, K. E., Maslov, O. A., Trush, E. A., Trush, V. A., Ovchynnikov, V. A., Shishkina, S. V., Amirkhanov, V. M.

(2009). Novel heteroligand complexes of Co (II), Cu (II), Ni (II) and Mn (II) formed by 2, 2'-dipyridyl or 1, 10-phenanthroline and phosphortriamide ligands: Synthesis and structure. *Polyhedron*, 28(13), 2661-2666.

[31] Gholivand, K., Oroujzadeh, N., Erben, M. F., & Della Védova, C. O. (2009). Synthesis, spectroscopy, computational study and prospective biological activity of two novel 1, 3, 2-diazaphospholidine-2, 4, 5-triones. *Polyhedron*, 28(3), 541-547.

[32] Schultz, C. (2003). Prodrugs of biologically active phosphate esters. *Bioorganic & medicinal chemistry*, 11(6), 885-898.

[33] Wu, L. Y., Do, J. C., Kazak, M., Page, H., Toriyabe, Y., Anderson, M. O., & Berkman, C. E. (2008). Phosphoramidate derivatives of hydroxysteroids as inhibitors of prostate-specific membrane antigen. *Bioorganic & medicinal chemistry letters*, 18(1), 281-284.

[34] Venkatachalam, T. K., Sarquis, M., Qazi, S., & Uckun, F. M. (2006). Effect of alkyl groups on the cellular hydrolysis of stavudine phosphoramidates. *Bioorganic & medicinal chemistry*, 14(18), 6420-6433.

[35] Gholivand, K., Farshadian, S., Hosseini, Z., Khajeh, K., & Akbari, N. (2010). Two novel diorganotin phosphonic diamides: syntheses, crystal structures, spectral properties and in vitro antibacterial studies. *Applied Organometallic Chemistry*, 24(10), 700-707.

[36] Gholivand, K., Dorosti, N., Ghaziany, F., Mirshahi, M., & Sarikhani, S. (2012). N-phosphinyl ureas: Synthesis, characterization, X-ray structure, and in vitro evaluation of antitumor activity. *Heteroatom Chemistry*, 23(1), 74-83.

[37] Gholivand, K., Alizadehgan, A. M., Mojahed, F., Dehghan, G., Mohammadirad, A., & Abdollahi, M. (2008). Some new carbacylamidophosphates as inhibitors of acetylcholinesterase and butyrylcholinesterase. *Zeitschrift für Naturforschung C*, 63(3-4), 241-250.

[38] Gholivand, K., Shariatinia, Z., Khajeh, K., & Naderimanesh, H. (2006). Syntheses and spectroscopic characterization of some phosphoramidates as reversible inhibitors of human acetylcholinesterase and determination of their potency. *Journal of enzyme inhibition and medicinal chemistry*, 21(1), 31-35.

[39] Oroujzadeh, N., Gholivand, K., & Jamalabadi, N. R. (2017). New carbacylamidophosphates containing nicotinamide: Synthesis, crystallography and antibacterial activity. *Polyhedron*, 122, 29-38.

[40] Caswell, K. K., Bender, C. M., & Murphy, C. J. (2003). Seedless, surfactantless wet chemical synthesis of silver nanowires. *Nano Letters*, 3(5), 667-669.

[41] Oroujzadeh, N., & Rezaei Jamalabadi, S. (2016). New nanocomposite of N-nicotinyl, N', N "-bis (tert-butyl) phosphorictriamide based on chitosan: Fabrication and antibacterial investigation. *Phosphorus, Sulfur, and Silicon and the Related Elements*, 191(11-12), 1572-1573.

[42] Srivastava, R., Tiwari, D. K., & Dutta, P. K. (2011). 4-(Ethoxycarbonyl) phenyl-1-amino-oxobutanoic acid–chitosan complex as a new matrix for silver nanocomposite film: Preparation, characterization and antibacterial activity. *International journal of biological macromolecules*, 49(5), 863-870.

[43] Wang, X., Du, Y., Yang, J., Wang, X., Shi, X., & Hu, Y. (2006). Preparation, characterization and antimicrobial activity of chitosan/layered silicate nanocomposites. *Polymer*, 47(19), 6738-6744.

24

Nanostructured Fe_2O_3/Al_2O_3 adsorbent for removal of As (V) from water

Faranak Akhlaghian[1*], Bubak Souri[2], Zahra Mohamadi[1]
[1]*Department of Chemical Engineering, Faculty of Engineering, University of Kurdistan, Sanandaj, Iran*
[2]*Department of Environmental Science, Faculty of Natural Resources, University of Kurdistan, Sanandaj, Iran*

Keywords:
Adsorption
Arsenate
Fe_2O_3/Al_2O_3
Immobilization
Water treatment

ABSTRACT

The presence of arsenate in drinking water causes adverse health effects including skin lesions, diabetes, cancer, damage to the nervous system, and cardiovascular diseases. Therefore, the removal of As (V) from water is necessary. In this work, nanostructured adsorbent Fe_2O_3/Al_2O_3 was synthesized via the sol-gel method and applied to remove arsenate from polluted waters. First, the Fe_2O_3 load of the adsorbent was optimized. The Fe_2O_3/Al_2O_3 adsorbent was characterized by means of XRF, XRD, ASAP, and SEM techniques. The effects of the operating conditions of the batch process of As (V) adsorption such as pH, adsorbent dose, contact time, and initial concentration of As (V) solution were studied, and optimized. The thermodynamic study of the process showed that arsenate adsorption was endothermic. The kinetic model corresponded to the pseudo-second-order model. The Langmuir adsorption isotherm was better fitted to the experimental data. The Fe_2O_3/Al_2O_3 adsorbent was immobilized on leca granules and applied for As (V) adsorption. The results showed that the immobilization of Fe_2O_3/Al_2O_3 on leca particles improved the As (V) removal efficiency.

1. Introduction

Arsenic enters our water sources through the leaching of soils and rocks, mining, smelting, disposal of industrial wastewater, and pesticides [1,2]. High concentrations of arsenic in drinking water is a serious problem in many countries such as India, Bangladesh, Taiwan, Mongolia, China, and Chile [3,4]. Exposure to arsenic causes diseases such as skin, lung, and bladder cancers, gastrointestinal disorders, and cardiovascular and cerebrovascular diseases [1,4,5]. The World Health Organization (WHO) recommends limiting arsenic concentration in drinking water to 10 µg/L [4]. There are several methods for removing arsenic from water such as adsorption, coagulation, precipitation, membrane, and ion exchange. Among these methods, adsorption is the most promising because of its low cost, simple operation, and non-harmful by products [6]. Many researches have focused on developing adsorbents to efficiency remove arsenic from water. Jeong et al. [7] compared iron and aluminum oxides as inexpensive

adsorbents for As (V) removal and found that Fe_2O_3 was a better adsorbent than Al_2O_3. Savina et al. [8] studied the removal of As (V) by applying iron nanoparticles embedded with macroporous polymer composites. Chen et al. [9] found that the high efficiency of As (V) adsorption on Ce-Fe bimetal oxide was related to its adsorbent mesoporous structure and abundant surface hydroxyl groups. Kong et al. [10] studied the role of an adsorbent of magnetic nanoscale Fe-Mn binary oxides loaded zeolites in the removal of arsenic from water. In most of the works, the adsorbents were in nano size powder forms which are difficult and expensive to separate from treated water [11]. The immobilization of adsorbent on substrate improves adsorption efficiency makes adsorbent separation from water easier in batch processes, and results in a lower pressure drop in column processes [11]. Therefore, in this work, the nanostructured Fe_2O_3/Al_2O_3 was synthesized via the sol-gel method and applied for the adsorption of As (V) from water. Also, Fe_2O_3/Al_2O_3 immobilized on leca granules was investigated for the adsorption of As (V) from water.

*Corresponding author
E-mail address: akhlaghianfk@gmail.com

2. Materials and methods

Aluminum isopropoxide (98%), iron (III) nitrate (98%), ethanol (98%), nitric acid (65%), sodium arsenate (98%), and polyethylene glycol with a molecular weight of 2000 g/gmol were purchased from Merck Company. The leca granules were purchased from Leca Company (Iran).

2.1. Synthesis of Fe_2O_3/Al_2O_3

In order to prepare the Fe_2O_3/Al_2O_3, double distilled water was added to aluminum isopropoxide, and hydrolyzed. The molar ratio of aluminum isopropoxide to water was 1:100. The mixture was stirred at a constant rate and heated to 85°C. Then, nitric acid was added to pepitize alumina sols. The molar ratio of water to acid was 1:0.07. Then, iron (III) nitrate was added. The mixture was stirred at 85°C under reflux for 24 h. The obtained gel was dried in an oven at 100°C for 12 h [12,13]. The dried gel was calcined at 400°C for 2 h. The synthesized particles were crushed, and sieved to the particles size of 60-90 µm.

2.2. Immobilization on leca particles

Leca granules with diameters of 4 to 10 mm were used as substrate. First, leca granules were cleaned for coating; so they were placed in a beaker containing nitric acid (10 wt.%) and exposed to ultrasonic waves for 30 min. Afterwards, the leca particles were rinsed with double distilled water and dried at 100°C in an oven for 24 h. The coating slurry was prepared by mixing distilled water (70 mL), polyethylene glycol (15 gr), nitric acid (1 g), and synthesized adsorbent (15 g). For 24 h, the mixture was stirred at a constant rate. The leca granules were coated by the dip coating method. They were immersed in the coating slurry, and then pulled out at a constant speed. The coated particles were dried at 100°C for 12 h [14]. Ultrasonic testing was used to examine the adhesion of the adsorbent to the substrate. A certain amount of coated granules was immersed in a beaker containing water, then it was exposed to ultrasonic waves for 30 min. The weight loss below 10% showed that the adsorbent particles were well adhered to the leca substrate [15]. The adsorbent was immobilized on the leca particles in this way.

2.3. Characterization

An X-ray fluorescence (XRF) spectrometer of Philips PW 2404 was used to determine the elemental composition of the adsorbent. The X-ray powder diffraction (XRD) analysis was performed using an X'Pert MPD Philips spectrometer with Co-k_α irradiation. The data were collected at 0.2°/s of scanning speed in the range of 10-80°. The specific surface area and pore volume were measured by nitrogen adsorption-desorption using an ASAP 2010 Micrometrics. The sample morphology was observed by a MIRA3 field emission scanning electron microscope from TESCAN.

2.4. Adsorption experiments of Fe_2O_3/Al_2O_3

The adsorbent dose of 1 g/L was added to 100 mL of arsenate synthetic wastewater with the concentration of 50 mg/L. The mixture was stirred at room temperature (25°C) for 12 h. Then, the mixture was centrifuged, and the arsenate concentration of the solution was determined by a Phoenix 986 atomic absorption spectrometer. The efficiency of the arsenate removal was calculated through Equation (1):

$$\text{Removal efficiency \%} = 100 \times \left(\frac{C_0 - C_f}{C_f} \right) \qquad (1)$$

where C_0, and C_f are the initial, and final concentration of arsenate in the water solution; respectively.

2.5. Adsorption studies

Adsorption experiments were performed to understand the behavior, nature, kinetics, and thermodynamics of the process. The effect of temperature on the adsorption of As (V) on Fe_2O_3/Al_2O_3 was examined. In Equation (2) [16,17]

$$K = \frac{mq_e}{C_e} \qquad (2)$$

K is the equilibrium constant; q_e (mg/g) is equilibrium adsorption capacity; C_e (mg/L) is the equilibrium concentration of As (V) in the solution; and m is the adsorbent mass.
In Equation (3) [16,17]:

$$\text{Log } K = \frac{\Delta S}{2.3R} - \frac{\Delta H}{2.3RT} \qquad (3)$$

R is the universal gas constant, and 8.314 J/mol.K; T (K) is temperature. K (equilibrium constant) was determined at different temperature using Equation (2). ΔH (enthalpy) and ΔS (entropy) can be determined from the slope (-ΔH/2.3R) and intercept (ΔS/R) of the linear plot of Log (K) versus 1/T.
From Equation (4) [16,17]:

$$\Delta G = \Delta H - T \Delta S \qquad (4)$$

Gibbs free energies were calculated at different temperatures [16,17].
The pseudo-first-order and pseudo-second-order models were employed to study the kinetics of the adsorption process. The pseudo-first-order (Equation (5)) and pseudo-second-order (Equation (6)) models are shown below [6,18]:

$$\text{Log}(Q_e - Q_t) = \text{Log} Q_e - \frac{k_1}{2.303} t \qquad (5)$$

$$\frac{t}{Q_t} = \frac{1}{k_2 Q_e^2} + \frac{t}{Q_e} \qquad (6)$$

where Q_e (mg/g) is equilibrium adsorption capacity; Q_t (mg/g) is adsorption capacity at any time t (min); k_1 and k_2 are rate constants of the pseudo-first-order and pseudo-second-order models; respectively. The Langmuir and Freundlich isotherm models were used for fitting of the experimental data. The Freundlich isotherm is expressed as follows [18]:

$$\text{Log}q = \frac{1}{n}\text{Log}C_e + \text{Log}K_F \qquad (7)$$

where n is a constant related to the adsorption energy; K_F K_F $(mg^{1-1/n}L^{1/n}/g)$ is a constant related to adsorption capacity.

The Langmuir isotherm model is shown below [18]:

$$\frac{1}{q} = \frac{1}{q_m b C_e} + \frac{1}{q_m} \qquad (8)$$

where q_m is the maximum adsorption capacity; b (L/g) is a constant related to the adsorption energy; q (mg/g) is the

As (V) concentration in the solid adsorbent; and C_e is the As (V) concentration in the solution (mg/L).

3. Results and discussion

3.1. Characterization

XRF spectrometry was used to determine the adsorbent composition. The composition of the optimized adsorbent was determined as $37.6\%Fe_2O_3/62.4\%Al_2O_3$. The X-ray diffraction patterns of the Al_2O_3 and $37.6\%Fe_2O_3/62.4\%Al_2O_3$ synthesized adsorbents are shown in Figure 1. The Al_2O_3 XRD pattern show the formation of amorphous alumina [19]. In the XRD pattern of Fe_2O_3/Al_2O_3, the peaks observed at 28.22°, 38.77°, 41.75°, 47.945°, 58.2°, 64.105°, 74.345°, and 76.005° were related to the formation of Fe_2O_3 with rhombohedral lattice (JCPDS File No. 13-0534) [20,21].

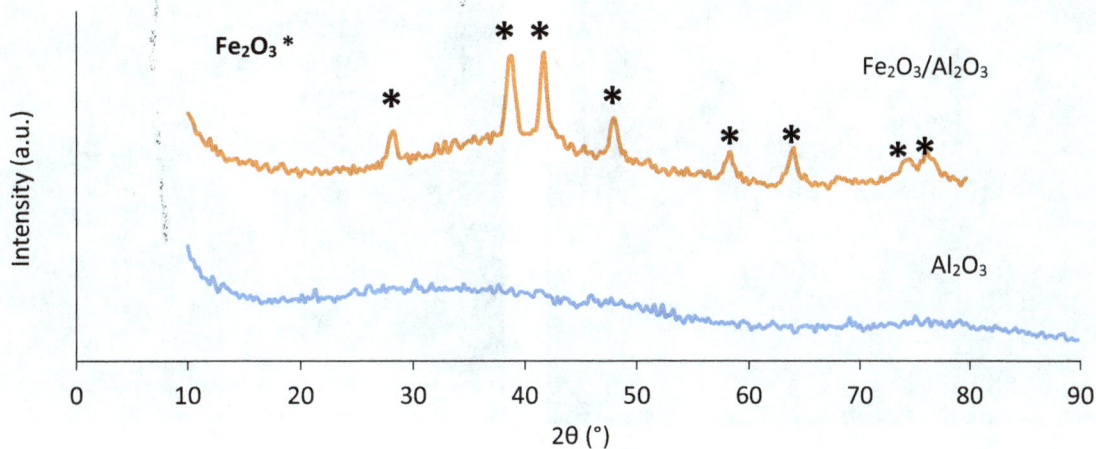

Fig.1. XRD pattern of Al_2O_3 and Fe_2O_3/Al_2O_3 adsorbents

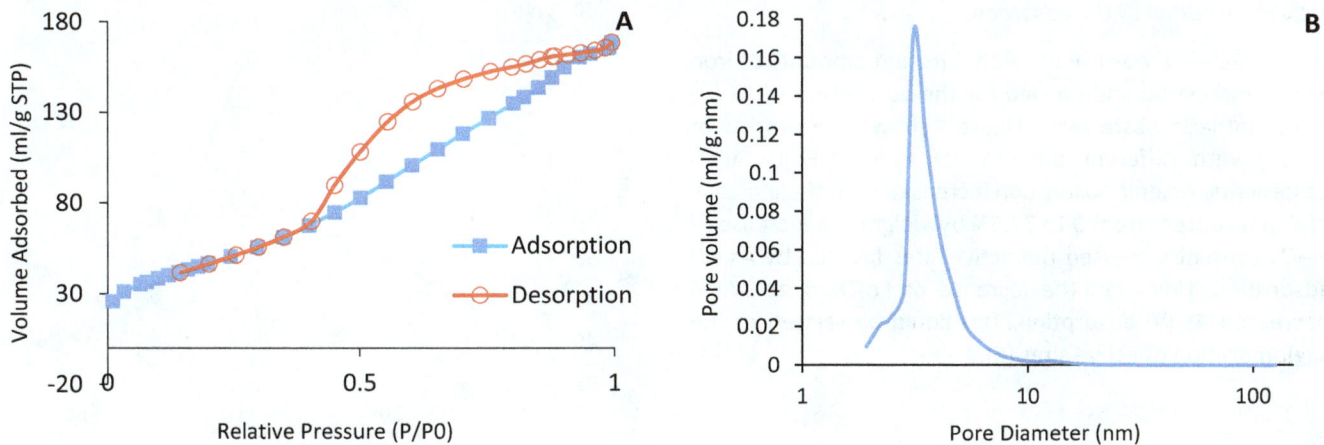

Fig. 2. (A) Liquid nitrogen adsorption/desorption isotherm of $37.6\%Fe_2O_3/62.4\%Al_2O_3$; (B) Pore size distribution of $37.6\%Fe_2O_3/62.4\%Al_2O_3$ based on BJH desorption model

Figure 2(A) displays nitrogen adsorption/desorption isotherms of the $37.6\%Fe_2O_3/62.4\%Al_2O_3$ adsorbent which is a mesopore of type IV according to the IUPAC

classification. The hysteresis loop is type H2, showing pores with large bodies and small mouths. Figure 2 (B) shows the pore size distribution in the range of 2-100 nm which is unimodal with a peak at 3.1 nm [22]. The specific surface area, total pore volume, and average pore diameter were determined as 269.151 cm^2/g, 0.260451 cm^3/g, and 3.9565 nm, respectively, based on the BJH desorption model. The

SEM images of the $37.6\%Fe_2O_3/62.4\%Al_2O_3$ adsorbent are shown in Figure 3 and dispersed nano size Fe_2O_3 particles are observed. The size of the Fe_2O_3 particles are less than 100 nm.

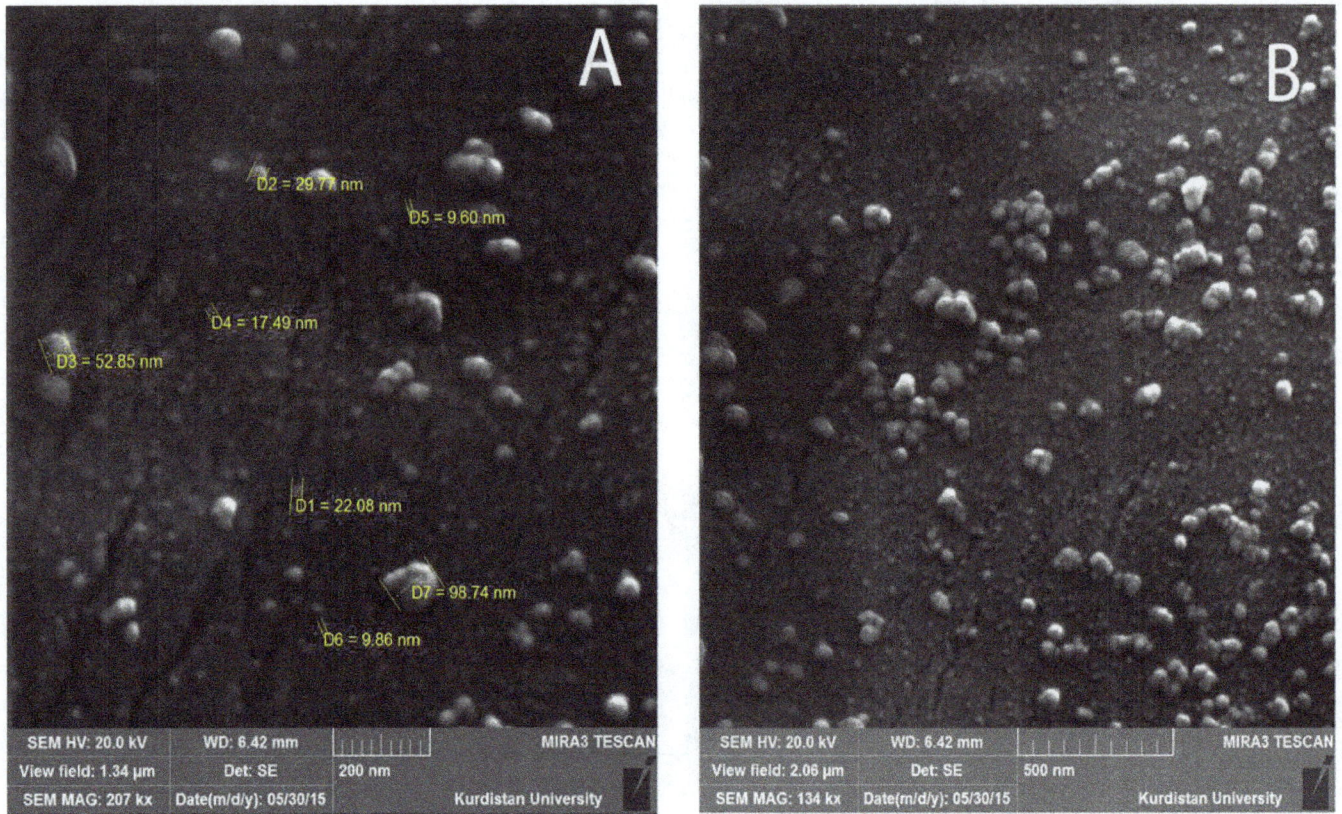

Fig. 3. SEM images of $37.6\%Fe_2O_3/62.4\%Al_2O_3$ adsorbent

3.2. *Optimization of the adsorbent*

The Fe_2O_3/Al_2O_3 adsorbents with different amounts of iron were synthesized and applied for the adsorption of As (V) from synthetic wastewater. Figure 4 shows the removal of As (V) with different amounts of iron in Fe_2O_3/Al_2O_3 adsorbents. Arsenic adsorption increased with the increase of Fe_2O_3 content from 0 to 37.6% by weight. An increase of Fe_2O_3 content increased the active sites favored by As (V) adsorption. However, the increase of Fe_2O_3 over 37.6% decreased As (V) adsorption. This could be related to the agglomeration of active sites [6].

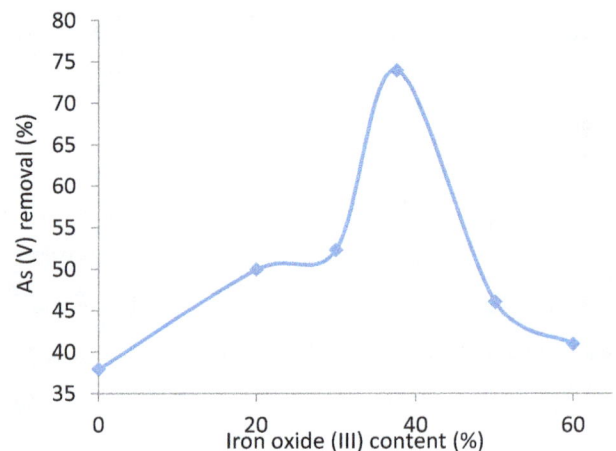

Fig. 4. The effect of Fe_2O_3 content on As (V) removal; operating conditions: As (V) initial concentration 50 mg/L, pH 6, adsorbent dose 1 g/L, contact time 12 h

3.3. Optimization of operating conditions

3.3.1. Effect of pH

The effect of pH on As (V) adsorption is shown in Figure 5 (A). In the pH range of 4 to 7, adsorption was nearly constant. In the pH of 7, adsorption was 72.61%. This amount dropped to 41.16% in the pH of 10. In the pH range of 2 to 10, As (V) occurs in the forms of $H_2AsO_4^-$ and $HAsO_4^{2-}$ [23]. The pH of zero point charge was 7. In pH<pHzpc, the adsorbent is positively charged because of the high concentration of protons in the solution and protonation. The attractive columbic force between the positively charged surface and negatively charged As (V) species led to adsorption. In pH>pHzpc, the adsorption decreased due to repulsive columbic force between the negatively charged surface and As (V) ions as well as the competition for adsorption between hydroxyl groups and As (V) species [23-25].

3.3.2. Effect of adsorbent dose

The adsorption increased with increasing of the adsorbent dose from 0.5 to 1 g/L (Figure 5 (B)). Increasing the available active site and specific surface area improved the adsorption. The adsorption remained nearly constant due to the agglomeration of adsorbent particles in the adsorbent dose more than 1 g/L [26].

3.3.3. Effect of contact time

In the first 120 min of the adsorption reaction, the rate was high, and then it became nearly constant (Figure 5(C)). In the first 120 min of adsorption, 64% of the As (V) was removed while in 720 min of adsorption, 74% of As (V) was removed. This high rate of adsorption in the first minutes is related to the large number of available surface sites. After sometime, the adsorption rate declined and finally reached equilibrium. The reason for the slow adsorption rate was the small number of active sites. At this stage, the adsorption reaction proceeded through the internal active sites of the adsorbent [27].

3.3.4. Effect of arsenate initial concentration

Figure 5 (D) shows that increasing arsenate initial concentration decreased adsorption. The saturation of the available sites with increasing of the arsenate concentration decreased the adsorption [27].

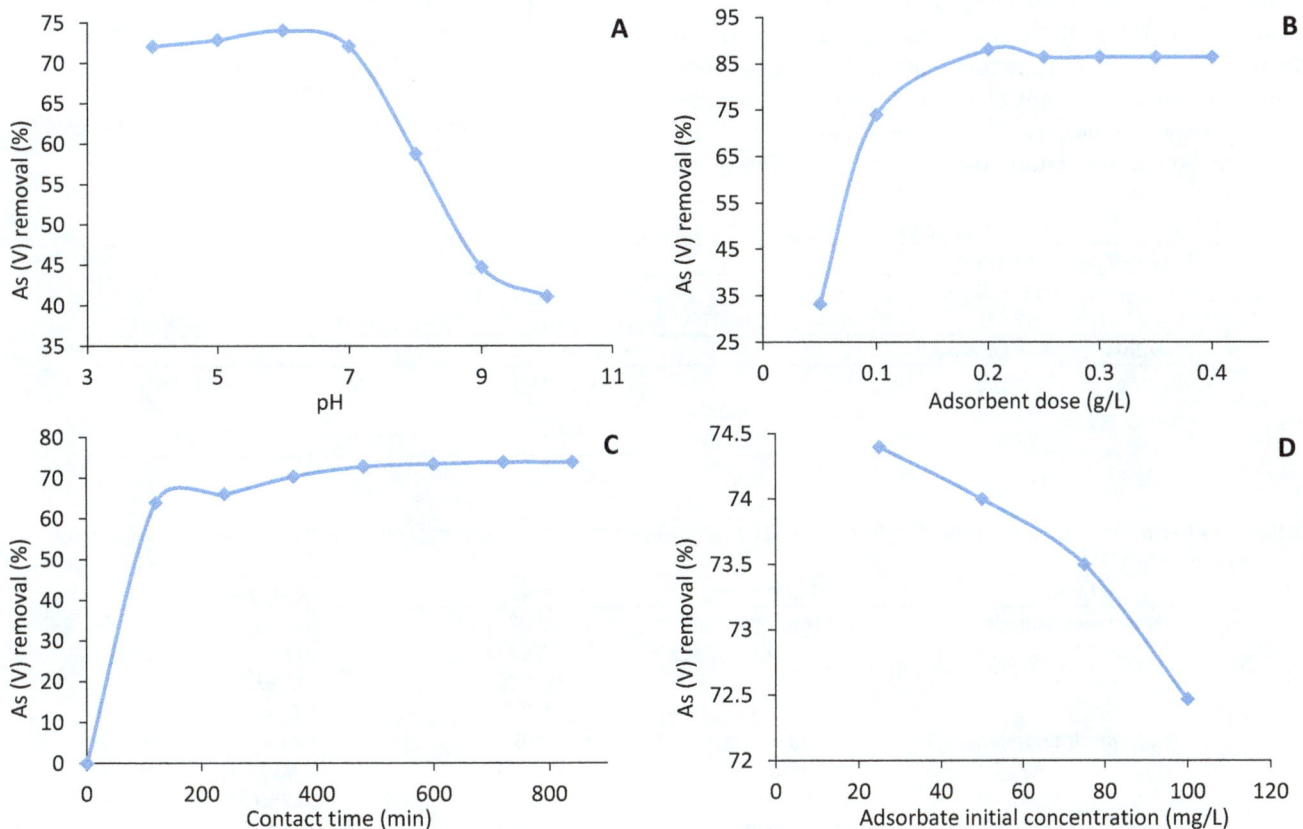

Fig. 5. Effects of operating conditions on As (V) removal (A) pH, (B) adsorbent dose, (C) contact time, (D) As (V) initial concentration. instead of; in the above figure, the operating conditions of As (V) initial concentration 50 mg/L, pH 6, adsorbent dose 1 g/L, and contact time 12 h were constant unless their effect was investigated

3.4. Thermodynamic, kinetics, and isotherms of adsorption

The results of the raising temperature showed slightly increased As (V) adsorption on $37.6\%Fe_2O_3/62.4\%Al_2O_3$. Equations (2) and (3) were used at different temperatures. The ΔH was calculated (+65.96 kJ/mol) showing it was an endothermic process. The ΔS was determined to be 228.87 J/mol presenting increased randomness at the interface of the solid solution for As (V) adsorption. Gibbs free energies were calculated at different temperatures by Equation (4) which is shown in Table 1.

Table 1. Thermodynamic data for As (V) adsorption by 37.4% $Fe_2O_3/62.6\%Al_2O_3$; operating condition: As (V) initial concentration 50 mg/L, pH 6, adsorbent dose 1g/L, contact time 12 h

T (K)	K	ΔG	ΔS (J/mol)	ΔH
298	2.97	-2.24	228.87	65.96
303	3.11	-3.39		
313	5.13	-5.13		
318	10.47	-10.47		

The values of Gibbs free energies were negative indicating the spontaneous process of As (V) adsorption [16,17]. The mechanism of As (V) adsorption on the $37.6\%Fe_2O_3/62.4\%Al_2O_3$ adsorbent was studied using Equations (5) and (6). The determination factor R^2 of the pseudo-second order model is larger than the pseudo-first-order model (Table 2), so it can be concluded that the kinetics obeyed the pseudo-second-order model and chemisorption is the controlling step of the adsorption [6,18]. The results of curve fitting are displayed in Figure (6). Table 3 shows the determination factor R^2 for fitting the

experimental data of As (V) adsorption on alumina to Freundlich and Langmuir isotherms. The larger determination factor R^2 shows that adsorption followed the Langmuir isotherm model. Similarly, Table 3 shows R^2s for the $37.6\%Fe_2O_3/62.4\%Al_2O_3$ adsorbent. The Langmuir isotherm model was better fitted to the experimental data due to the larger R^2 compared with the Freundlich model. For alumina maximum adsorption capacity, q_m was 40.65 mg/g and b was 0.0275 L/mg (Table 3). For the adsorbent $37.6\%Fe_2O_3/62.4\%Al_2O_3$, q_m was 74.6 mg/g and b was 0.0392 L/mg. The larger adsorption capacity indicates that $37.6\%Fe_2O_3/62.4\%Al_2O_3$ is a more efficient adsorbent for the removal of As (V) from water compared with alumina.

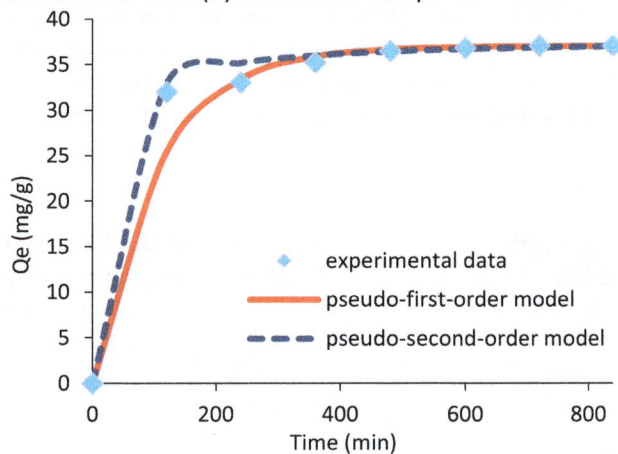

Fig. 6. Kinetics of As (V) adsorption on $37.6\%Fe_2O_3/62.4\%Al_2O_3$ adsorbent

Table 2. Kinetic parameters for As (V) adsorption by $37.6\%Fe_2O_3/62.4\%Al_2O_3$ adsorbent

Pseudo-first-order model				
Q_e (mg/g)	k_1 (min^{-1})	Q_1 (mg/g)	R^2	Variance
37.01	0.0097	36.017	0.9468	10.87
Pseudo-second-order model				
Q_e (mg/g)	k_2 (g/mg.min)	Q_2 (mg/g)	R^2	Variance
37.01	0.0015	37.7	0.9989	1.53

Table 3. Isotherm model parameter for As (V) by Al_2O_3, $37.6\%/Fe_2O_3/62.4\%Al_2O_3$; operating conditions: pH 6, adsorbent dose 1g/L, contact time 12 h

Model	Parameter	Al_2O_3	$37.6\%/Fe_2O_3/62.4\%Al_2O_3$
Freundlich equation	K_F (mg$^{1-1/n}$L$^{1/n}$/g)	2.97	2.87
	n	1.93	0.992
	R^2	0.9679	0.997
	Variance	0.093	0.0743
Langmuir equation	q_m (mg/g)	40.65	74.6
	b (L/mg)	0.0275	0.0392
	R^2	0.9769	0.999
	Variance	0.0067	0.0052

3.5. Adsorption of immobilized adsorbent

The leca was coated by the $37.6\%Fe_2O_3/62.4\%Al_2O_3$ adsorbent and applied for As (V) adsorption. The coating

layer load on leca particles was 4% by weight (Figure 7). The calculation of the immobilized adsorbent dose according to its load is given in the caption of Figure 7. The adsorption of As (V) improved using immobilized adsorbent on leca granules due to increasing available surface area. In the batch process, the immobilized adsorbent on the leca granules separated more easily from water. In the column process, the leca granules increased the porosity of the bed and decreased the pressure drop.

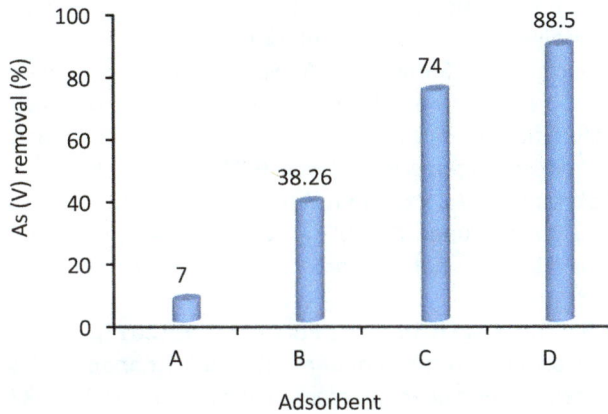

Fig. 7. As (V) adsorption on different adsorbents (A) leca, (B) alumina, (C) 37.6%Fe$_2$O$_3$/Al$_2$O$_3$, (D) coated leca; operating conditions: As (V) initial condition 50 mg/L, pH 6, contact time 12 h, adsorbent dose (A) 50 g/L, (B) 1 g/L, (C) 1g/L, (D) 50 g/L (50 g of coated leca × (4 g coating layer)/ (100 g of coated leca) × 1/2 = 1 g of 37.6%Fe$_2$O$_3$/ 62.4%Al$_2$O$_3$), the ratio of 37.6%Fe$_2$O$_3$/62.4%Al$_2$O$_3$ to polyethylene glycol+ 37.6%Fe$_2$O$_3$/62.4%Al$_2$O$_3$ in coating slurry was 1/2, so in our calculation we multiplied the result by 1/2.

Table 4 shows the results of fitting Langmuir and Freundlich isotherms to the experimental data for the immobilization of 37.6%Fe$_2$O$_3$/62.4%Al$_2$O$_3$ on leca particles (coated leca). The larger R^2 reveals that Langmuir isotherm model fitted the experimental data better. The results of this work are compared with other literature works in Table 5. It is clear that the best adsorption capacity belonged to 37.6%Fe$_2$O$_3$/62.4%Al$_2$O$_3$ on leca. The adsorption capacity of 37.6%Fe$_2$O$_3$/62.4%Al$_2$O$_3$ was also good. These results suggest that the Fe-Al binary metal adsorbent prepared with the proposed method as well as coating it on a substrate like leca can be very efficient for the removal of As (V) from water.

Table 4. Isotherm model parameter for As (V) adsorption by coated leca; operating conditions: pH 6, adsorbent dose 1g/L, contact time 12 h

Model	Parameter	coated leca
Freundlich equation	K$_F$ (mg$^{1-1/n}$L$^{1/n}$/g)	2.094
	n	0.678
	R^2	0.776
	Variance	0.429
Langmuir equation	q$_m$ (mg/g)	125
	b (L/mg)	0.1
	R^2	0.999
	Variance	0.0313

Table 5. Comparisons of As (V) adsorption capacities of different adsorbents

No.	Adsorbent	Adsorption capacity (mg g^{-1})	Reference
1	Al$_2$O$_3$	0.17	[7]
2	Fe$_2$O$_3$	0.66	[7]
3	Iron oxide particles-embedded macroporous polymers	91.74	[9]
4	Fe-Al double hydrous oxide	24.1	[11]
5	Cryogel embedded with Fe-Al double hydrous oxide	24.6	[11]
6	Ni-Fe binary oxide	90.1	[29]
7	Iron-ziconia coated sand	45.05	[28]
8	Al$_2$O$_3$	40.65	present work
10	37.6%Fe$_2$O$_3$/62.4%Al$_2$O$_3$	74.6	present work
11	37.6%Fe$_2$O$_3$/62.4%Al$_2$O$_3$ coated on leca	125	Present work

Conclusions

The Fe$_2$O$_3$/Al$_2$O$_3$ adsorbent was synthesized using aluminum isopropoxide and iron (III) nitrate as precursors via the sol-gel method. The synthesized adsorbent was used for the adsorption of As (V) from water. The load of iron oxide of the adsorbent was optimized and the XRF results determined the composition of the adsorbent 37.6%Fe$_2$O$_3$/62.4%Al$_2$O$_3$. The XRD results showed the formation of hematite Fe$_2$O$_3$ and amorphous Al$_2$O$_3$. The SEM images depicted nano size iron oxide particles. The kinetics studies revealed that As (V) adsorption of 37.6%Fe$_2$O$_3$/62.4%Al$_2$O$_3$ obeyed the pseudo-second-order model. The Langmuir isotherm model fitted the experimental data better than Freundlich model. The adsorption capacity of pure Al$_2$O$_3$ and 37.6%Fe$_2$O$_3$/62.4%Al$_2$O$_3$ were determined to be 40.65 and 74.6 mg/g, respectively. The adsorption capacity of 37.6%Fe$_2$O$_3$/62.4%Al$_2$O$_3$ coated on leca was determined to be 125 mg/g which revealed its higher efficiency in respect to uncoated adsorbents. A comparison of the adsorption capacity of the nanostructured 37.6%Fe$_2$O$_3$/62.4%Al$_2$O$_3$ adsorbent with those available in the literature revealed that this adsorbent is promising.

Acknowledgments

The financial support of the University of Kurdistan is gratefully acknowledged.

References

[1] Shevade, S., Ford, R. G. (2004). Use of synthetic zeolites for arsenate removal from pollutant water. *Water research*, *38*(14), 3197-3204.

[2] Reza, R., Singh, G. (2010). Heavy metal contamination and its indexing approach for river water. *International journal of environmental science and technology*, *7*(4), 785-792.

[3] Kundu, S., Kavalakatt, S. S., Pal, A., Ghosh, S. K., Mandal, M., Pal, T. (2004). Removal of arsenic using hardened paste of Portland cement: batch adsorption and column study. *Water research*, *38*(17), 3780-3790.

[4] Sigdel, A., Park, J., Kwak, H., Park, P. K. (2016). Arsenic removal from aqueous solutions by adsorption onto hydrous iron oxide-impregnated alginate beads. *Journal of industrial and engineering chemistry*, *35*, 277-286.

[5] Sigdel, A., Park, J., Kwak, H., Park, P. K. (2016). Arsenic removal from aqueous solutions by adsorption onto hydrous iron oxide-impregnated alginate beads. *Journal of industrial and engineering chemistry*, *35*, 277-286.

[6] Han, C., Zhang, L., Chen, H., Shan, X., Li, X., Zhu, W., Luo, Y. (2016). Removal As (V) by sulfated mesoporous Fe–Al bimetallic adsorbent: Adsorption performance and uptake mechanism. *Journal of environmental chemical engineering*, *4*(1), 711-718.

[7] Jeong, Y., Fan, M., Singh, S., Chuang, C. L., Saha, B., Van Leeuwen, J. H. (2007). Evaluation of iron oxide and aluminum oxide as potential arsenic (V) adsorbents. *Chemical engineering and processing: Process intensification*, *46*(10), 1030-1039.

[8] Savina, I. N., English, C. J., Whitby, R. L., Zheng, Y., Leistner, A., Mikhalovsky, S. V., Cundy, A. B. (2011). High efficiency removal of dissolved As (III) using iron nanoparticle-embedded macroporous polymer composites. *Journal of hazardous materials*, *192*(3), 1002-1008.

[9] Chen, B., Zhu, Z., Guo, Y., Qiu, Y., Zhao, J. (2013). Facile synthesis of mesoporous Ce–Fe bimetal oxide and its enhanced adsorption of arsenate from aqueous solutions. *Journal of colloid and interface science*, *398*, 142-151.

[10] Kong, S., Wang, Y., Hu, Q., Olusegun, A. K. (2014). Magnetic nanoscale Fe–Mn binary oxides loaded zeolite for arsenic removal from synthetic groundwater. *Colloids and surfaces A: Physicochemical and engineering aspects*, *457*, 220-227.

[11] Kumar, P. S., Önnby, L., Kirsebom, H. (2013). Arsenite adsorption on cryogels embedded with iron-aluminium double hydrous oxides: Possible polishing step for smelting wastewater. *Journal of hazardous materials*, *250*, 469-476.

[12] Chen, H., Bednarova, L., Besser, R. S., Lee, W. Y. (2005). Surface-selective infiltration of thin-film catalyst into microchannel reactors. *Applied catalysis A: General*, *286*(2), 186-195.

[13] Germani, G., Alphonse, P., Courty, M., Schuurman, Y., Mirodatos, C. (2005). Platinum/ceria/alumina catalysts on microstructures for carbon monoxide conversion. *Catalysis today*, *110*(1), 114-120.

[14] Germani, G., Stefanescu, A., Schuurman, Y., Van Veen, A. C. (2007). Preparation and characterization of porous alumina-based catalyst coatings in microchannels. *Chemical engineering science*, *62*(18), 5084-5091.

[15] Stefanescu, A., Van Veen, A. C., Mirodatos, C., Beziat, J. C., Duval-Brunel, E. (2007). Wall coating optimization for microchannel reactors. *Catalysis today*, *125*(1), 16-23.

[16] Goswami, A., Raul, P. K., Purkait, M. K. (2012). Arsenic adsorption using copper (II) oxide nanoparticles. *Chemical engineering research and design*, *90*(9), 1387-1396.

[17] Xiong, C., Li, Y., Wang, G., Fang, L., Zhou, S., Yao, C., Zhu, Y. (2015). Selective removal of Hg (II) with polyacrylonitrile-2-amino-1, 3, 4-thiadiazole chelating resin: batch and column study. *Chemical engineering journal*, *259*, 257-265.

[18] Wu, K., Liu, T., Xue, W., Wang, X. (2012). Arsenic (III) oxidation/adsorption behaviors on a new bimetal adsorbent of Mn-oxide-doped Al oxide. *Chemical engineering journal*, *192*, 343-349.

[19] Amini, G., Najafpour, G. D., Rabiee, S. M., Ghoreyshi, A. A. (2013). Synthesis and Characterization of Amorphous Nano-Alumina Powders with High Surface Area for Biodiesel Production. *Chemical engineering and technology*, *36*(10), 1708-1712.

[20] Biabani-Ravandi, A., Rezaei, M., Fattah, Z. (2013). Catalytic performance of Ag/Fe$_2$O$_3$ for the low temperature oxidation of carbon monoxide. *Chemical engineering journal*, *219*, 124-130.

[21] Golsefidi, M.A., Abbasi, F., Abrodi, M., Abbasi, Z., Yazarlou, F. (2016). Synthesis characterization and photocatalytic activity of Fe$_2$O$_3$-TiO$_2$ nanoparticles and nanocomposite. *Journal of nanostructures, 6* (1), 61-66.

[22] Leofanti, G., Padovan, M., Tozzola, G., Venturelli, B. (1998). Surface area and pore texture of catalysts. *Catalysis today*, *41*(1), 207-219.

[23] Sahiner, N., Ozay, O., Aktas, N., Blake, D. A., John, V. T. (2011). Arsenic (V) removal with modifiable bulk and nano p(4-vinylpyridine)-based hydrogels: The effect of hydrogel sizes and quarternization agents. *Desalination*, *279*(1), 344-352.

[24] Zheng, Y. M., Zou, S. W., Nanayakkara, K. N., Matsuura, T., Chen, J. P. (2011). Adsorptive removal of arsenic from aqueous solution by a PVDF/zirconia blend flat sheet membrane. *Journal of membrane science*, *374*(1), 1-11.

[25] Kumar, A. S. K., Jiang, S. J. (2016). Chitosan-functionalized graphene oxide: A novel adsorbent an efficient adsorption of arsenic from aqueous solution. *Journal of environmental chemical engineering, 4*(2), 1698-1713.

[26] Lisha, K. P., Maliyekkal, S. M., Pradeep, T. (2010). Manganese dioxide nanowhiskers: a potential adsorbent for the removal of Hg (II) from water. *Chemical engineering journal, 160*(2), 432-439.

[27] Nath, B. K., Chaliha, C., Kalita, E., Kalita, M. C. (2016). Synthesis and characterization of ZnO: CeO 2: nanocellulose: PANI bionanocomposite. A bimodal agent for arsenic adsorption and antibacterial action. *Carbohydrate polymers, 148*, 397-405.

25

Effect of anion interaction on the removal efficiency of nanofilters for potable water treatment

Peyman Mahmoodi, Mehrdad Farhadian[*], Ali Reza Solaimany Nazar, Reihaneh Bashiri

Department of Chemical Engineering, Faculty of Enginerring ,University of Isfahan, Isfahan, Iran

Keywords:
Box–Behnken design
Hexavalent chromium
Ion interaction
Nanofiltration
Nitrate
Potable water treatment

ABSTRACT

The interaction between the ions and the charge of membranes can affect the efficiency of pollutant removal. The present study investigated the removal efficiency of hexavalent chromium and nitrate ions from both actual and synthetic contaminated water via two different commercial spiral wound polyamide nanofilters. In addition, the interaction of ions under different experimental conditions was investigated by using a Box-Behnken design (BBD). The Box–Behnken design optimized the contributing factors which included pH (5-9), the initial concentration of Cr (VI) (0.05-5 mg/L) and the initial concentration of nitrate (40-160 mg/L). The maximum removal efficiency of both Cr (VI) and nitrate was achieved at a pH of 9.0, as 99 % and 90 % for the Iranian nanofilter (NF-I) and 98 % and 82 % for the Korean nanofilter (NF-K), respectively. The results also indicated that as the initial concentration of Cr (VI) increased, the removal efficiency was enhanced while the removal efficiency of nitrate decreased according to the pH. However, by increasing the initial concentration of nitrate, the removal efficiency of both the Cr (VI) and nitrate increased. For actual water samples at an optimum pressure of 0.6 Mpa (NF-K) and 0.8 Mpa (NF-I), the removal efficiency of Cr(VI) and nitrate obtained was 95% and 76 % for the NF-K and 97 % and 86 % for the NF-I, respectively.

1. Introduction

Recently, the extensive application of nitrogen-rich fertilizers as well as utilizing municipal wastewater in the agriculture industry has resulted in severe water contamination by nitrate [1], turning many of the wells into stand-by ones [2]. Since the excessive concentration of nitrate causes various health problems and diseases such as methemoglobinemia [3], many researchers have focused on the removal of nitrate from water resources. On the other hand, the presence of heavy metals in water and wastewater has always been a matter of concern for environmentalists, who promote practical methods in an attempt to remove them. Of the various water polluting ions, hexavalent chromium is the most frequently found toxic agent in water and wastewater, originating mainly from industries such as mining, tanning, cement,

photography, metal manufacturing and electroplating [4]. This carcinogenic element can cause a number of illnesses affecting the kidneys, lungs, and liver [5]. Therefore, developing a novel pragmatic approach to remove these contaminating ions is of crucial importance. The World Health Organization (WHO) has reported a maximum contamination level (MCL) of 45 and 0.05 as mg/L for nitrate and Cr (VI), respectively [6]. However, the amount of these ions is usually higher. In Iran, there is also an increased potential for the presence of both nitrate and hexavalent chromium in water resources due to agricultural and industrial activities. Although a number of methods such as membrane filtration [7], adsorption [8], ion exchange [9], electrodialysis [10] and biosorption [11-12] are available for ion removal from contaminated water, most of them are unable to thoroughly eliminate such impurities. Furthermore, these methods suffer from several

*Corresponding author
E-mail address: m.farhadian@eng.ui.ac.ir

disadvantages such as requiring expensive equipment, monitoring systems and energy as well as the management of the produced toxic sludge [13]. Nanofiltration(NF) is not only an effective method to remove both the Cr (VI) (ionic radius for $HCrO_4$= 0.242 nm) [14] and nitrate (ionic radius= 0.189 nm) [15], it is also cost-effective and environmentally friendly. The separation of ionic species by the nanofilter membrane depends on both the charge and size effects [16]. Besides, nanofiltration can simultaneously remove a broad range of other impurities such as salts, dyes, viruses, bacteria and parasites. Yu Y., et al. [17] investigated the effect of ion concentration and natural organic matter on arsenic removal by nanofiltration. Recently, Li K., et al. [18] used NF to reuse wastewater by operational optimization and analysis of the membrane fouling. Since the removal of nitrate and hexavalent chromium from water has not been studied extensively, this study aims to compare the efficiency of two different nanofilter membranes in removing nitrate and hexavalent chromium from both actual and synthetic drinking water. In addition, the interaction of ions under different experimental conditions including the pH and the initial concentration of nitrate and hexavalent chromium was investigated by using a Box-Behnken design. The removal of both nitrate and chromium were first tested by nanofilters using synthetic water samples. Then, the experiments were repeated for the actual water samples.

2. Materials and methods

All the chemicals used in this study were supplied by the Merck Company, Germany, and included potassium dichromate ($K_2Cr_2O_7$), potassium nitrate (KNO_3), sulfuric acid, hydrochloric acid, and sodium hydroxide. The nitrate and hexavalent chromium solutions were also prepared using distilled water. Additionally, the polyamide spiral wound nano-filter membranes used in the experiments were selected from two different brands, one developed by the Noshirvani University of Technology, Iran (NF-I), and the other by the TFC Company, Korea (NF-K). Their characteristics are defined in Table 1. The actual water

samples were taken from the city of Isfahan's potable water (Table 2).

Table 1. Physicochemical characteristics of the commercial polyamide nanofilter membranes

Specification	Allowed range	
	NF-I	NF-K
Maximum operating pressure (bar)	20	9
Maximum operating temperature (ºC)	50	-
pH range	3-12	2-11
Active surface (m²)	0.35	0.35
Isoelectric point	4.6	4.5
Surface charge	Negative	Negative

Table 2. Isfahan's potable water characteristics

Characteristics	Measured value
K^+ (mg/L)	1
Na^+ (mg/L)	11
Ca^{2+} (mg/L)	38
Mg^{2+} (mg/L)	13
HCO_3^- (mg/L)	173
NO_3^- (mg/L)	10
Cl^- (mg/L)	15
SO_4^{2-} (mg/L)	19
pH	8.1
EC^*(µS/cm)	310
TDS (mg/L)	205

*Electric conductivity

2.1. Experimental set-up

A schematic picture of the apparatus is illustrated in Figure 1. The setup included a temperature control device and two diaphragm pumps with a capacity of 1.6 liters per minute at a maximum outlet pressure of 8.5 bars.

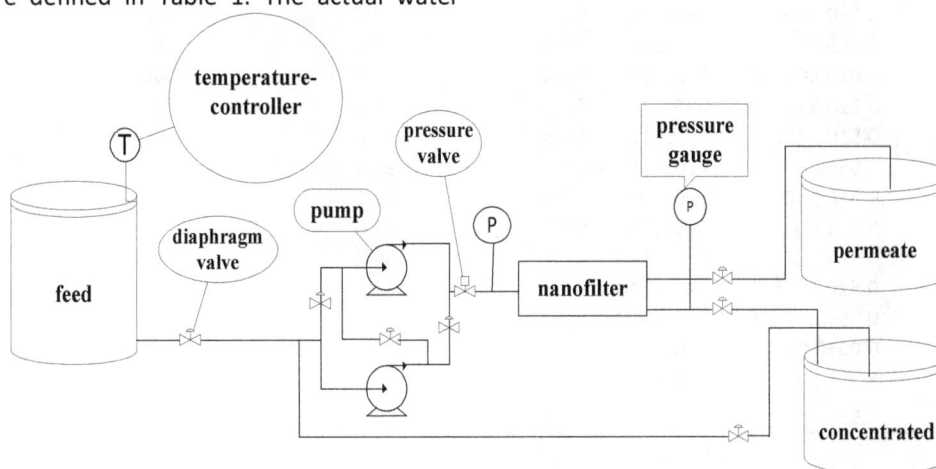

Fig. 1. Schematic picture of the experimental setup.

2.2. Experimental procedure

The experiments were conducted in 15, 30, 45 and 60 minute intervals after the startup time. The sampling analysis indicated that the system had the maximum removal efficiency at the 15 minute interval after the startup. In fact, by increasing the time interval, the removal efficiency remained constant. In order to achieve an efficiency rate of 75 %, the pressure was adjusted using the pressure valves on permeate and the concentrate stream. In addition, to avoid both measurement and human errors, the sampling and all the experiments were triplicated and duplicated, respectively. The data were introduced as the mean value to the Design Expert software. The feed concentration was also measured to obtain more accuracy in calculating the contaminant removal efficiency. During the experiments, the feed temperature and optimum pressure were kept constant at 20±1 °C and 0.8 (NF-I) and 0.6 Mpa (NF-K), respectively. According to the standard methods of 4500B and 3111B [19], the nitrate and chromium concentrations were determined using UV spectrophotometry (Jasco V-570, Japan) and atomic absorption spectrophotometry (Philips PU-9100, Netherlands), respectively. While the initial concentration of Cr (VI) was designated as 0.05, 0.5 and 5 mg/L, the nitrate concentration was selected as 40, 80 and 160 mg/L at three pH values of 5, 7 and 9. The removal efficiency of Cr (VI) and nitrate by the nanofilter membrane were determined by Eq. 1 [20]:

$$R\% = [1 - (\frac{C_p}{C_0})] \times 100 \tag{1}$$

where R represents the removal percentage and C_p and C_0 are the concentrations of the pollutant in the permeate and the feed water, respectively.

2.3. Design of experiment (DOE)

The response surface method (RSM) is an effective method for response optimization [21]. Therefore, it was employed in this study to maximize the removal of nitrate and hexavalent chromium. In order to determine the effective variables of the experiments and develop a RSM for optimization, the main factors mentioned above were further studied via BBD which is an accurate method in the case of water and wastewater treatment processes. Eq. 2 was used to calculate the number of experiments:

$$N = 2K \times (k - 1) + C \tag{2}$$

where k and C_0 denote the number of factors and central points, respectively [22]. On the other hand, Eq. 3 represented a second order polynomial [21]:

$$Y = b_0 + b_1X_1 + b_2X_2 + b_3X_3 + b_{11}X_1^2 + b_{22}X_2^2 + \tag{3}$$
$$b_{33}X_3^2 + b_{12}X_1X_2 + b_{13}X_1X_3 + b_{23}X_2X_3$$

where Y, [X_1, X_2, X_3] and [b_0, b_i, b_{ij}] represent the responses, coded parameters and the estimated model coefficients, respectively. It is worth mentioning that the response can be analyzed by applying contour diagrams. The contour line plot is two dimensional, illustrating fixed responses. The RSM used the contour plots to analyze the results. According to the number of factors and by employing the RSM method via Design Expert software (8.0.1), it was determined that 15 experiment runs should be conducted.

3. Results and discussion

The levels of independent variables according to the BBD method and the rejection percent of hexavalent chromium (Y_1) and nitrate (Y_2) responses for all the experiments done by both membranes are presented in Table 3.

Table 3. Cr (VI) and nitrate experimental design (conditions and responses) for NF-I and NF-K.

Run	Nitrate concen. (mg/L)	Cr (VI) concen. (mg/L)	pH	NF-I		NF-K	
				Cr (VI) Removal Y1 (%)	Nitrate Removal	Cr (VI) Removal Y1 (%)	Nitrate Removal
1	80±3 (0)	0.5±0.1 (0)	7±0.1 (0)	93.2	76.3	90.4	72.2
2	40±2 (-1)	0.5±0.1 (0)	5±0.1 (-1)	86.4	73	87.4	67.8
3	40±2 (-1)	0.5±0.1 (0)	9±0.1 (1)	95.2	89.7	93.2	81.5
4	40±2 (-1)	5±0.5 (1)	7±0.1 (0)	93.5	75.4	91.2	70.3
5	40±2 (-1)	0.05±0.02 (-1)	7±0.1 (0)	91.4	83.1	88.7	75.9
6	160±5 (1)	0.5±0.1 (0)	5±0.1 (-1)	89	65.4	87.4	66.7
7	160±5 (1)	0.5±0.1 (0)	9±0.1 (1)	98.8	86.9	96.8	77.1
8	80±3 (0)	0.05±0.02 (-1)	9±0.1 (1)	97.2	90.4	96.2	80.6
9	80±3 (0)	0.5±0.1 (0)	7±0.1 (0)	92.6	79.1	90.4	72.2
10	80±3 (0)	5±0.5 (1)	9±0.1 (1)	98.5	86	97.9	77.8
11	80±3 (0)	5±0.5 (1)	5±0.1 (-1)	91	66.2	91.4	64
12	80±3 (0)	0.5±0.1 (0)	7±0.1 (0)	93.8	78	91.5	71.6
13	80±3 (0)	0.05±0.02 (-1)	5±0.1 (-1)	86.2	77.5	83.6	72
14	160±5 (1)	0.05±0.02 (-1)	7±0.1 (0)	93.6	77.6	92.2	76.6
15	160±5 (1)	5±0.5 (1)	7±0.1 (0)	95	70.8	94.6	66.9
AWS*	80±3	0.5±0.1	8±0.1	97.3	86	95	76.4

*Actual Water Samples

3.1. ANOVA study

The results of the analysis of variance (ANOVA) for Cr(VI) and nitrate removal are presented in Table 4. Since the confidence level was taken as 95%, the effect of any factor was only significant if its P-value was less than 0.05. This meant that there was only a 5% probability of error to consider a non-significant factor as significant. The greater F-value showed a greater effect of the factor on the response. The F-value was defined as the ratio of the mean square of regression (MRR) to the mean square of error (MRe) (F = MRR/MRe). This implied that the linear effects of pH(X_2), Cr (VI) concentration(X_1), nitrate concentration (X_3) and interactive effect of pH and Cr (VI) concentration were more significant. Table 4 also indicates that the interactive effects of pH and nitrate concentration as well as Cr (VI) and nitrate concentration had no significant influence on the removal efficiency of Cr (VI) and nitrate. It is worth mentioning that the interactive effects between Cr (VI) concentration and pH on the response can be adjusted by pH. In fact, as the pH increased from 5 to 9, the concentration of the chromium ion which was in the form of $HcrO_4^-$ shifted to other forms like CrO_4^{2-} and $Cr_2O_7^{2-}$ [23-24]. Therefore, their electrical charge and ionic radius

changed, thus affecting the contaminants removal efficiency. The mathematical model based on actual values for hexavalent chromium and nitrate removal percentages are expressed by Eqs. 4 and 5 for the NF-I and Eq. 6 and Eq. 7 for the NF-K, respectively:

$$R_1(\%) = 93.2 + 1.25X_1 + 4.6375X_2 + 1.2875X_3 \quad (4)$$
$$- 0.875X_1X_2$$

$$R_2(\%) = 77.8 - 3.775X_1 + 8.862X_2 - 2.5625X_3 \quad (5)$$
$$+ 1.725X_1X_2 + 2.125X_2^2$$

$$R_1(\%) = 91.3 + 1.825X_1 + 4.925X_2 + 1.925X_3 \quad (6)$$
$$- 1.525X_1X_2 + 1.3375X_1^2$$

$$R_2(\%) = 72.27 - 3.2625X_1 + 5.8125X_2 \quad (7)$$
$$- 1.025X_3 + 1.30X_1X_2$$

The regression parameter R^2 was applied to determine the agreement in comparison of the experimental responses to the ones estimated by the BBD method. The R^2 value for Eq. 4 and Eq. 5 was found to be 0.9910 and 0.9906, respectively, and 0.9840 and 0.9874 for Eq. 6 and Eq. 7, respectively. Thus, there were good agreements between the experimental and the predicted removal efficiency (Figure 2).

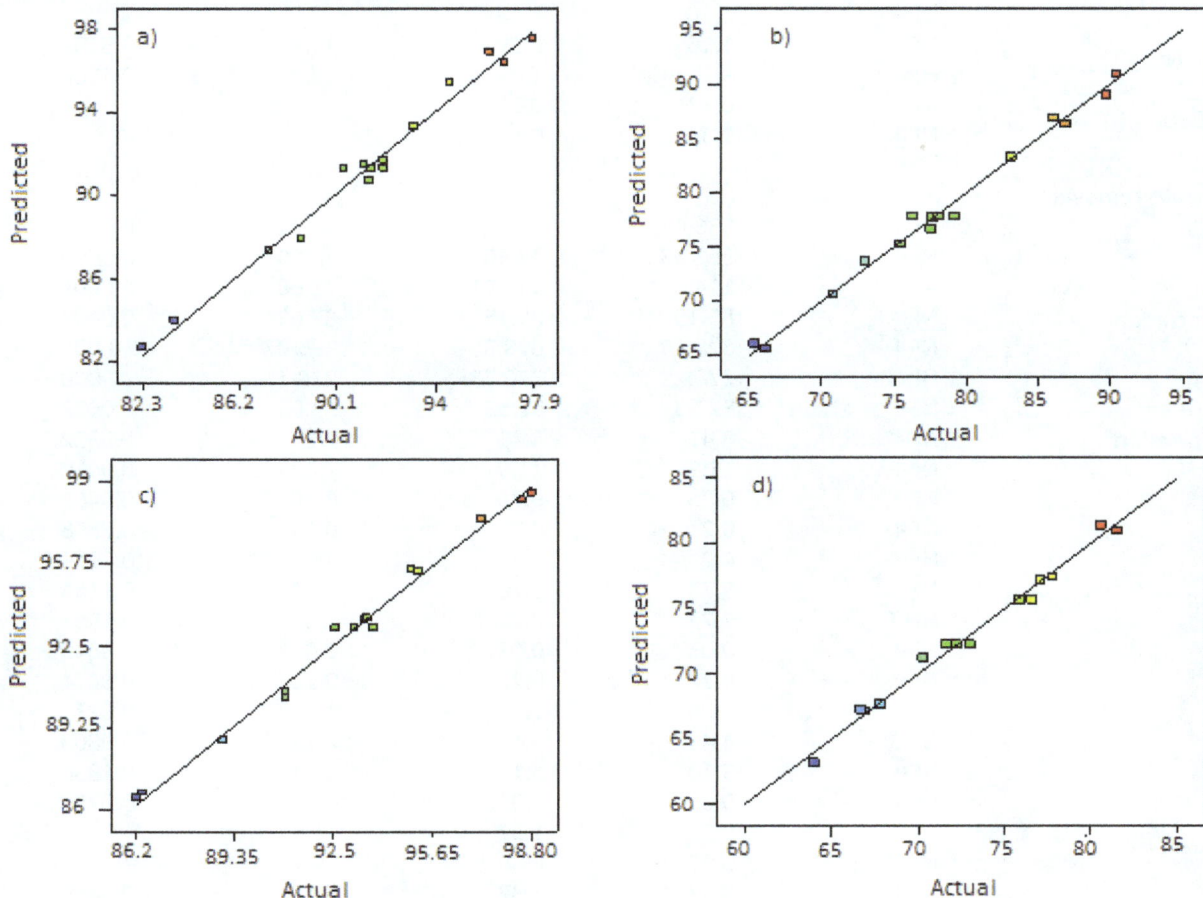

Fig. 2. Regression parameter of nanofilters a) Cr(VI) removal with NF-I; b) Ni removal with NF-I; c) Cr (VI) removal with NF-K; d) Ni removal with NF-K.

Table 4. ANOVA results of Cr (VI) (A) and nitrate (B) removal for NF-I and NF-K.

Sours	d.f.	Membrane type	Adj MS	Seq SS	F-value	P-value
(A) Cr (VI) removal						
Model	9	NF-I	203.74	22.64	60.29	0.0001
		NF-K	271.98	30.22	34.03	0.0006
X_1-Cr (VI) concentration	1	NF-I	12.5	12.5	33.29	0.0022
		NF-K	26.64	26.64	30.01	0.0028
X_2-pH	1	NF-I	172.05	172.05	458/19	<0.0001
		NF-K	194.05	194.05	218.52	<0.0001
X_3-Nitrate concentration	1	NF-I	13.26	13.26	35.32	0.0019
		NF-K	29.64	29.64	33.38	0.0022
X_1X_2	1	NF-I	3.06	3.06	8.16	0.0356
		NF-K	9.3	9.3	10.48	0.023
X_1X_3	1	NF-I	0.3	0.3	0.81	0.4106
		NF-K	0.0025	0.0025	0.0028	0.9597
X_2X_3	1	NF-I	0.25	0.25	0.67	0.4516
		NF-K	0.56	0.56	0.63	0.4622
X_1X_1	1	NF-I	0.83	0.83	2.22	0.1965
		NF-K	6.61	6.61	7.44	0.0414
X_2X_2	1	NF-I	0.75	0.75	1.99	0.2173
		NF-K	0.49	0.49	0.55	0.4930
X_3X_3	1	NF-I	0.59	0.59	1.57	0.2652
		NF-K	3.79	3.79	4.26	0.0939
Residual error		NF-I	0.38	1.88	-	-
		NF-K	1.46	7.28	-	-
Lack of Fit	3	NF-I	1.16	0.39	1.07	0.5159
		NF-K	3.1	1.03	1.54	0.4166
Pure Error	2	NF-I	0.72	0.36	-	-
		NF-K	1.34	0.67	-	-
(B) Nitrate removal						
Model	9	NF-I	836.08	92.9	58.41	0.0002
		NF-K	382.13	42.46	43.69	0.0003
X_1-Cr (VI) concentration	1	NF-I	114.01	114.01	71.68	0.0004
		NF-K	85.15	85.15	87.62	0.0002
X_2-pH	1	NF-I	628.35	628.35	395.07	<0.0001
		NF-K	270.28	270.28	278.11	<0.0001
X_3-Nitrate concentration	1	NF-I	52.53	52.53	33.03	0.0022
		NF-K	8.41	8.41	8.65	0.0322
X_1X_2	1	NF-I	11.9	11.9	7.48	0.0410
		NF-K	6.76	6.76	6.96	0.0461
X_1X_3	1	NF-I	0.2	0.2	0.13	0.7358
		NF-K	4.2	4.2	4.32	0.0921
X_2X_3	1	NF-I	5.76	5.76	3.62	0.1154
		NF-K	2.72	2.72	2.8	0.155
X_1X_1	1	NF-I	0.037	0.037	0.023	0.8849
		NF-K	0.22	0.22	0.22	0.6574
X_2X_2	1	NF-I	16.67	16.67	10.48	0.023
		NF-K	4.4	4.4	4.53	0.0866
X_3X_3	1	NF-I	5.10	5.10	3.21	0.1334
		NF-K	0.026	0.026	0.026	0.8773
Residual error		NF-I	0.27	1.37	-	-
		NF-K	0.97	4.86	-	-
Lack of Fit	3	NF-I	3.97	1.32	0.67	0.6469
		NF-K	3.87	1.29	0.97	0.2885
Pure Error	2	NF-I	3.98	1.99	-	-
		NF-K	0.99	0.49	-	-

3.2. Effect of pH on Cr (VI) and nitrate removal

The results indicated that by increasing the pH, the removal efficiency of Cr (VI) and nitrate by both of the membranes was enhanced (Figure 3 a and b). However, the pH had a greater influence on the removal efficiency of Cr (VI) than that of the nitrate. At a nitrate concentration of 80 mg/L and when the pH was increased from 5 to 9, the removal efficiency of Cr (VI) showed a 9 % and 10 % rise for NF-K and NF-I, respectively, and the removal efficiency of nitrate increased about 14 % and 19 % for the NF-K and NF-I, respectively. This can be explained by the fact that the Cr

(VI) ions were in the form of dichromate at around a pH of 5, while chromate and dichromate ions were formed at pH values higher than 7 [23-24]. Hence, by increasing the pH from 5 to 9, the Cr (VI) ions transformed from monovalent into the divalent form which increased the removal efficiency. Additionally, since the selected pH range was above the membrane isoelectric point, any increase in the pH level multiplied the membrane negative charge. Therefore, the repulsion force between the anions and the negatively charged membrane surface intensified and consequently, raised the removal efficiency of both the Cr (VI) and nitrate.

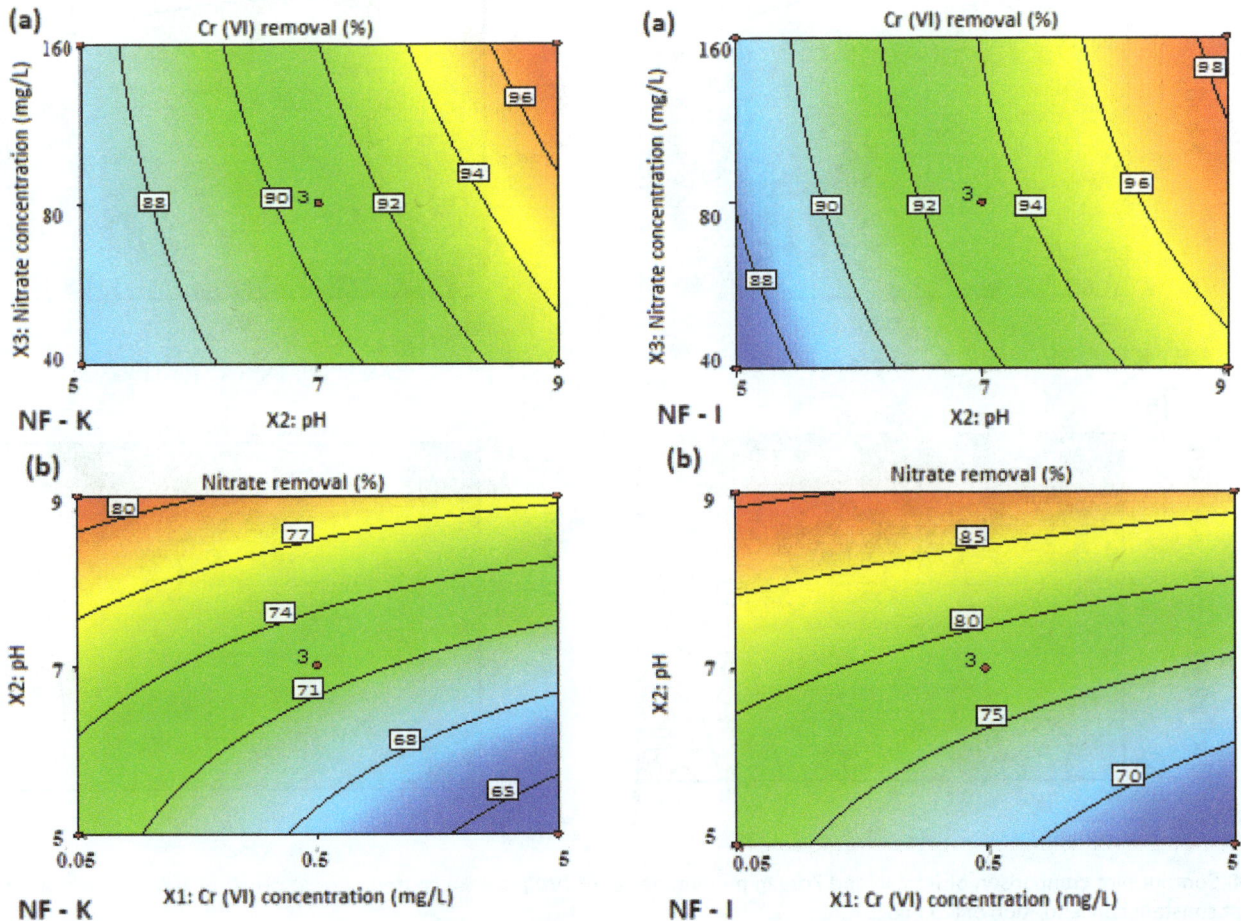

Fig. 3. Contour plot comparison of Iranian and Korean membranes a) Cr (VI) removal contour plot at a constant concentration of 0.5 mg/L; b) Nitrate removal contour plot at a constant concentration of 80 mg/L.

3.3. Effect of initial nitrate concentration on Cr (VI) and nitrate removal

Considering Figure 4a and b, it is evident that the more the nitrate concentration increased, the more the removal efficiency dropped. However, the trend was reversed for the removal efficiency of Cr (VI). At a pH=7 and a Cr (VI) concentration of 0.5 mg/L, as the nitrate concentration increased from 40 mg/L to 160 mg/L, the removal efficiency of Cr (VI) was enhanced up to 4 % and 2 % for the NF-K and NF-I, respectively. On the other hand, this enhancement in

the removal efficiency of the nitrate varied between 1.5 % and 0.5 % for the NF-K and NF-I, respectively. In fact, when the nitrate concentration increased, the cation (K$^+$) concentration was raised too. These cations were absorbed on the membrane surface and since the membrane itself had the negative charge, a shield formed on it. Thus, the newly formed cation layer on the membrane surface weakened the repulsion force between the negatively charged membrane and the anions which further facilitated the interaction of Cr (VI) and nitrate anions with the membrane surface. Because both the ion size and the electrical charge of Cr (VI) were larger compared to nitrate ions, they could hardly pass the membrane pores and as a result, the removal efficiency of Cr (VI) increased. Similar results have also been obtained by several authors [15,25,26]. On the other hand, the nitrate ions could cross the pores of the membrane more easily due to their smaller size and caused a drop in the removal efficiency of nitrate. However, in this condition, the reduction rate of the

efficiency depended on the pH of the feed solution. For instance, at a pH of about 5, the Cr (VI) ions were in the divalent form and were larger in size in comparison to the nitrates. Therefore, the removal efficiency of nitrate decreased while the removal efficiency of chromium increased. On the contrary, at pH values higher than 7, the Cr (VI) ions were in the divalent form and caused a shift in the removal efficiency of Cr (VI).

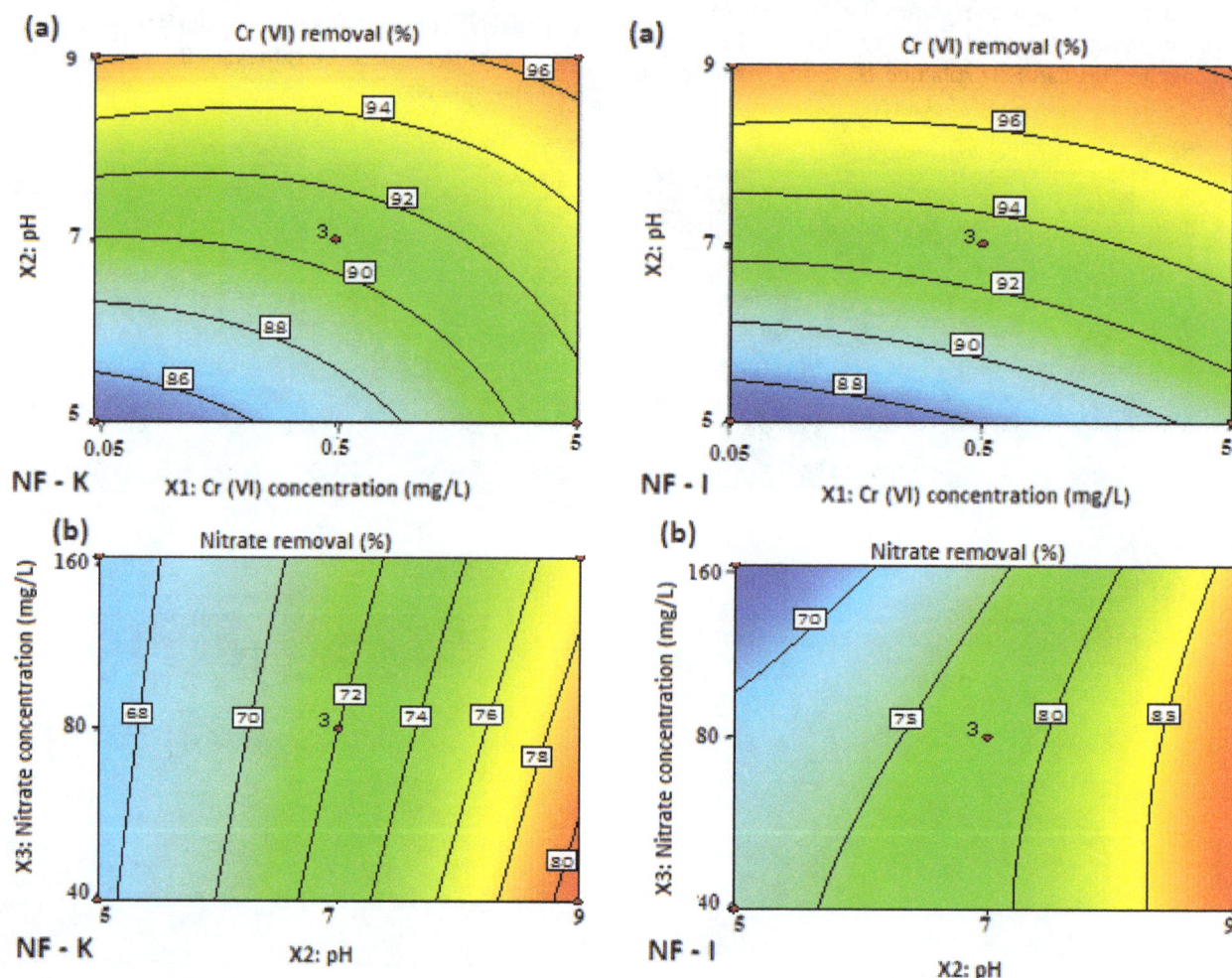

Fig. 4. Contour plot comparison of Iranian and Korean membranes a) Cr (VI) removal contour plot at pH=7; b) Nitrate removal contour plot at constant concentration of 0.5 mg/L.

3.4. Effect of initial Cr (VI) concentration on chromium and nitrate removal

According to Figure 3b and 4a, as the Cr (VI) concentration increased, the removal efficiency of Cr (VI) was enhanced. In contrast, the nitrate was removed less efficiently. At constant pH of 7 and a nitrate concentration of 80 mg/L, when the Cr (VI) concentration increased from 0 to 5 mg/L, the removal efficiency of Cr(VI) for the NF-K and NF-I rose about 3.5% and 2.5 %, respectively. However, the removal efficiency of nitrate for the NF-K and NF-I was reduced up to 6 % and 7%, respectively. The explanation is similar to the facts mentioned before. Indeed, by increasing the concentration of Cr (VI), the concentration of cations (K[+])

increased too, which further resulted in the formation of a cation layer on the membrane surface and weakness of the repulsion force between the negatively charged membrane surface and the anions. Therefore, the dichromate and nitrate anions could get closer to the membrane surface. At a pH of around 5, since the size of the nitrate ions were smaller than the Cr (VI) ions, they could cross the membrane pores more easily; consequently, the removal efficiency of nitrate decreased while the Cr (VI) was removed more efficiently. At pH values higher than 7, since Cr (VI) ions were in divalent form and had a larger ionic radius and electrical charge compared to nitrates, their removal efficiency increased. In this condition, in order to maintain the electrical balance between the two sides of the membrane, more nitrate ions crossed the membrane pores and the nitrate removal efficiency was further increased. Similar results were observed in several other works [15,25,26, 27].

3.5. Interaction of Cr (VI) and nitrate in actual water samples

The results of nanofiltration on actual water samples at an optimum pressure of 0.6 and 0.8 Mpa for NF-K and NF-I, respectively, are presented in Table 3. However, the water characteristics are specified in Table 2. According to Table 3, the NF-K could remove Cr (VI) and nitrate by 95 % and 76.4 %, respectively, while the NF-I was more successful and yielded about 97.3 % and 86 % for them, respectively. This could be explained by the fact that the NF-I had smaller pores or more exclusion effect and could remove anions with higher rates. The effect of other contributing factors on the actual water samples were the same as indicated for the synthetic water.

4. Conclusions

In this study, the removal efficiency of hexavalent chromium and nitrate ions by means of two different commercial spiral wound polyamide nanofilters was investigated for both synthetic and actual water specimens. The results are summarized as follows:
1- Both of the commercial polyamide spiral wound nano-filters can effectively remove Cr (VI) and nitrate from the contaminated water.
2- The interaction between the ions and charge of membranes influenced the pollutant removal efficiency. However, the NF-I showed a slightly better efficiency compared to the NF-K due to its smaller pore size or more exclusion effect.
3- The removal efficiency of nitrate and Cr (VI) depended greatly on pH, their initial concentration, and the interaction between the concentration of Cr (VI) and the pH.
4- The removal efficiency was obtained in the range of 86.2 to 98.8% for Cr (VI) and 65.4 to 90.4% for nitrate. The results showed that the presence of chromium ions in the solution reduced the amount of nitrate removal from water.
5- The Box–Behnken design was employed to develop a mathematical model for predicting the removal efficiency of Cr (VI) and nitrate by the nanofiltration process.

Acknowledgements

The authors would like to thank the Environmental Research Institute at the University of Isfahan for their support of this research.

References

[1] Mohseni-Bandpi, A., Elliott, D. J., Zazouli, M. A. (2013). Biological nitrate removal processes from drinking water supply-a review. *Journal of environmental health science and engineering*, *11*(1), 35.

[2] Epsztein, R., Nir, O., Lahav, O., Green, M. (2015). Selective nitrate removal from groundwater using a hybrid nanofiltration–reverse osmosis filtration scheme. *Chemical engineering journal*, *279*, 372-378.

[3] Richard, A. M., Diaz, J. H., Kaye, A. D. (2014). Reexamining the risks of drinking-water nitrates on public health. *The Ochsner journal*, *14*(3), 392-398.

[4] Baig, U., Rao, R. A. K., Khan, A. A., Sanagi, M. M., Gondal, M. A. (2015). Removal of carcinogenic hexavalent chromium from aqueous solutions using newly synthesized and characterized polypyrrole–titanium (IV) phosphate nanocomposite. *Chemical engineering journal*, *280*, 494-504.

[5] Romero-Gonzalez, J., Peralta-Videa, J. R., Rodriguez, E., Ramirez, S. L., Gardea-Torresdey, J. L. (2005). Determination of thermodynamic parameters of Cr (VI) adsorption from aqueous solution onto Agave lechuguilla biomass. *The journal of chemical thermodynamics*, *37*(4), 343-347.

[6] World Health Organization. (2006). *Guidelines for the safe use of wastewater, excreta and greywater* (Vol. 1). World Health Organization.

[7] Chakravarti, A. K., Chowdhury, S. B., Chakrabarty, S., Chakrabarty, T., Mukherjee, D. C. (1995). Liquid membrane multiple emulsion process of chromium (VI) separation from waste waters. *Colloids and surfaces A: Physicochemical and engineering aspects*, *103*(1), 59-71.

[8] Calace, N., Di Muro, A., Nardi, E., Petronio, B. M., Pietroletti, M. (2002). Adsorption isotherms for describing heavy-metal retention in paper mill sludges. *Industrial and engineering chemistry research*, *41*(22), 5491-5497.

[9] Tiravanti, G., Petruzzelli, D., Passino, R. (1997). Pretreatment of tannery wastewaters by an ion exchange process for Cr (III) removal and recovery. *Water science and technology*, *36*(2-3), 197-207.

[10] Ayyasamy, P. M., Rajakumar, S., Sathishkumar, M., Swaminathan, K., Shanthi, K., Lakshmanaperumalsamy, P., Lee, S. (2009). Nitrate removal from synthetic medium and groundwater with aquatic macrophytes. *Desalination*, *242*(1-3), 286-296.

[11] Aksu, Z., Kutsal, T. (1990). A comparative study for biosorption characteristics of heavy metal ions with C. vulgaris. *Environmental technology*, *11*(10), 979-987.

[12] Aksu, Z., Özer, D., Ekiz, H. I., Kutsal, T., Çaglar, A. (1996). Investigation of biosorption of chromium (VI) on Cladophora crispata in two-staged batch reactor. *Environmental technology*, *17*(2), 215-220.

[13] Bansal, M., Singh, D., Garg, V. K. (2009). A comparative study for the removal of hexavalent chromium from aqueous solution by agriculture wastes' carbons. *Journal of hazardous materials*, *171*(1), 83-92.

[14] Sharma, D. C., Forster, C. F. (1994). The treatment of chromium wastewaters using the sorptive potential of leaf mould. *Bioresource technology, 49*(1), 31-40.

[15] Paugam, L., Taha, S., Dorange, G., Jaouen, P., Quéméneur, F. (2004). Mechanism of nitrate ions transfer in nanofiltration depending on pressure, pH, concentration and medium composition. *Journal of membrane science, 231*(1), 37-46.

[16] Schaep, J., Van der Bruggen, B., Vandecasteele, C., Wilms, D. (1998). Retention mechanisms in nanofiltration. In *chemistry for the protection of the environment 3* (pp. 117-125). Springer US.

[17] Yu, Y., Zhao, C., Wang, Y., Fan, W., Luan, Z. (2013). Effects of ion concentration and natural organic matter on arsenic (V) removal by nanofiltration under different transmembrane pressures. *Journal of environmental sciences, 25*(2), 302-307.

[18] Li, K., Wang, J., Liu, J., Wei, Y., Chen, M. (2016). Advanced treatment of municipal wastewater by nanofiltration: Operational optimization and membrane fouling analysis. *Journal of environmental sciences, 43*, 106-117.

[19] American public health association, American water works association, water pollution control federation, water environment federation. (1915). *Standard methods for the examination of water and wastewater* (Vol. 2). American public health association.

[20] Chakraborty, S., Dasgupta, J., Farooq, U., Sikder, J., Drioli, E., Curcio, S. (2014). Experimental analysis, modeling and optimization of chromium (VI) removal from aqueous solutions by polymer-enhanced ultrafiltration. *Journal of membrane science, 456*, 139-154.

[21] Myers, R. H., Montgomery, D. C., Anderson-Cook, C. M. (2016). *Response surface methodology: Process and product optimization using designed experiments.* John Wiley and Sons.

[22] Betianu, C., Caliman, F. A., Gavrilescu, M., Cretescu, I., Cojocaru, C., Poulios, I. (2008). Response surface methodology applied for Orange II photocatalytic degradation in TiO₂ aqueous suspensions .*Journal of chemical technology and biotechnology, 83*(11), 1454-1465.

[23] Hafiane, A., Lemordant, D., Dhahbi, M. (2000). Removal of hexavalent chromium by nanofiltration. *Desalination, 130*(3), 305-312.

[24] Taleb-Ahmed, M., Taha, S., Maachi, R., Dorange, G. (2002). The influence of physico-chemistry on the retention of chromium ions during nanofiltration. *Desalination, 145*(1), 103-108.

[25] Santafé-Moros, A., Gozálvez-Zafrilla, J. M., Lora-García, J. (2005). Performance of commercial nanofiltration membranes in the removal of nitrate ions. *Desalination, 185*(1-3), 281-287.

[26] Mahmoodi, P., Hosseinzadeh Borazjani, H., Farhadian, M., Solaimany Nazar, A. R. (2015). Remediation of contaminated water from nitrate and diazinon by nanofiltration process. *Desalination and Water Treatment, 53*(11), 2948-2953.

[27] Mahmoodi, P., Farhadian, M., Solaimany Nazar, A. R., Noroozi, A. (2014). Interaction between diazinon and nitrate pollutant through membrane technology. *Journal of Applied Research in Water and Wastewater, 1*(1), 18-25.

PERMISSIONS

All chapters in this book were first published in AET, by Iranian Research Organization for Science and Technology (IROST); hereby published with permission under the Creative Commons Attribution License or equivalent. Every chapter published in this book has been scrutinized by our experts. Their significance has been extensively debated. The topics covered herein carry significant findings which will fuel the growth of the discipline. They may even be implemented as practical applications or may be referred to as a beginning point for another development.

The contributors of this book come from diverse backgrounds, making this book a truly international effort. This book will bring forth new frontiers with its revolutionizing research information and detailed analysis of the nascent developments around the world.

We would like to thank all the contributing authors for lending their expertise to make the book truly unique. They have played a crucial role in the development of this book. Without their invaluable contributions this book wouldn't have been possible. They have made vital efforts to compile up to date information on the varied aspects of this subject to make this book a valuable addition to the collection of many professionals and students.

This book was conceptualized with the vision of imparting up-to-date information and advanced data in this field. To ensure the same, a matchless editorial board was set up. Every individual on the board went through rigorous rounds of assessment to prove their worth. After which they invested a large part of their time researching and compiling the most relevant data for our readers.

The editorial board has been involved in producing this book since its inception. They have spent rigorous hours researching and exploring the diverse topics which have resulted in the successful publishing of this book. They have passed on their knowledge of decades through this book. To expedite this challenging task, the publisher supported the team at every step. A small team of assistant editors was also appointed to further simplify the editing procedure and attain best results for the readers.

Apart from the editorial board, the designing team has also invested a significant amount of their time in understanding the subject and creating the most relevant covers. They scrutinized every image to scout for the most suitable representation of the subject and create an appropriate cover for the book.

The publishing team has been an ardent support to the editorial, designing and production team. Their endless efforts to recruit the best for this project, has resulted in the accomplishment of this book. They are a veteran in the field of academics and their pool of knowledge is as vast as their experience in printing. Their expertise and guidance has proved useful at every step. Their uncompromising quality standards have made this book an exceptional effort. Their encouragement from time to time has been an inspiration for everyone.

The publisher and the editorial board hope that this book will prove to be a valuable piece of knowledge for researchers, students, practitioners and scholars across the globe.

LIST OF CONTRIBUTORS

Mojtaba Ahmadi, Mohmmad Hamed Ardakani and ALi Akbar Zinatizadeh
Chemical Engineering Department, Faculty of Engineering, Razi University, Kermanshah, Iran

Araz Tofigh Kouzekonani and Majid Mahdavian
Department of Chemical Engineering, Quchan University of Advanced Technology, Quchan, Iran

Masoumeh Ghorbani and Seyyed Mostafa Nowee
Department of Chemical Engineering, Ferdowsi University of Mashhad, Mashhad, Iran

Mohammad J. Nosratpour, Morteza Sadeghi and Saied Ghesmati
Department of Chemical Engineering, Isfahan University of Technology, Isfahan 84156-83111, Iran

Keikhosro Karimi
Department of Chemical Engineering, Isfahan University of Technology, Isfahan 84156-83111, Iran Industrial Biotechnology Group, Institute of Biotechnology and Bioengineering, Isfahan University of Technology, Isfahan 84156-83111, Iran

Gehan E.Sharaf El-Deen
Radioactive Waste Management Department, Hot Laboratories Center, Atomic Energy Authority, Egypt

Mahboubeh Saeidi and Marzie Komeili
Department of Chemistry, Faculty of Science, Vali-e-Asr University of Rafsanjan, Iran

Atena Naeimi
Faculty of Science, Department of Chemistry, University of Jiroft, Jiroft, Iran

Jamshid Behin and Negin Farhadian
Department of Chemical Engineering, Faculty of Engineering, Razi University, Kermanshah, Iran

Farzaneh Saadati and Narjes Keramati
Faculty of Nanotechnology, Semnan University, Semnan, Iran

Mohsen Mehdipour Ghazi
Faculty of Chemical, Petroleum and Gas Engineering, Semnan University, Semnan, Iran

Nayereh Yahyaei and Pedram Mohebbi
Department of Chemical Engineering, Shahrood Branch, Islamic Azad University, Shahrood, Iran

Javad Mousavi and Mehdi Parvini
Department of Chemical Engineering, Gas and Petroleum, Semnan University, Semnan, Iran

Saeed Almohammadi and Masoomeh Mirzaei
Department of Chemical Engineering, Mahshahr Branch, Islamic Azad University, Mahshahr, Iran

Kourosh Razmgar and Zahra Beagom Mokhtari Hosseini
Department of Chemical Engineering, Hakim Sabzevari University, Sabzevar, Iran

Vahide Elhami
Department of Chemical Engineering, Faculty of Chemical and Petroleum Engineering, University of Tabriz, Tabriz, Iran

Afzal Karimi
Faculty of Advanced Technologies in Medicine, Iran University of Medical Sciences, Tehran, Iran

Hossein Lotfi and Mohsen Nademi
Department of Chemical Engineering, North Tehran Branch, Islamic Azad University, Tehran, Iran

Mohsen Mansouri
Department of Chemical Engineerin, Ilam University, Ilam, Iran

Mohammad Ebrahim Olya
Department of Environmental Research, Institute for Color Science and Technology, Tehran, Iran

Mohammad Ali Salehi and Nasrin Hakimghiasi
Department of Chemical Engineering, University of Guilan, Rasht, Iran

Zohreh Alimohammadi, Habibollah Younesi and Nader Bahramifar
Department of Environmental Science, Faculty of Natural Resources, Tarbiat Modares University, Noor, Iran

Jamal Alikhani, Jalal Shayegan and Ali Akbari
Department of Chemical and Petroleum Engineering, Sharif University of Technology, P.O. Box 11155-9465, Azadi Ave., Tehran, Iran

Index